Would a Mar

From Edward's detailed childhood ...
Liverpool through to his early career ... apprentice and aircraft engineer at Speke Airport and beyond, this well researched book takes you deep into the minutiae of Edward's daily life.

An unusual contemporary history 'Would a Man Return' combines the complexity of hangar life with the vicarious thrill of seeing the aircraft and other machinery of yesteryear as first recorded through the enthusiastic eyes of a young, knowledgeable Edward. This book brings the forties and fifties to life.

Author's Note

This is not really a book of memoirs, nor is it written with delusions of grandeur, it is simply a literary account of how a boy grew up. Through the learning curves of childhood and schooling, through adolescence and an apprenticeship training, eventually to reach manhood, it seeks to invoke the emotions of readers. It highlights the complexities of human thought and human relationships. Against a strong aviation backdrop, the story is descriptive of places, people and environments. It is sometimes matter-of-fact, often humorous, and occasionally tragic.

Most certainly the chapters describe impressions of youth, long since consigned to the past. The location and environment of Liverpool's original airport at Speke are well documented in Phil Butler's eminently interesting book "Liverpool Airport - An illustrated history". I read his account only after I had written of my own experiences as an apprentice during the years of 1953 thru 1955. Mr. Butler's book serves well to confirm the relative accuracy of my observations in adolescence, recalled and set down after the passing of almost half a century.

I'm pleased to say that my memory hasn't played too many tricks on me.

Ted Weinel.
Denia 2011.

Note: Although many of the characters in 'Would a Man Return', as an autobiographical work, are based on real people, the portrayal of those characters in the book is fictional and their names, except where integral to the plot, have been altered to protect their privacy. The events, character motivations and interpretations are those of the author and entirely fictional, however, the information about systems, aircraft and events are factual, based on the author's life.

First published in Great Britain in 2012 by U P Publications Ltd
25 Bedford Street, Peterborough, UK PE1 4DN

A CIP Catalogue record of this book is available from the British Library

FIRST PAPERBACK EDITION

ISBN 978-1908135094

Printed in England by The Lightning Source Group

FIRST KINDLE EDITION

ISBN 978-1908135247

www.uppublications.ltd.uk
www.edwardweinel.com

Would a Man Return

Edward H. Weinel

U P Publications
2012

Part 1

Beginnings in a Time of War

1

The family moved to Liverpool in 1940. Edward's first impressions of childhood, and of his surroundings, were only just beginning to take tangible shape. At barely two years old, embraced in a loving household, he knew nothing of the troubled war-torn world beyond.

France was about to fall to Hitler's Nazi regime and the British Expeditionary Force was in full retreat to Dunkirk. The blitz had yet to come, with the bombing of London and many other cities and ports of the land. The Battle of Britain was about to begin. Ill prepared for war, the nation and its dominions braced for the onslaught. The British people stood alone against the armoured might of Hitler's Third Reich with less than thirty miles of the English Channel separating the belligerent, combatant nations.

Perhaps it was a darkened space, illuminated by the flame of a single candle, that was Edward's earliest recollection – or was it of his gentle mother's face, serene in the dim light of the candle's flickering glow? Was it the warmth and softness of the mattress where he lay, next to his elder brother Freddy, in the narrow confines of the space beneath the staircase? Was it the heavy drone of aircraft above, the howl and shriek of falling bombs, the crash of impact not far away and the hollow bang of a defence gun down the road? Was it the urgent, heart-pounding descent in his father's arms, down the stairs to their refuge? ...Or was it simply his blue, hooded dressing gown, thrown about him as their parents snatched Edward and his brother from their beds at the sound of the air-raid siren's mournful warning.

He well remembered how the house shook one night, when bombs fell on two homes along the road to Hale, but he couldn't recall fear, only the sight of it perhaps, reflected in his mother's face: fear for her children's safety.

On such perilous nights, the two brothers lay side-by-side

beneath the slope of the staircase. A heavy wooden book-case that once belonged to their grandfather, then filled with their father's books, closed the triangular opening; leaving space enough at the highest end for them to pass within. It was an improvised shelter of sorts, snug, but safe in the strongest part of the house. Their mother always lit a candle, perching it in a saucer on top of the bookcase. The gleam of its flame, flickering as wafts of draught fanned its delicate light, would throw vaguely moving shadows and images dancing upon the walls of their sheltering space.

Edith would sit upon a chair just inside the sanctuary, next to the opening left by the bookcase. From there, she could see quickly to her children's needs or run to the front door at the sounding of the all-clear siren, to let in their father; his ARP, fire-watch duty done.

Night-time air raids were frequent on the cities of their land: Coventry, London, Birmingham, Manchester, where ever Britain's concentrated populations lived and industry produced armaments and munitions for the war effort. Just as everywhere else in the realm, the call to civil duty touched all non-combatants. Husbands and fathers kept fire-watch and undertook voluntary civil defence duties, on those nights when bombs fell on Liverpool.

How many mothers fretted over the well-being of their children and their men-folk on duty, out in the darkness? This was war-work of another kind. There were many sleepless nights for parents and children alike, when the war came to them in the dark. They showed no lights. They huddled in their homes or in brick and concrete shelters, the growl and drone of aircraft high above their rooftops. They waited with bated breath to hear the shriek of falling bombs. If the bombs fell near, they waited again: perhaps for death.

Yet, to the brothers in childhood, it was exciting, like an adventure, being snatched up from sleep and dreams, when 'Jerry' came to drop his deadly cargoes of terror. Conceivably, the boys even enjoyed the hours spent in their warm space beneath the stairs. Sometimes, on those awful nights, to relieve his loved ones' fear in the midst of mayhem, their father would call through the front-door letterbox, to let his family know that he was near at hand and safe. The boys were lucky to have both parents close in such times. They felt secure, even in the thunder of conflict. At daybreak, the birds always sang again.

During those uncertain times of war, Fred and Edith tried to make home-life as normal as they were able for their boys and although holidays were distant memories of peace-time, before the war, the family sometimes made a day-trip to New Brighton, by steam-ferry across the Mersey. On one such occasion the family had taken deck-chairs and a picnic basket onto the rocky shore.

They picked a smooth rise, one of many sandy hummocks uncovered by the sea's tidal retreat, and settled down to enjoy the sunshine.

At some point during the afternoon (so his parents told him in later years) Edward became agitated; there was something wrong. They were taking the sun and building sandcastles when, because of Edward's obvious distraction, Edith realised that people were shouting from the shore-line and from the esplanade, trying to draw the attention of several family groups sitting and playing on the sands. The tide had turned.

On that part of the coast, close to the river's estuary-mouth, the waters of a rising tide rapidly covered the beaches as the sea returned. The family found themselves surrounded, cut off, and in danger of being caught up in the swift currents and swirling rollers. They abandoned their chairs and grabbing towels, clothes and basket, they made for the stone and concrete walls of the esplanade. Fred took Edward upon his shoulders whilst Edith half swam and half waded with Freddy to reach safety.

Edward's agitation had indeed warned them of the danger, although he was unable to articulate his concerns verbally; he was too young, but it was lucky for the family that his tearful fretting had alerted them to the hazard of their situation. The episode scared them all, not least Edward, though it was a different kind of fear from that of the bombings. It was a touchable and obvious threat to life, one that was probably the root of Edward's latent fear of water and swimming in later years.

With a birth date in March, astrology placed Edward's existence under the sign of Pisces; bestowing him, they said, with a loving, pacific nature. Of stoic character with somewhat philosophical and artistic leanings he was a dreamer, not given to decision making. Even so, one day close to Christmas, as he stood upon a kitchen chair to watch his mother mix the ingredients for a fruitcake, there was a brief moment in which he recorded what was, probably, the first significant decision he'd ever made. By then he was all of four years old.

As Edward spooned up the sweet scrapings from the mixing-bowl, his mother asked him when he would like to start school. He considered the question deeply before committing himself to an answer. His brother Freddy, eight years old, had been at the local Primary for several years already but Edward hadn't given too much thought to the idea of going to school himself until his mother mentioned it. It wasn't that he discounted the notion completely, but faced with the direct question, he discovered that he really had no great desire to go.

Accordingly, his response was emphatic – quite decisive, for one born under the sign of Pisces. "Not until after Christmas!" His pronouncement was assertive, much to his mother's amusement. Of course, he would have to go, eventually, that he knew; but if there was a choice, then 'after Christmas' was his. After-all, he thought, the logic crossing his mind clearly at the time, he couldn't allow the worries of going to school to interfere with preparations for the festive season, could he? Not because he had said so, but only because that was the way of it, Edward started at Infant's school in the following year, 1943, after Christmas and before his fifth birthday in March.

Emotions of that first day were not so traumatic as such moments can be for some children. As it happened, Edward was already quite familiar with the school. He'd seen it many times before, from the seat of a perambulator or push-chair, and later, walking with their mother, on the several-times-daily trip to deliver and retrieve Freddy. Nevertheless, on the day of his own induction, the place seemed very much larger to his young eyes: less than awesome but quite definitely bigger.

With its lofty, echoing halls, Kingsthorne Road County Primary was an impressive, modern structure. It stood tidily in its own grounds, on the edge of Hunts Cross, with a high wooden fence separating it from the surrounding fields and from the approach road that gave its name to the school. The long, two-storied building, built of neatly pointed, light-brown, rough-cast bricks, supported a flat roof where a square chimney squatted to carry smoke and hot fumes from Caretaker Cave's infernal machines (water-heating furnaces) that lurked and rumbled in the mysterious, subterranean depths of the forbidden basement.

Mothers arrived each weekday morning to deposit their children at the school before nine o'clock, when the Head Mistress rang a hand-bell for assembly with inflexibly timed precision. Teachers of the infants' classes receiving the children at their respective classrooms shushed and fussed over their noisy charges, whilst the juniors, gathering in the playgrounds, lined-up in response to a teacher's whistle, blown as the bell rang inside. Thus mustered, ready for morning prayers, each class would then march, along with their teachers, to the large assembly-hall at one end of the building.

Infant classes filing in first, to stand in fidgeting lines across the wide expanse of floor at the front, close to the stage, found it hard to keep still and quiet as the older children, the juniors, arriving from the play-grounds, formed-up behind to fill the hall. It was a disciplined and precise exercise, conducted daily in near total silence, except for the whispered remonstrations of teachers and the tramp of many shoes upon the hard, wood-block floors.

The assembly would wait then, in a silence broken only by sporadic, muted coughs and the shuffling of impatient feet, for the arrival of the Head Mistress. Her appearance upon the stage, hurrying as always, dressed in her official-looking skirt and jacket of clerical grey or sometimes black, clutching a sheaf of papers, and with a clearly enunciated, "Good morning school," received a well rehearsed, almost choral response, as the children came to order, "Good-morning-Miss-Scott, good-morning-all."

At this cue, one of the teachers, usually the tall, elderly, kindly, bespectacled, golden-toothed Miss Lewis, by then seated at the upright piano to one side of the stage, would strike up the introductory chords for the first hymn.

The sound of children's voices lifted in praise of the Lord, would drift sweetly in song, if not all-together in tune, the way that only young voices can, trying hard to follow Miss Lewis's exuberant pounding upon the keys, as she too sang heartily. Then, a moment for prayers, another for announcements, a second hymn and the children would disperse in orderly groups for classes. Edward enjoyed morning assemblies. He loved to march; but more, he loved to sing.

At the front of the school building, jutting, half-circular or 'D' shaped protrusions, one at each extremity, enclosed wide staircases connecting the ground floor with the upper-storey corridor. The stairs supported metal banisters with handrails of smooth, polished wood; often used (against the rules) as slides by some of the older juniors. Many a blazer, cardigan or shirt bore witness to this exciting, though strictly illegal, method of descent.

The wooden rails were at just the right height to mount with one outstretched leg and one buttock cheek. An arm crooked over the metal banister capping, alongside, gave support for the sliding plunge to the bottom of the stair-flight. Friction under the armpit and beneath the seat was intense, and many-a-sleeve, not to mention knickers and short trousers, ended up with a shined nap or, more often, well ventilated by holes that inevitably appeared. The holes were perhaps a mystery to thoughtful mothers who sewed them up or darned them whilst their children slept.

The Infant's department occupied the lower floor of the school and the Juniors', the upper storey. Long corridors running the whole length of the building on the front side, between the two stair-towers, gave access to classrooms, built in-line along the other side, to face the rear. Tall, metal-framed windows allowed plenty of natural light into the corridors and classrooms.

The windows fascinated Edward. Each metal framed glazing contained several 'swing-over' panes for ventilation. They had latches with rings, he recalled, that needed a pole with a hook at the top, to operate them. Every classroom had its own pole, standing in a corner, ready for use, and children in the upper classes would vie for selection as pole-monitor, during the warmer months, to open and close the windows at the teacher's bidding.

The school's, unpolished, wooden, parquet floors were smooth and finely finished. Together with his new classmates, Edward soon discovered that the long, hardwood surfaces of the broad corridors were perfect for sliding. Perhaps, indeed, they were there for that very purpose, providing lively lads and lasses with a short-lived moment of diversion at their new-found sport. A short run before snapping leather-soled shoes into a balanced, side-ways-standing, arm-flailing glide, gave brief outlet to healthy exuberance.

Wary teachers soon put a stop to the wild experiments, for the umpteenth time, as they surely did with every new intake of high spirited youngsters.

The metal stair-banisters, window frames and the wooden doors, all bore shiny, blue-green, turquoise paintwork, complimenting the cream-coloured walls. In the corridors, hand painted pictures of imps, pixies and elves, seated upon branches amid leafy foliage, hung at intervals on the walls between the windows to gaze, with mischievously smiling countenances, upon the children as they passed.

Annexes at both ends of the main building accommodated entrance doors, toilets and the cage-like cloakrooms where the children deposited their hats and caps, topcoats and mackintoshes. They hung their out-door garments upon numbered, wrought-iron pegs, set in rows, alternating high and low on black-painted, squared-wire dividers. Built in the mid-nineteen-thirties, it was altogether a well-appointed, functional and modern school: a model of the latest in educational facilities.

Outside, large, flag-paved playgrounds gave the children ample space in which to recreate and participate in the popular outdoor games that youngsters played in those days: relieve-oh, piecrust, tag and ball games of every sort. The playgrounds also hosted seasonal activities – skipping, conkers, marbles, kite flying, and not least, the thrills and many a spill from sliding on the icy, flat, concrete paving flags, when winter's first frosts and snows appeared.

An enormous playing field at one end of the juniors' yard provided the grassed area essential for organised sports – football and cricket for the boys, rounders and netball for the girls.

The school's curriculum listed gardening as a subject for the upper classes. The study of vegetable plant life, and husbandry thereto, went hand in hand with a need for all to know about growing food in those wartime days. Excellent results were plain to see beneath the windows of downstairs classrooms, in a cultivated area behind the school. All manner of vegetables grew in that ample plot, exemplifying strong horticultural activity.

Soon after the outbreak of war, the authorities built a number of brick and concrete air-raid shelters, siting them on the flanks of the playgrounds. The shelters provided a measure of protection for the children and staff, should an air-attack on nearby factories occur. Fortunately, daylight air raids were not commonplace then, over Liverpool, so during Edward's time the school's shelters never saw use. Nevertheless, they provided ideal spots for play, at hide-and-seek or variations of goodies-and-baddies like Cops-and-Robbers, and Cowboys-and-Indians. More ominously, and because of their isolation from the eagle eyes of the yard monitors, the shelters or rather the spaces between and behind them, became unsavoury locations for bullying and 'bashings-up.'

Such were the impressions of school-life to an observant youngster of those times. Edward's world had expanded. Whoever would have thought there was a war on?

Miss Sharp was Edward's first teacher at Kingsthorne. She was younger than were most educators of that time. Short of stature and robust, she wore glasses with thick lenses: floral-print dresses in the summer, tweeds in the winter and, always, sensible shoes. She smiled and laughed often during her teaching, and when she addressed her class, Miss Sharp used the children's given names. Edward was therefore, Teddy: one of two so named in her class. He liked that because that was what his family called him at home. In the beginning he therefore felt very comfortable at school.

Miss Sharp taught letters and numbers, and how to write them, on paper with a pencil. Some of her class, including Edward, already knew the alphabet and numbers up to ten. As a diversion between serious lessons, the children learned to work with *Plasticine* and coloured crayons, their artistic teacher guiding unaccustomed hands in the ways of self-expression. She was good at drawing and painting and demonstrated many things by illustration on the blackboard. Indeed, Miss Sharp had painted the watercolour pictures of elves and pixies that hung between the windows in the corridors. Edward often gazed upon them in admiration, wishing that one day, perhaps, he might draw and paint as well as she did. It was a lasting impression that was to have impact and influence upon him, throughout his life.

In class, the children frequently sang: sometimes together and sometimes individually. It was something Edward quite enjoyed, at first. However, his enthusiastic and unselfconscious pleasure for performing before an audience ceased quite abruptly. His outlook and perspectives changed because of an embarrassing incident that produced a reticence or shyness in his character thereafter.

It happened in this way...

Over the months, Miss Sharp trained her pupils well in classroom protocols, not the least of which was the raising of one's hand to catch the teacher's eye, to ask a question courteously or seek advice. “If you wish to speak, put your hand up,” she told them. “If you need to leave the room (to use the toilet for example), then raise your hand.”

One unfortunate day, Edward needed to use the toilet quite desperately. He tried to hold it until playtime, but he couldn't. He put up his hand to catch the teacher's eye. “What is it Teddy?” She enquired, looking up at the spontaneous movement in her class, as Edward invoked the hand-up rule.

“Please, Miss, may I leave the room?” He asked, qualifying his reason. “I need to go to the toilet.”

“No ...you can wait until playtime, it's not long to wait,” commanded Miss Sharp. Edward was bursting and he put up his hand again. “Teddy, I told you before, you must wait until playtime. There are only a few minutes before the bell rings.” Her rebuke came even before he could open his mouth to speak. He could feel the eyes of the whole class upon him. His audience of classmates was glaring at him with disdain and disbelief. Nevertheless he could wait no longer. He made his decision. It was imperative, even in the face of censure.

Miss Sharp had turned her attention to the blackboard. Edward stood up from his desk and strode forward to the front of the room and the classroom door, resolute only in the need to relieve himself. Sensing his movement, Miss Sharp turned again to face the class, firmly telling Edward to stop and not to go any farther. At this point, nature got the best of him, compelling him to wet himself in front of the class.

How his peers giggled and sniggered. They grimaced, held their noses and pointed with their fingers, as the puddle spread around Edward's feet. Miss Sharp stood next to her desk in horror, transfixed for a moment. Alone and ashamed, he stood in front of her class, unable to stem the flow of nature. Too late to help his plight, the bell rang for the mid-morning break.

Miss Sharp hurriedly dismissed her class for playtime before she came to where Edward stood in misery, with the front of his short trousers, his socks and shoes, all soaking wet. She was very apologetic and almost in tears herself, as she took the shame-filled boy along to the nursing room to get him dried up.

For quite some time after that, his class-mates called him wee-wee Weinel. He'd lost the sympathetic and kindly audience, before whom he'd often raised his voice in song, solo and unabashed. After the wetting episode, Edward never sang again in front of the class, nor did he any longer look forward to, or enjoy, the prospect of recitation, reading or any other form of activity that required him to face an audience. His embarrassment was profound. He didn't like it, nor did he like the reaction of the other children. He vowed inwardly that it would never happen again. For him, thenceforward, school took on rather a negative and hostile character.

2

Home-life, however, remained a constant sanctuary of love, companionship and understanding among his family, in spite of the war and in spite of school. It upset his parents when Edward explained to them about the wetting incident in class, but next day, before his mother could register her complaint at the occurrence, Miss Sharp made a particular point of apologising for it. Though it did not make Edward feel any better about the shame he felt.
Home for the Weinel family was a rented house on Higher Road, the main highway out of Hunts Cross, running through the districts of Halewood, Hale, and on towards Warrington.

The house was one of many, semi-detached dwellings lining both sides of the main road for many miles out into the countryside. Farther along there were bungalows and a place where horses lived when they were too old to work. It was rather like an old-folks home and had a sign 'The Horses Rest' set close to the fence where the footpath ran. Sometimes the family would walk that way to see the horses, taking much delight in rubbing the noses of those aging animals that were able to hobble across their field, at the sight of people at the fence.

They were a friendly mob, old farm horses mostly: Shires with huge frames and great hoofs with hairy fetlocks, which Edward thought were to keep the horses' ankles warm. Memories of those moments, in later life, always reminded Edward of his early love for the rural countenance of the district – the green fields, the farming community and the hamlets, then still untouched by city expansion and industrialisation.

Just like all the houses lining both sides of Higher Road, that in which the Weinels lived had big wooden, double gates opening from the road onto a broad concrete driveway leading to even bigger gates at the side of the house, and beyond, to a large garage at the back. Neglected, green, sun-blistered paintwork, without lustre, dull and peeling in places, matched the window frames and outside doors of the house.

One of the front-gate posts bore the number 100 in large, weather-rusted characters, whilst one of the two gates displayed the name "MARONDA" upon a neatly set plaque, next to another saying "NO PEDLARS or HAWKERS." Edward thought it a silly sign because his mother often said that many of such travelling people couldn't or wouldn't read it anyway. The gypsies still came knocking at the door, selling pegs, as did the rag-and-bone man with his donkey cart.

Built in the late twenties or early thirties, the spacious houses reflected architectural styles and taste peculiar to that period in the rural North West of England. Most had bay or bow windows at the front, surmounted by gabled roofs of tile or slate. All had chimneys to serve fireplaces in both living rooms and upstairs bedrooms. The Weinel home was no exception.

Unlike many latterly built, semi-detached dwellings, the front door, hallway and staircase of the house were next to the central dividing wall, separating their home from that of their neighbour's. The staircase climbed from a dismally dark hallway towards the rear of the house, turning once at the top to reach a small landing with doors to a bathroom and three bedrooms on the upper floor. Three doors off the hallway downstairs served a small kitchen and two large rooms.

A little-used front room or parlour, with elegant bow windows, faced the front garden and road beyond while a living room at the back of the house, next to the kitchen, overlooked the spacious back garden through angular, square, bay windows. Gummed, brown-paper strips, stuck diagonally to the inside surfaces of all the panes to prevent, or at least reduce, splintering under explosive concussion, formed neat diamond patterns across the glass. It was wartime.

The family used the back room mostly because it was brighter in the summer months and warmed in winter by an enormous wrought iron, enamelled fireplace; resplendent with its oven, hob and back-boiler. The boys' mother, Edith, spent hours each week with black lead and brushes to keep the metal gleaming. A brass artillery shell-case from the Great War, containing a fire-poker, brush and tongs, stood to one side of the hearth, together with a copper coalscuttle. The ritual of fireplace cleaning and fire setting was a daily chore in the cold months of winter: essential for warmth in the house and for hot water at the taps. How well Edward remembered the sound of the little hand-shovel his father used, to clean out the cinders. The scraping and tapping, amplified through the chimney, sounded clearly at a small fireplace in the bedroom above, where the two brothers slept.

"Yut-soot, Yut-soot-soot, Yut-soot-soot," it went; a comforting sound, Edward's verbal imitation of which, often brought amused smiles to the faces of his parents, once they'd discovered the answer to their son's perplexed question. "What is it that goes Yut-soot-soot, Yut-soot-soot, in the chimney every morning?"

A very mixed variety of furnishings filled the living room below, providing all the basic comforts of home. Many pieces had belonged to the boys' grandparents. Edward loved the room dearly. An aging, brown, patterned carpet covered most of the wooden floor, but where it didn't reach, sheets of beige linoleum hid the boards from sight. Several oak dining-chairs and a matching, round-topped, gate-legged table stood against one wall, whilst a couple of easy chairs and an enormous sofa or settee surrounded a well worn, but still fluffy, hearth-rug of blue, black and white in front of the fire-place. The settee's thick, brown, velvet cushions made it a comfortable refuge close to the warmth of the hearth, although the rounded heads of the bright, brass pins securing the brown hide of its shiny upholstery at its foot always seemed to catch unwary anklebones in passing.

In that generously large room, many objects of adornment decorated the walls, windowsills and shelves to gaze down upon a young family thrown again into the constraints of wartime. Reminders of the Boer Wars, two large brass dishes from South African bazaars hung from picture-rails on either side of the fireplace. They nestled with polished brightness in the alcoves formed by the chimney breast while 'Old Nick's' blemished, oval mirror above the mantle shelf, gave reflected depth to the family's most-used room.

The boys sometimes stood upon chairs to gaze into the depths of the mirror's back-to-front image of their living room; or just to make faces at their own reflections, capering in the silvered glass. It was a pass-time for which their mother scolded them, more than once, though not unkindly, telling them that one day they'd see 'Old Nick' looking over their shoulders if they persisted in grimacing at their own images for too long. Edward didn't really believe his mother, but he and Freddy gave up the narcissistic practice, just to be on the safe side. Old Nick wasn't someone they really wanted to catch sight of. Edward often wondered who Old Nick was. Perhaps he was the bogeyman who hid beneath the beds of naughty boys.

Many small, brass ornaments lined the mantle-shelf above the fireplace: several candle-sticks, a tiny windmill, an eighteenth century fisherman standing with his rods, a small bell in the form of a crinoline-lady, perhaps once used by a titled lady to summon her maid or bid entrance to callers. The bell stood in modest silence then, between the fisherman and a smiling Buddha wrought in brass and seated upon a chair of carved ebony. These, along with a copper kettle (one of the very first electric ones) and an equally handsome copper hot-water jug, all held cherished memories for Edith and Fred. Memories of their own childhood and of their own parents long since taken or passed away ...grandfathers who had died in the service of their country, far away, in inglorious conflict, and grandmothers who hadn't lived long enough, afterwards, to see their children married or to know their children's children.

'Wireless' played a vital part in family life during those long, wartime years of the nineteen-forties, bringing news along with humour and music to cheer the people on the home-front. To improve reception of his big *Cossor* wireless, the boys' father had strung an aerial that drooped in a sweeping curve from their house to a tethering ring, mounted on a tall flagpole at the farther end of the garden.

The wireless-set, 'Dad's pride and joy', stood upon on a square, *Chippendale* gaming-table in one corner of the room, next to the bay windows where blackout curtains hung behind floral drapes that Edith had made on her Singer sewing-machine. The living room was snug and comfortable. It contained all that the family needed to meet the daily routines. It was bright, cheerful and functional. For Edward, it was a room filled with happiness and joy.

The family rarely used the front room except at Christmas-time, or on special occasions like birthday parties, or if relatives came to stay. Well-furnished though it was, its door remained closed for the most part, for it was such a cold, dank room, with rising-damp in the walls. Sometimes, in the summer months, the boys used it as a playroom, but the sun rarely cast its beams through the bow windows shaded by leafy foliage and the fencing of the garden, in front.

Every year, several days before Christmas, their father would light a fire in the elegant, pink-tiled, arched fireplace, to give the room a good airing to dry out the damp in the North-facing wall.

Thus prepared, everyone participated in decorating – bedecking the lonely parlour with holly from the garden and cheerful paper-chains, they'd made from coloured crepe-paper, ready for their Yule-tide celebrations.

What wonderful times they had, too. The family seated together for Christmas lunch at the big, dark-wood table, set with fresh white linen and the best silver, and again for afternoon tea with delicate sandwiches, jellies and rich Christmas cake. Later, they'd have mince pies around the fire, and play silly games before bedtime. Edward felt that they honoured the sad room they so little used. Perhaps they made it happy. He hoped they had. But it was an ironical quirk of war-time local administration that later in the war, long after the air-raids had stopped, the authorities finally delivered a heavy, steel-topped, cage-like, indoor air-raid shelter.

It took up some of the front-room space next to the large dining table, but it served no useful purpose, except to provide an interesting play device for the boys, on rainy days in summer. So long as the shelter remained, the room saw no more use for their Christmas celebrations; a saddening thought for Edward for he knew the room needed people, laughter, life and vitality. It wasn't until after the war, when men came to dismantle and remove the shelter, that Maronda's front room could smile again at occasional human occupation.

Edith's younger sister and her husband arrived, taking a week's holiday away from London one summer, during those early wartime years. The front room became a guest's bedroom, aired and lovingly prepared for the visitors. To the boys, it was an exciting occurrence, introducing them to Aunty Biddy and Uncle Chris, and providing an opportunity for them to put on a hastily prepared theatrical show. Edward sang a song and Freddy recited a poem. As a finale, the boys sat back to back on the stage they had constructed from an old mattress set between two deck chairs. Inspiration came from the war, the battle of Britain, *Spitfire*s and aerial conflict that dominated the radio news.

They flew imaginary planes against the 'Jerries'; imitating the hum of engines and the rat-tat-tatting of guns as they chased their invisible enemies in blue skies, high above the English countryside. The boys leaned this way and that as they hauled their machines through tight turns, steep dives and climbs, calling out to each other on an imaginary RT. Their audience appreciated this last act, more especially Uncle Chris because he had served with Royal Air Force attached to the British Expeditionary Force, when Hitler attacked France. The visitors stayed only for a week, before returning to their home in Brixton.

The Weinels′ brick-built house stood well back from the road; its front and side facades pebble-dashed with flint stone chippings reached from the eaves down to a line level with the ground floor window sills. Edward and his elder brother discovered the wonders of flint in later years, making sparks fly with the pebbles they prised from the walls.

Curved bow windows, rising from behind three holly bushes that dominated the fenced front-garden and capped with a triangular gable, added a certain elegance to the front of the house whilst, at the back, angular, bay windows roofed with lead, jutted from the living room to provide a complete view of the ample garden and open spaces of farm-land beyond.

High, wood-plank fencing enclosed the large well-developed garden at the back, where a broad expanse of lawn, flanked with black currant bushes and rose-trellises that hid outhouses beyond a long garage, gave the family a very private garden to enjoy. It was altogether a comfortable place: a quiet, middle class dwelling that the Weinels called home for some fifteen years. Certainly, they weren't well off, quite the contrary, but they were content in those pleasant surroundings.

Higher Road began at Hunt's Cross, on the edge of Speke, at the boundary separating Liverpool Corporation Authority from that of the Lancashire County Council's. The postal address therefore reflected Halewood, Near Liverpool, as the district. The main road connecting Speke and South Liverpool with the industrial Northwest of England, carried traffic to and from Warrington, Widnes, Runcorn and Manchester. In those times, the highway extended through the farmlands of Halewood and Hale, passing villages and countryside along its meandering way eastwards.

From an early age Edward knew that he and his family were strangers in the region, indeed they were almost foreigners, having moved up to Liverpool from the South in 1940, with his father's posting. His brother's birthplace was Surrey. Edward was born four years later at Hornchurch, but he had no recollections of Essex nor of life in the South, being just two years old when the family moved. Freddy was six and started his schooling almost immediately after arrival in the Halewood district.

Perhaps because of their Southern accent, acquired naturally from their parents, local children often asked the brother's about their family's origins. The boys always answered by telling their inquisitors that they came from London. This immediately branded them as evacuees. Apparently, during that early part of the war, many Northern children thought that anyone who came from the South must fall into the category of being an evacuee, something to be despised in the strange workings of immature minds. It was not true, of course.

The Weinel family moved from the Home Counties because of Fred's wartime job and his posting to an important aircraft factory at Liverpool. Yet, the intimidating accusation stuck. For Edward's part, the hypothesis that he was an evacuated child only served his classroom peers to forge a new name for him at school – supplanting that which they had given him earlier. For a while thereafter, until the cruel novelty wore off at least, his heartless classmates re-named him EvacuWeeWee Weinel.

It vexed him, but he learned to ignore the jibes. Although the name-calling pained him, Edward had developed a protective cloak and outwardly remained calm and casual about it, as he was with most things by then. His new-found nonchalance brought him to note several times with Miss Scott, the Head Mistress at Kingsthorne.

Miss Scott was an upright lady with a severe manner. Her shortness of stature belied a stern, strong-willed and authoritative character. She ruled with iron discipline to match her iron-grey hair, and it was not unknown for her to use the cane against outright disobedience when displayed in upper-junior grades. Her commands demanded instant compliance.

When Edward's mother had occasion to visit the school office one day, Miss Scott revealed concern at what she considered was an over-casual manner on Edward's part. She sought his mother's help. It seemed that when the children of Edward's class had occasion to march past their Head Mistress, after assembly or at play time, for example, or when she met up with him alone, on a solitary errand in the corridors, Miss Scott frequently found it necessary to correct him regarding his comportment.

Her concern centred upon his hands; one or the other, or both, would invariably be in his trouser pockets. Patently, she thought it indecorous. His mother's retort to the notion implied that the pocketed hand was a parody of his father's mannerism, but in spite of that, she sewed up the offending trouser pockets in a vain attempt to foil the habit. It really didn't matter that much to Edward; he simply transferred his hand to a blazer pocket, with his thumb protruding at the front. This was equally comfortable and he felt no less unworried and grown up, as he believed the characteristic displayed.

Apart from that singular criticism, Miss Scott really did have a kindly side. It came through at the reading tests that she personally conducted at the end of each term. As the last named pupil on the class register, Edward was always last in the line of nervous children waiting for the call to her office. When his turn finally came she would bid him enter at his timid tapping upon her office door. She would seat him, and with a kindly smile, ask him how he felt about his progress. She would carefully clean his glasses while he answered, before opening the Beacon reader book from which he should read.

Edward couldn't claim mastery of his early reading ability, he was slow to pick up the art but somehow he always managed to struggle through. Miss Scott was gentle and patient – perhaps understanding the nervousness her children suffered in the awesome presence of their Head Mistress. Their ordeal over, and as a reward for trying hard, she always gave her relieved pupils two sweets from the glass dish, close at hand, on her desk.

On those rare occasions when Edward found himself in Miss Scott's room for reading tests, he discreetly looked about him in a vain attempt to catch a glimpse of the dreaded stick of castigation, but never once did he see the cane. Other boys who had tasted it, told him that she kept it locked away in a cupboard, a secret place, from whence it would emerge only infrequently for application. Whilst Edward did not necessarily agree with caning, he did believe in fundamental discipline at school, as well in the home. His father often spoke of it as an essential ingredient to decent social conduct in later life.

At home, the Weinels taught their sons the importance of good manners, courtesy and respect for others. They encouraged the use of simple words like 'please' and 'thank you' or phrases such as 'May I have' instead of 'I want' or 'give me,' as a matter of common courtesy. Requests were better and more acceptable than demands. After all, civility costs nothing and guarantee's civil or kindly response.

The boys' father always lifted his hat when greeting women, or ladies, in the street. Similarly, he expected his sons to tip the peeks of their school-caps as a sign of respect. Likewise, when they occasioned to pass women on the footpath, courtesy required the brothers to take the outside position, walking next to the road and allowing the lady to pass on the inside. When they were old enough, their Dad explained that they should offer to give up their bus seats for ladies who boarded to find standing-room only. The reward was always a delighted smile and profound thanks. They felt grown up and responsible. In showing respect, they earned respect.

A kiss for their mother, before going to bed and on getting up in the morning, expressed the boys' love, affection and appreciation for the central person in their household. Teaching was unnecessary for this, of course. It came quite naturally, although if the boys forgot in the excitement of some unusual activity, their father sometimes had to remind them. Perhaps they were lucky to have such a loving, caring, patient and protective mother; lucky too, in having a father of similar character, stern though he was at times.

Nevertheless, it wasn't all ‘goody-two-shoes’ in the Weinel household, by any means, and although Edward could count on one hand the number of spankings he'd had, he didn't believe they did him any harm. For sure they put him straight on the subjects of right and wrong.

3

Before the move from the South to Liverpool, the brothers both suffered with Whooping cough. This left Edward with serious eye weakness and astigmatism, for which corrective glasses became an encumbrance for the rest of his waking life. At the time, opticians thought that the more seriously weakened eye might possibly recover its strength by covering the stronger one with a patch. Edward's first eyeglasses, therefore, (starting at the age of two) had one lens blanked over with sticky tape: the pink-coloured stuff, used for bandages or to hold lint in place over cuts and grazes. The therapy was partially successful until the optician died in a motoring accident and another specialist took over the practice.

Perhaps the new fellow lacked experience, for when the time came to use both eyes with a modified prescription, the lenses were mixed up, left for right and vice versa. Edward's complaints about the difficulties he then suffered drew only sympathetic platitudes until his vehement insistence, over several months, occasioned another unscheduled visit to the optician. The specialist immediately identified the problem with some embarrassment but, even with new lenses correctly fitted, Edward had lost much of the previous corrective work to his eyes. As a result, thereafter, he used spectacles permanently.

He never thought that glasses were that much of a restriction however, although he supposed they were constraining in some respects, especially at sport. On the other hand, since normal vision was unknown to him, he felt decidedly 'undressed' and definitely at a disadvantage with his spectacles removed.

For most people, wearing eyeglasses is a compromise with the natural deterioration of eyesight, which they would not normally suffer until much later in life. Edward, however, was content and felt neither rejection nor embarrassment through wearing lenses as an aid to his sight. He tolerated the occasional jibe of 'four-eyes'. It didn't irritate him greatly. Through his framed lenses, he believed that he could probably see better than most people anyway. Four eyes are better than two.

Moreover, in the face of aggressive behaviour, sometimes directed towards him in later life, Edward found some advantage in taking the wind out of a would-be assailant's fighting sails, by simply removing his glasses meaningfully, and placing them out of harm's way. Such action very often succeeded in defusing a close-to-fisticuffs situation. It was, however, a bluff that was never called and Edward sometimes wondered if he could have backed it up without the lenses.

His first encounter with snow, as a wearer of glasses, was a disaster. The soft, cold precipitation had come down thickly and silently overnight and the two boys woke to see a deep, white carpet covering the whole vista outside their bedroom window. The sky was heavy with grey cloud and the snow glistened in purest white upon the ground. Their excitement grew as their father stepped out after breakfast, wading through the hip-high snow to find his shovel in the shed. He cleared the pathways from the kitchen door to the coal shed, throwing the snow to one side, on top of the thick white carpet that covered the lawn. It was like walking in a canyon when the boys took their first tentative steps outside, to investigate the cold, white phenomenon.

Wearing warm, thick clothing and wellington-boots, balaclava helmets, scarves and woollen mittens, they marched delightedly out into a world of Arctic cold. It struck Edward how quiet it was out there. Not a sound carried. The soft whiteness surrounding them, muffled their utterances, deadening and absorbing all sound. They heard no bird-song. Silence reigned in near totality.

Whilst his brother, Freddy, helped to clear a path to the garage doors, Edward borrowed his father's little coal-shovel to experience his first dig in snow. Another impression struck him ...how light, how un-heavy a shovelful of snow was. Tentatively, he threw his first diggings gleefully into the air and seconds later, learned the lesson of what-goes-up, must-come-down. The snow returned in a shovel-shaped lump to deposit itself upon his upturned face, as he watched its aerial progress.

Some of the snow lodged thickly and tightly behind his glasses. It was wet and very cold. He could not see and he stumbled about, temporarily sightless and completely bemused. The result of his experiment frightened him for a moment. Edward shouted fearfully; until his Dad came to the rescue, with unhidden mirth at his son's self-inflicted predicament. With comforting words, he gently removed the spectacles to wipe away the snow from the distraught boy's face and eyes, before polishing the lenses with a handkerchief he'd fished out of a pocket.

Snowballing was fun, but it soon became evident that knitted gloves didn't take kindly to the wetness produced by squeezing handfuls of snow into throw-able missiles. Ears, hands and feet quickly became numb and tingled with the cold, prompting a retreat to take warmth from the fire indoors, and to gaze in continued wonderment at the thick white carpet that reached almost as high as the windowsills.

Edward remembered thinking how unkind such deep snow must be for animals, both wild and domestic. Where did the wild ones go in winter; did cats and dogs have to stay indoors until the snow had melted? When he asked his Mother about his concerns she told him about hibernation, and how many of the wild-life creatures slept in snug places, burrows, hollow trees and warm spaces in the farmer's barn, right through the winter months. As for cats and dogs that lived with people, well, they slept a lot too, though not in a state of hibernation. Her answers allayed Edward's youthful fears and sensitivities for the welfare of creatures in Mother Nature's sometimes-unkind world.

Perhaps his concern for animals made his parents think, for one day in the following spring, when winter's snow and grey skies had long since gone, the boys' Dad arrived home as usual from work, at lunchtime. The brothers raced, as always, to greet him at the side-gate. They noticed that their father had his raincoat buttoned to the top. They thought it strange, since the sun shone warmly and there was little wind. Even stranger was the bulge, so obviously evident beneath the closed lapels.

Fred Weinel hailed his sons with a beaming smile as he propped his bike against the wall of the house, as if nothing was amiss. The boys were desperate to know what it was that he had under his coat and followed him to the open kitchen door, at the back. Fred strode in with a kiss for Edith and his usual brief greeting, as he stooped to take off his cycle-clips. It was then that the bulge moved.

The brothers were beside themselves with unspoken curiosity. What was it that their father had under his coat? Still smiling, he slowly and very carefully unbuttoned the front of his Gabardine. In a moment, the tiny ears and head of a kitten appeared: black and fluffy, its eyes still half closed. Tiny claws plucked at the open lapels as the diminutive creature struggled into the light, as if re-living birth. Fred had decided that it was high time the boys had a pet to look after, perhaps prompted by Edward's clear concern for animals, expressed during the preceding winter months.

"What shall we call him? He's as black as soot."

"What about Dinky?" Edith's suggestion appealed. "Years ago, we used to have a cat just like him and his name was Dinky."

And so it was that Dinky came to them, the first of several cats that lived within their family circle over the years. The older Edward, in retrospect, could not remember his parents' home without one. With a smile at the recollection, he recalled quite distinctly how he told his mother, with a child's instinct and logic, that if Dinky were to venture out into the snow, he would never get lost. His coat was so black it would be easy to see him against the whiteness.

Furthermore, he told her that she shouldn't worry anyway because Dinky would always come home again to sleep, curled up by the fire. For his display of faith in nature and for his words of comfort for what he presumed would concern his mother greatly, Edward earned a mother's cuddle. Such was the closeness of Edith to her sons.

As a family, the Weinels took much pleasure in the regular bus journeys they made into the City of Liverpool. Certainly, the monthly excursions were always popular with the boys. They travelled by means of the only bus service that operated as far out as Halewood in those days. The Crossville bus company's rural terminus lay at New Hut Lane, some five or six bus stops distant from the house. Their only link with the city, apart from infrequent, local passenger trains through the village station, the green double-deckers passed at hourly intervals.

Very conveniently, their home lay midway between two bus stops, both of which were within sight of the front gates. This gave them a distinct advantage if they were late catching the hourly schedule. If the bus was already at the Wood Road bus stop, before passing the house, then they could dash up the road to the next stop in front of the Rigby's cottage, where the bus halted again after passing their house. It didn't matter which bus stop they used since the fare was the same from either. The service operated to the very heart of Liverpool and shared its riverside terminus in the city with Corporation buses and electric tramcars at the Pier Head in front of the famous Liver building with its clock tower and huge Liver birds at the pinnacles. The family usually alighted at the foot of Mount Pleasant, in front of John Lewis's, opposite the Adelphi Hotel.

To Edward, the city's busy streets were very different from the serenity of rural Halewood. It was exciting to be in the City, with rumbling tramcars grinding and clanking upon their metal wheels. They rolled on steel rails set in the square-cobbled paving of broad thoroughfares. Everywhere, tall buildings, towered above the roadways to throw long shadows into treeless avenues, where traffic moved slowly between the pavements, halting in procession at traffic lights and the orange beacons of pedestrian crossings.

It was a different world, with streets full of shop windows reflecting crowds that moved upon flag-stoned pavements: peddlers selling matches and trinkets, newspaper sellers shouting headlines and people hurrying by, unseeing, intent upon their way through the clamour.

At his eye level, as a child walking with his parents through the throngs, Edward saw the city from between legs and wheels. Everything was taller than him in those vast, concrete and stone surroundings, yet he revelled in the bustle.

Spaces between the buildings along the way, ragged gaps where daylight shone brightly, plainly showed where bombs had fallen. Damaged buildings reared like broken teeth between the bare, brick walls of their neighbours. Windows gaped in tumbled walls, scorched and burned timbers lodged at crooked angles and some basements were open to the sky.

Edward often thought how strange it was, to see inside bombed-out apartments and shops where people used to live or work. Possessions were exposed for all to see: a shred of wallpaper flapping in the breeze, a drape of charred curtain, a picture still hanging from its hook at a crazy angle from a jagged, bomb-blasted wall: a bedstead standing where it had for many-a-year upon a few square feet of flooring, jutting out at three floors up, apparently without support. He thought how sad it was for people to lose a home in such a way. Did the people lose their lives too?

The eight-mile bus journey into the city usually took forty-five minutes, terminus to terminus, by way of the Woolton route. The boys' favourite seats were at the front, upstairs. From there they had a clear, unobstructed view of the route and of the interesting places they passed. Edward always felt detached sitting in the front seats of the bus, on the upper deck. He found strange interest in the effect of foreshortening when viewing objects and people on the sidewalk, from so high above their heads: a phenomenon that intrigued him.

The Crossville buses had 'bench' type seats upstairs, seating four people abreast. Access was from a sunken aisle on the right hand side of the deck. Edward liked to sit by the side window on the left, with Freddy next to him. Their mother would sit next to Freddy, leaving the aisle seat for their Dad, where he could pay the conductor and smoke his cigarettes.

On one of these excursions Edward vaguely recalled trying to draw his mother's attention to what he called a ‘Churchill’. His mother recounted the whole incident to Edward in later years. The bus had drawn up to the bus stop, in front of a convent, opposite a large hospital for the war-wounded, on the outskirts of the city, when Edward, watching the line of people below move along to board the bus, made his astounding discovery. “Look Mum! There's a Churchill!” Edward exclaimed, straining his nose against the glass window to look downward to the footpath below.

"A Churchill, whatever does the child mean?" His mother solicited in astonishment.

"There's a Churchill getting on the bus," insisted her youngest son. From their seats, neither Freddy nor parents were able to see the subject of Edward's observation, boarding the bus, downstairs. They shook their heads and shrugged at Edward's utterances and they said no more about his bizarre name for the mysterious sighting.

The forgotten incident re-emerged on another day, when his mother and he were on a shopping trip, on foot, walking into the village of Hunt's Cross. Edward excitedly drew attention to the subject of his notice, passing by on the other side of the road.

"There's a 'Churchill,' Mum." He volunteered, pointing with his finger to emphasise the fact. His mother scolded him for pointing, but laughed as she corrected the child's error.

"No, Teddy, she's a Nun, dear." Then, mystified by her son's assertion, she asked him, "Wherever did you get the word Churchill?" To explain this, it seems that the constant references made to Winston Churchill over the radio, coupled with Edward's five-year-old understanding that Nuns were people of the church, his childish logic invented a name for these silent women with covered heads and billowing gowns. To Edward, 'Churchill' seemed like a suitable appellation and until corrected, he didn't know the word Nun. The older Edward often wondered what the famous man himself would have said, to hear his name applied in such a way.

As a child, it wasn't normal for Edward to express anger. Some might say that he was a placid youngster, compliant and easy to please. However, there were several occasions in his younger life when his temper flared and he had to give battle. The earliest episode revolved around a toy. His elder brother teased him over ownership and held the toy at arm's length high above his head, well out of Edward's reach. All his pleading made no headway. They wrestled silently for a while until Edward began to see red.

His elder brother was taller, and stronger. Edward's anger grew until, finally, his sufferance snapped and he buried his teeth in the soft flesh of Freddy's chest. His brother certainly shouted and Edward did get his toy back, but such an ugly quarrel that ended only after the drawing of blood, failed to impress their Dad. They both got a telling off after the dust had settled and treatment applied to the wound.

For the most part, the brothers got on well together. They were very close in reality. They shared the large bedroom at the back of the house in Higher Road. It was a fine room, facing approximately southward. It was furnished with two single beds, a couple of chairs and a chest of drawers for their clothes. An airing-cupboard for linen stood in one corner, between the three-paned casement window and a small, wrought-iron fireplace set in the Western wall.

Edith kept her sewing machine in that bedroom too, set beneath the window because it was a bright and spacious room, ideal for her sewing activities while the lads were at school. The window overlooked the back garden, the farmland and, with its broad vista, wide-open spaces and countryside beyond. That sunny room was the venue of childhood creativity. It was the laboratory of youth and the birthplace for so many of the boys' ideas and plans.

4

At weekends, it was always their Dad who woke the boys from their night's sleep. He would swish the blackout curtains back to let in the morning light, pulling them wide as he loudly whistled a tune that never failed to bring the two brothers, blinking and stretching to the surface. The tune was either a rendering of 'This Old Man' or that silly, Laurel and Hardy signature that reminded Edward of a man walking with two left feet. Whichever one, it was always off-key, for their father was tone deaf and music wasn't really an attribute he aspired to, but it was a good way to get the boys out of bed.

Sometimes, if they were reluctant to rise, he would follow up his curtain-swishing and whistling routine with a comic parody of a Sergeant Major's voice. "Come on you lot. Wakey-wakey, rise and shine. The sun's scorchin' your bloomin' eyes out, are you going to lie there all day, you lazy lot: ups-a-daisy lads."

The boys would tumble from their beds, urged on, no doubt, by the smell of cooking drifting up the stairs along with the sound of clattering plates in the kitchen. They'd scamper down to the living room with kisses for their mother, to breakfast on scrambled eggs and a rasher of bacon – a pleasant Saturday-morning-change from the usual porridge or corn-flakes served on week-days. The weekend eggs were the powered, dehydrated variety that came in waxed, boxes, displaying an American flag; real, fresh eggs were few-and-far-between.

On such days, with no school for the boys and their father with a whole weekend off, it was quite usual for the three of them to take a Saturday-morning-shopping walk. Over breakfast, they would discuss the route for their outing and by nine o'clock they'd be on their way to the village of Hunt's Cross, or sometimes, in the other direction to buy vegetables at the glazed nursery on the corner of Old Hut Lane, opposite Mrs. Bonner's cafeteria.

At the end of each month, there were bills to pay and a visit to the bank was a necessary chore. The boys' father used a branch of Barclays some way out of the village, on the road to the industrial area of Speke, and not far from where he worked as an A.I.D aircraft examiner at *Rootes* 'Shadow' Aircraft factory where *Bristol Blenheim* and later, *Handley-Page Halifax* bombers were built.

The factory was close to the aerodrome, from which the finished aircraft flew, for delivery to waiting R.A.F bomber squadrons. The boys often saw the big planes, engines roaring in the climb after takeoff, their out-bound courses generally taking them over the farmland of Halewood, behind the Weinel home. Such sightings and sounds were strong reminders of the war; it was never very far away, but at least when the boys saw those thundering aeroplanes, they knew by the markings that the bombers were friendly.

The brothers would shout and wave as the planes climbed for altitude, but from that height, of course, the crews probably couldn't see the prancing, waving children far below but, to the boys, it was the sentiment that counted. Frequently too, they saw fighter-aircraft zooming noisily at low level across the fields or passing high overhead, almost invisible to the naked eye, their existence marked by a flash of reflected sunlight or by their droning engines, the sound trickling down to the land from lonely patrols in the wide skies. Even from quite an early age, the boys could recognise *Spitfires, Hurricanes,* American *Thunderbolts* and the like, flying from Speke, or from one of the many other satellite airfields in the region.

It was, of course, quite natural that the boys should have a leaning towards aviation, because of their father's RAF background and his wartime job in aircraft production at *Rootes*. Freddy probably had more interest than Edward at the time, for he was old enough to remember the pre-war flying-displays at Biggin Hill and Hornchurch, when their Dad served with the Royal Air Force.

Whilst aeroplanes intrigued both boys, little did the younger Edward realise that his own future would revolve about a life in the aviation industry, and a career that would one day propel him abroad, in due course, to live and work in foreign places. Until then however, his childhood ambitions related only to a small world of play, learning, youthful spare-time activities, birthdays and Christmases among family at home and Saturday morning walks to the village.

On such occasions, Fred Weinel and his two sons would set out before nine. The boys loved to be out in the fresh air and sunshine. They'd hop and skip, side-by-side with their father as he strode briskly, one hand buried in a trouser pocket, a broad smile on his face and his brow shaded beneath the brim of his best Trilby hat.

From their home, it took only fifteen minutes to reach the small parade of shops at the edge of the village. There, partially hidden by bushes and shrubs growing in two small plots contained by low wooden fences, the shops stood back from the road with a paved area in front. There were three, an electrical repairer's shop, a grocer's and Miss Duncan's sub-post office & tuck shop at the end of the row.

The boys often posted their parents' letters at the familiar, red-painted pillar box set next to a pair of telephone kiosks and a bus stop, where buses halted to pick up people bound for the city. Across the road, Mrs. Morris's Dairy-shop lay behind a thick, broad-leafed hedgerow on the corner of Macket's Lane. The Post Office and the lane marked the official City Boundary at the edge of Liverpool's nineteenth district.

At that point, no more than half-a-mile from the village of Hunt's Cross, the road widened into dual carriageways, but because of the war that section remained unfinished, leaving only one side completed with its concrete road-surface, cycle track and flag-paved footpath. The other side, beyond the central divider, old and uneven tarmacadam still paved the surface, following the original course of the once narrow road out of the village. Farther along, the road traced the edge of a broad, grassy, open space in front of Kingsthorne Road Primary school, where the boys attended on weekdays. The boys frequently used well-worn tracks across one corner of the field, as a short cut to reach the gates of their school.

It had been a sports-field until the war came, and the authorities fenced off most of it with wooden stakes supporting square-mesh and barbed wire. The compound thus formed served as a storage ground for utility war materials, hidden under camouflage nets. Six-wheeled trucks bearing the white star of the American army were frequent visitors there, loading or off-loading cargoes of tentage, Jerry cans or huge wooden crates containing Lord-knows-what, as soldiers armed with rifles or Tommy-guns patrolled the perimeters. The two youngsters questioned everything they saw and somehow their Dad always managed to find the answers to his sons' inquisitive chatter.

The focal point of Hunt's Cross was the ‘circle’ (their mother's name for the big roundabout) at the centre of the village, where two major roads crossed. In earlier times, one main road from Liverpool's outer districts came in from Woolton, to pass through on its route towards the village of Speke and thence to Hale. Another came from Garston and Allerton, crossing the village, bound for Halewood and Warrington.

Later, Liverpool's pre-war development plans of the nineteen-thirties extended the city boundaries to absorb Hunt's Cross, bringing it within the district of Speke. Modern tramways and dual carriageway roads had appeared, driven through the county-side to the South of Woolton to join with Speke, even then designated for industrial development in the future.

If, at the end of a month on their Saturday excursion, it was their intention to visit the bank, they would turn left at the circle and head down Woodend Avenue towards Speke. Passing the estates of housing, built before the war to accommodate factory workers and their families, the road was very wide, having two lanes flanked by service roads and broad, flagged footpaths. A central reservation separating the two traffic lanes should have carried electric tramcars to and from the industrial zone, but there were no rails because of the redirection of all metals into the war effort, for munitions. Work to finish the tramway system ceased in 1939 and the years of austerity, after the war, prevented completion of those wonderful ideas to connect the outskirts and industrial areas with the city and its famous seaport. Sadly, post war re-development took a different route, when the notion of extending the high-speed-tram system gave way to cheaper diesel fuel and buses.

Edward could remember one weekend in particular, if only for the clarity of its image in his recollections. Perhaps it was the weather, so fine and clear, or perhaps a fresh awakening to his environment at that point in his early life, which prompted retention of its memory. They were on their way to the bank. Fred Weinel and his sons had made good time out of the village, reaching the first of the factories, only ten-minutes on from the circle at Hunt's Cross.

As they walked, Fred spoke of recent air raids on Liverpool that had left Speke untouched for once. The city itself, however, and its enormous dock area had taken a pounding. Invariably such enemy activity occurred at night but daylight raids were an ever-present possibility. For this reason the walls and roofs of factories and office buildings, bore camouflage paintwork of broad, green and brown undulating shapes, to reduce their otherwise conspicuous visibility from the air. This observation was very clear in Edward's mind's-eye. Here and there, sandbagged ARP posts were plain to see, prompting from the boys, more youthful questions about the air raids.

Like many other civilians in the district, the boys' father had volunteered as an ARP Warden, with duties at night whenever there was an air-raid on, making sure no lights were showing in the locality, and helping with fire-fighting and rescue, should the need arise. He had a gas mask in a khaki bag and a stirrup pump, with what seemed a patently short length of narrow bore hosepipe attached to it, a bucket and a tin hat with the letters ARP stencilled on the front. Everyone carried gas masks. The boys kept theirs in square cardboard boxes with a cord to sling over their shoulders.

Only light traffic moved on the wide dual carriageways. Cyclists swooped along on the service roads and people on foot hurried about their business, or stopped to chat with neighbours as they passed on the side-walks.

The boys had grown impatient for more interesting things to see; fences, factory walls and streets of houses had become boring. Sea gulls wheeled and circled high above, calling and crying plaintively, drawing the attention of the two brothers who, as always, were alert for the unusual. This was unusual, and they stopped for a moment to observe the aerial behaviour of those coastal monarchs. Their Dad said that the gulls, circling in-land that way, were a sure sign of a change in the weather. Rain could be on the way.

They moved on as the distant sound of running aircraft engines reached their ears. They could hear it over the noise of traffic on the roads and bird-song in the trees that grew on fenced-off waste ground alongside the road. More questions – and the boys' father reminded them that they'd be able to see the airfield from the top of the bridge, ahead of them.

The road narrowed to a single carriageway as the gently curving gradient of a brick-built bridge rose before them; gently arching the roadway to carry traffic over four main line railway tracks that served Liverpool. North, East and South they went, connecting this major port city with all parts of England. The trio topped the bridge and halted to watch a goods-train pass; with all its fuss and bother, steam and smoke, rattling and thundering as hundreds of tons of war materials and freight went speeding inland. Meanwhile, over the rumble of trucks passing beneath their feet, the far-away sound of aircraft engines rose to a crescendo, held for a moment, and fell back again.

Perhaps their father knew by the sound that it would happen, and if so, he couldn't have timed their arrival at the crest of the bridge any better for, as they watched the train recede and finally disappear around a distant bend, an aeroplane took off from the airfield not far away: a *Halifax* on air test. They watched its progress from their vantage point on the rise of the bridge, the four engines roaring as it lifted from the aerodrome's runway.

The plane droned into the sunny morning air, wheels tucking up and engines throttling back to climbing speed as it passed almost above the man and two children who gazed upwards from the pavement of the bridge. Its markings were plain to see, as were the silhouettes of the pilots in its glazed cockpit that glinted briefly in the strong, morning sunlight.

Shading their eyes against the sun's glare, they watched the plane until it reached cruising height and levelled off, turning gently to fly eastwards. The undulating, airborne sound of the aero-engines faded, as sight of its shape and form diminished into the distant haze. Something awoke in Edward's mind, he couldn't put his finger on what it was but the passing of that aircraft, flying so close, left a mark upon his memory. It was like a bookmark, saving a page to be re-read many times over.

It was an exciting diversion, but with no more trains or planes to watch, the three turned back off the bridge and walked the short distance to the bank's entrance on the corner of Milner's factory. This was the first of several factories lining a side road, at the beginning of Speke's industrial zone. Milner's specialised in manufacturing safes and strong boxes before the war, but had turned over to munitions work for the duration. The factory built tanks for the nation's hard pressed, fighting armies. It occurred to Edward's young mind that the transition from making safes to tanks wasn't such a big step to take: both being designed to protect the contents. Tanks were merely mobile safes, though not for keeping money in.

Beyond the perimeter fence, and through a gap in the factory's sliding doors, they could see long lines of military tanks, in all stages of construction. The boys' father spoke again of the war, telling his sons that the tanks would go to the front when the time came. Perhaps they'd go to Montgomery's campaigns in North Africa or maybe to the European continent itself, if the allies could ever get a foothold in France or Italy: against Jerry. "But they should keep the doors closed," he said, "anybody could see what they're making in there."

To Edward it was all academic, for at his age, he had no real idea of what the war was all about anyway. All he knew was that 'Jerry' was the enemy who sometimes came over at night, to drop bombs. He and his brother sat quietly in the bank's tiny front office while their Dad went to the counter. In whispers, the boys discussed the tanks and wondered if the war would ever finish. Would it go on long enough for them to join up? If it did, perhaps they'd get to drive one of the tanks they had seen just a moment or two ago. Freddy didn't fancy that – he preferred aeroplanes.

Their Dad took them back to the village after his bank business. He wanted to buy cigarettes and pay for his newspapers, his sons' comics, and their mother's weekly magazine at Davis's paper shop. Passing the big roundabout at Hunt's Cross again, they took the Woolton road, leading to the village station and beyond, where they could turn right into Stuart Avenue, just after the Church of St. Hilda's. This brought them out directly in front of their objective, the parade of shops in Macket's Lane.

The paper shop was last on the row of shops lining the farthest side of the lane, next to another railway cutting. The boys knew the lane well since the Weinel family shopped there frequently, and they sometimes had their hair cut in O'Rouke's barbershop, halfway along.

The tall, thin, bespectacled barber had a cupboard at floor-level in his shop; just in sight of the pump-up chair. Hidden inside, there was a Teddy bear or some-such toy. Mr. O'Rouke would spring the cupboard open with his foot when least expected, startling his younger customers with a loud squeak through his lips, as he snipped away at a standard 'short-back-and-sides.' He was always full of silly jokes and pranks to play on his youthful clients. Edward did not like him very much.

Edith didn't like the man much either – not much anyway since the last time she'd taken the boys to O'Rouke's. After depositing her sons at the barber's shop, the Edith had slipped off to do some shopping in the lane, leaving Edward in the barber's chair and Freddy seated with the other customers, awaiting his turn. By the time she got back, Freddy and Edward had just changed places. Mr. O'Rouke's over-jocular manner soon turned to defensive embarrassment as Edith discovered that he'd cut off a distinctive forelock of curling hair, which, until a few moments before, had graced Edward's brow. The barber's short-back-and-sides had extended to the top and front, too.

She scolded him roundly, for she really loved that curly lock, and it never grew again in the same way. The incident didn't distress Edward very much. He had no liking for the curl himself anyway; it had been the subject of ridicule at school, from time to time. However, perhaps its removal was the beginning of the end for his hair because, over the years, he slowly lost the rest.

The Weinels often visited McAllister's, the chandler's shop next to O'Rouke's, to buy household things and soap. Carbolic seemed to be the only soap available in those far off days, but Edith always made a point of asking, hopefully, if they had perfumed soap. She always got the same answer.

"Gerr-off Missis. Perfumed soap? Toilet-soap? The's a war-on y'know. Sorry luv, on'y got Carbolic." The chandler would announce in a broad Liverpool accent, with a shake of his head. It was a mirthless phrase but often humorously imitated by Edith, in a moment of merriment. She thought it was hilarious. "On'y got Kaabolik!"

Sometimes, they bought vegetables and fruit at Hollingworth's, the greengrocer's shop, from what limited selection there was during those difficult times. The shop was a gloomy place inside and smelt of earth, old potatoes, turnips and cabbage. Its dismal atmosphere reminded Edward of a wicked wizard's cave and he preferred to stay by the doorway when the green-smocked, hand-wringing wizard himself emerged from the dark interior, to serve his waiting customers. Edward often wondered where the wizard kept his pointed hat, with stars and moons and whizz-bang signs all over it; for he never saw the man wearing it.

Apart from a chemist's, with its large, bulbous leech-jars filled with coloured water and set high on shelves in the window, the remainder of the shops in the lane amounted to a butcher's, a baker's, and finally a rather posh gentleman's outfitters at the farthest end. The only shop missing was a candlestick maker, although to make up for that deficiency, the mysteries of Mr. Caplin's dental surgery lurked behind a brass plaque, on a red door, set between two of the shops in the row. Edward's brother was a regular visitor to Mr. Caplin's torture chamber.

The row of shops occupied two buildings set in a single line and separated by an unpaved passage or alleyway leading to gardens and garages at the back. Most of the shop owners lived above their business premises in first-floor flats. Their front doors were set next to the shop-front windows, at pavement level, just like Mr. Caplin's surgery door. A flagged footpath along the front of the shops almost needed a stepladder to mount, so high were the kerbstones. It was just as well really, since the lane's unfinished roadway, being much lower and full of potholes, frequently flooded with deep puddles when it rained.

At the beginning of the parade, next to the paper shop, the railway ran in a cutting below street level, at right angles to the lane. The lane crossed by way of a stone built bridge, carrying a cobbled roadway over the lines. Cyclists often used the bridge's sidewalks of crushed ashes because the footpaths were smoother than the cobbles of the road. Pedestrians therefore preferred to walk on the road itself, as a favour to the two-wheeled traffic. It was safe enough, since few people could run cars because of fuel shortages and rationing.

On that day, as they always did, the trio stopped on the little bridge for a moment, to see if there were any trains in sight. The boys reached to look over the stonework: Freddy on tiptoe and their father lifting his younger son under the armpits. Edward was too short to see over the top, even on tiptoe. From there, they could then see up the line towards the station they had passed earlier, with its ticket-office at road level, and the covered access staircases down to the platforms in the cutting, next to the road-bridge built of red bricks and steelwork.

On other occasions, at the Macket's Lane bridge, they'd clung to the parapet stone-work excitedly, caught in billowing clouds of smoke and steam that enveloped them like thick, warm fog as trains passed beneath. It was exciting, but Edward discovered that smoky steam-clouds left his glasses severely impaired with sooty grime after the billows had cleared. That day there were no trains and they moved off chattering about the difference between engine drivers and firemen.

Edward observed that railways seemed to surround the village on all sides. Train driving therefore seemed like a logical step for a young lad's ambitions, when he was old enough. Fred had often drawn the boys' attention to the shunting-engines that plied upon the raised sidings along the embankments of Halewood. The steam locomotives, with saddle or pannier water-tanks, big round boiler-heads and wheels joined by connecting rods, worked there frequently. “Listen to what the engines are saying, boys,” their father would say, as they watched the small engines struggling to pull enormous trains of coal-trucks that seemed much too long for them to manage.

As smoke belched forcefully from the funnels and steam spewed in white plumes, alternately from the wheel-driving cylinders, the boy's father would speak in time with the engine's hissing and chuffing. “I think I can, I think I can, I think I can.” Then, with the speed increasing slowly until, the engine would seem to say, “I thought I could, I thought I could, I thought I could,” before blowing a triumphant shriek upon the whistle: such dear memories.

Macket's Lane continued from the cobble-stoned bridge, to intersect with the main road farther on, emerging opposite Miss Duncan's Post Office. The main road, of course, was the highway upon which the Weinels lived. They had come full circle; halting briefly at Morris's dairy-shop, to pay for the milk that Mrs. Morris's son, Harry, delivered around the district each morning. A simple country lad, Harry always arrived on time, whatever the weather, and never without a smile and a cheery greeting, making his deliveries on an errand-boy's bike that carried two enormous baskets and the dairy's nameplate fixed to the frame.

With their final chore done at the dairy shop, they turned again into Higher Road, to make the last, plodding mile or so homeward, along the broad footpath where the dense foliage of Copper Beeches threw their dark midday shadows. Passing fine houses, gardens and gateways, the doctor's house with its surgery at the side, the boys walked wearily, clinging to their father's hands, until they reached the curve in the road at Boundary Drive. From there they could see their home, half a mile on, beyond Tom Lunt's orchard. Leg-tired and hungry, but elated with their outing, they were ready for lunch.

5

To the two brothers, the war seemed very far away on days like that. Little did they really understand of the dangers, the rationing, the clothing coupons, the savings stamps, the fear of invasion and the ever-present reason for blackout at night. However, one stark reminder of war was the roadblock that stood almost in front of their house.

Built at the beginning of hostilities because of the feared invasion, it closed the road down to single line traffic, forcing vehicles to pass through a chicane. The impediment consisted of two, square, camouflage-painted, concrete pillboxes pierced with several small apertures for weapons use. Zigzag lines of sharply pointed, concrete tank-traps that stood in rows like miniature pyramids on either hand, closed the flanks and footpaths to anything bigger than a bike, on both sides of the road.

Sometimes the Home Guard used this place for practice at repelling the would-be invaders. Men in ill-fitting, khaki uniforms, dashing about with rifles and fixed bayonets, frequently delighted the boys with their antics when the brothers watched the noisy proceedings from an upstairs window. Some rushed here and there with cudgels or pitchforks, to fight off others bent on attacking the post, emerging through thick smoke like demons, accompanied by the loud bangs of thunder-flash-grenades. Snipers with rifles crouched among the dragon-tooth traps, firing blanks, as Corporals and Sergeants bellowed their orders.

The pointed tank traps extended from the road, into front gardens, between houses and on into the fields beyond. Behind Mrs. Paine's house, two doors down from the Weinels, the line of traps intersected a canyon-like ditch running along the rear of the neighbouring gardens. The ditch was both wide and deep and, being designed to impede the approach of invading vehicles, tanks and infantry, it had shear sides, shored-up with timber and planking.

It was always muddy and wet at the bottom – quite a dangerous place for children, as Edward learned to his cost one day. He lost his balance while crossing a plank that the two brothers had earlier placed across the mud, as an aid to traversing it. Edward's misplaced foot and leg sank almost to the knee in the soft, wet, sticky clay. Try as he might he could not pull it out. There was too much suction. He'd heard about quicksand sucking people to their doom, and his fear grew as he struggled to pull his leg from of the glutinous ooze. Freddy went running back to help when he heard the tearful shouts. Never again!

The tank-trap ditch extended along the edge of Lunty's field, but only as far as a neighbour's back garden. The boys discovered an easy way to scramble through a thick privet hedge at the bottom of their own garden. This gave them access into their green world of play. Their beloved fields belonged to Tom Lunt's small farm, but the lads claimed them as theirs. The grassy spaces were dear to them and in some way they did belong to the brothers, for there, when weather permitted, they spent many of their free hours. They played-out grand adventures and they fought and won great battles against imaginary foes.

The brothers mapped the fields and gave names to the dips, hollows, trees and streams. Siskin Marsh and Tri-Marsh were hollows in the field that lay beyond their garden. They were not marshes at all in reality: just damp hollows. They gave one hollow the name of Tri-Marsh. At its lowest point, any observation of the boys' presence was from one direction only: the farmer's orchard. From the other three sides no one could see them. The 'Elephant' tree at the gateway to the next field, they named for its peculiar bark, reminiscent of an elephant's skin. Probably, it was an Elm.

Farther along, in a far corner of the field, grew a small oak tree. It thrived close to a brook in a shallow hollow filled with smooth stones and gravel that might once have been the wider stretch of a faster running stream. The tree's broad branches were in easy reach and its foliage hid their presence when the boys climbed into the leafy limbs. They called it the 'Helicopter' tree. When seated on a main branch amongst the greenery, they were several feet above the stony pit below. They supposed it was rather like being airborne, perhaps in an imaginary helicopter, as the branch swayed under their weight and tilted with the summer breezes.

Some years later, it was in this tree that Edward took a pellet in his posterior, fired from an air-gun. The pellet was one of several that pinged around the boys one day, when a gang of youths stumbled upon their presence in the tree. The gang came from a new council-house estate known as Laxton Road, which then filled the open space in front of Kingsthorne primary school.

Before the pellets flew, Freddy whispered insistently that they should keep strict silence in their perch; and that they should draw their legs up out of sight. In compliance with this, Edward followed his brother's lead, gritting his teeth as the first pellets flew. The gang might not have seen the two boys in the tree, who can say but, certainly, they had sighted the blue material of Edward's shorts. The fellow with the air-rifle took aim and fired.

The impact of the pellet hurt but Edward uttered not a sound. He told his brother later that he didn't mind admitting to being scared out of his wits. Fortunately, a shout from the distant farmyard gate scattered the Laxton Road mob and they retreated at a run, down the track behind the hedgerow, from whence they had come. The farmer saw them off, but he had not seen the two lads in the tree. They breathed freely again.

In the same corner of the field, not far from the Helicopter tree, and well in sight of the Weinels' house, pre-war builders had left a pile of house bricks. The boys used some of these, and several sheets of corrugated iron, to build a low 'den' from where they could secretly spy upon imaginary enemies if they dared approach.

The brothers often observed real people too, farm labourers and ramblers, as they passed by on the farmer's cart track. Unseen and unheard they were. Judicious use of removable bricks in the walls enabled the boys to open small apertures through which to look out in any direction. Overgrowth and branches of a hedgerow, where juicy blackberries grew and birds nested, provided good cover to hide the hut and its entrance opening at the rear.

Beyond the next field, planted often as not with root crops or kale and cabbage, a fence of barbed wire prevented access to an unusual railway bridge known to the boys as 'Table Bridge'. They could see it from their bedroom window, silhouetted on the skyline. Except for their smoke and plumes of steam, which billowed from the high embankments of the cutting, trains passing beneath remained unseen. The name for the bridge derived from an unusual structure built on its span: a concrete, table-like adjunct with four perpendicular legs. Beneath its flat roof, the army had mounted an artillery piece as a protection against enemy assault upon the industrial area, by way of the railway. In later years, the boys discovered, with adventuring children's delight, the secrets of its concrete pillars and its dark bunkers.

After their morning's walk into the village and a light lunch, Edith suggested a picnic excursion to Bluebell Wood. The place was only two or three fields distant, beyond the railway, following the public footpaths. The pathways no longer bore the signs and pointers normally affixed to guide ramblers and country walkers.

As with all road signs, the authorities had removed them for the duration of the war, on the premise that invading armies or foreign spies would have a harder time finding their way without signs.

Recovered from their morning route march and with the picnic-basket packed, the family strolled through the countryside, bathed in warm afternoon sunshine. The footpath started at a stile next to the end house of the row, then went into the fields of Tom Lunt's farm. This was the boys' usual way into the fields to play. The path led on behind a hedgerow to bring them out on the farthest edge of Mr. Lunt's first field. Passing the distant pond with willows all around, where the farmer's ducks and geese bobbed and squabbled, the path continued on to a five-bar gate where the elephant-tree grew, its roots close to the brook running to one side of a pasture that the boys knew as the 'cow field'.

The way through was muddy and usually trampled by the farmer's cattle, which often waited there at milking time but, that day, the herd grazed another pasture. The way was clear. They climbed another wooden stile next to the gate and moved on into the broad field, following the long, age-hewn cart track to Old Hut Lane. They skirted another farmyard, owned by one of Tom Lunt's brothers, crossed a narrow, stone bridge over the railway, and went on through the countryside towards Hale and Ogglet, where their objective lay, on the river side of Halewood.

Named for its abundant carpet of blue flowers beneath the trees, the small wood was a favourite place to picnic. The war seemed very distant with the rustle of leaves in gentle air and bird song all about. How tasty, too, were the sandwiches, rock cakes and tea.

Afterwards, they played a ball game with an improvised bat for French cricket.

"Don't let the ball touch your legs!"

"Catch it if you can Teddy."

"There, you're out Freddy, it's Mum's turn to bat."

It was a silly game but they had fun. Everyone participated and the little woodland place echoed with their laughter. It was a family together and happy. After the exertions of their game, they lay in the warmth, upon the sweet-smelling grasses; a contented family, relaxing and chatting until the lowering sun in a clear evening sky cast lengthening shadows, prompted them to take a final cup of tea before moving homeward again in the twilight, in time for an evening's entertainment with the wireless.

In those times, wireless was an important feature of most households. Not only was it a link with the war news, but Tommy Handley, Arthur Askey, Kenneth Horne with Richard Murdoch and Sam Costa entertained with their wartime-related humour and situation comedy. The boys listened daily, with ears glued to the big *Cossor* radio set, to hear 'Children's Hour' with Uncle Mac. This was a favourite after-school activity at tea time. They thrived upon the epics of Norman and Henry Bones, basing many ideas for play upon the adventures of the radio's young, intrepid detectives. With less interest, they listened to hear the BBC Home Service for news of the war's progress but with little understanding of its importance.

There were times, too, when one of their parents would read stories to them from books given for Christmas or birthdays. On those evenings when their Dad had time, they'd ask him if he would read a story. He never refused, and the boys would sit upon the arms of his big easy chair, their arms around his shoulders, as they listened intently to his voice.

He was a good story-reader and he shared in the boys' excitement as the stories unfolded. It was always such a wonderful way to end the day, before bedtime.

Sundays were also pleasant and happy days; days to help in the garden, for in those austere times of rationing most people who were able, set aside a section of their gardens to cultivate fruits, salad and vegetables. Cabbage, cauliflower and potatoes grew well in the good soil at the farthest end of their garden. The boys had a small patch where they planted carrots and radish. Under direction, they used soot from the chimney, sprinkled liberally and dug into the soil, to help the growth along. Lettuce needed glass, so their Dad dealt with that. Mint, chives and parsley thrived and shallots grew thickly next to the compost heap behind the coal shed. Rhubarb growing in a shady corner and several blackcurrant bushes produced abundantly for tarts, pies and preserves.

On Sunday afternoons the brothers attended Sunday school at the little red-brick church of Saint Hilda's, in the village. They enjoyed walking to and from the church, for it provided an opportunity for them to discuss their projects and plans to fill their spare-time. Such conversations were extensions of their imaginations and developed their ingenuity. How they talked, how grand their ambitions were: to be soldiers, to be airmen, to be brave in the face of greater odds. The King, St. George and The Flag! Down with Hitler and Goering: down with Mussolini and down with Tojo! Propaganda had done its job. The boys' sense of loyalty to their country at war, was clear.

When the Americans came into the fighting, they occupied a number of bases in Britain. One such was at Burtonwood, not far from Liverpool. The American service men passed frequently in their huge, six-wheeler-trucks, convoying war materials and soldiers to and from Liverpool's seaport. Strange though it was, the brothers had remarkably little time or respect for those smartly uniformed cousins from the other side of the Atlantic, in spite of their necessary presence in Britain. Perhaps it was the patronising way the soldiers seemed to treat the local population. Sweets and spearmint chewing-gum, thrown from the windows of their passing trucks by grinning soldiers, often lay where they fell.

Conversely, when bus-loads of Italian prisoners came into the district, to dig drainage ditches in Mr. Lunt's fields, there was an empathy with them. It seemed to Edward that these were poor souls in a foreign land, unwilling soldiers perhaps, and prisoners now, to a conquering army, far away from their own families. Yet they were hearty men who smiled readily and seemed so friendly. They had nothing to give but their labour and good humour even in defeat and capture.

The Summer holidays always seemed so long and sunny. The weeks and months at school, during term, passed quickly enough but Edward's thoughts were always afloat with ideas – for after-school and weekend activities, and for how to fill the summer holidays. Freddy was the leader. Indeed, he was Edward's mentor, his protector and, most of all, his friend. Other friends the boys had, of course, classmates, sons and daughters of neighbours and families living in the district but to Edward's young mind, none was equal to his brother. Kind and considerate, Freddy always looked after his younger brother, with a care that Edward never forgot.

Freddy was four years ahead at school. He was invariably well to the top of class in test results yet he was never pretentious about his abilities. He strove constantly for the best, wrestling with difficulties until he won, just like their Dad in that respect: stubborn in his pursuit of one hundred percent.

Edward remembered his brother's concern for him, soon after moving up at school, from the infants to the juniors. The juniors used a different playground and the appearance of new faces amongst the older children generated interest. A bully, who specialised in the intimidation of newly up-graded pupils, brought Edward tears one day. Freddy dealt with the ruffian behind the air raid shelters, where he persuaded the boy not to do it again.

The brothers often played with friends after school and at weekends but always, it seemed to Edward, on terms weighted in favour of the Weinel boys. Freddy enjoyed organising things and leading. Edward always felt that the others wanted a leader anyway. Besides, his brother was older than the rest, so it seemed logical that he should be in charge of things.

Although Freddy was of placid nature for the most part, he did sometimes get into the occasional scrap with other kids, on a defensive basis. One such occasion came about when a lad of Freddy's age insisted on disrupting the proceedings of a game based upon two groups playing at soldiers. An unsavoury boy, Gordon, wasn't a person with whom they normally fraternised. He was too aggressive and had a bad reputation in the district. In questioning Freddy's leadership, an argument ensued. Gordon became angry and lashed out, thinking, perhaps that he could change the situation by wrestling Freddy to the ground. The resulting fight scared Edward out of his wits, as he watched the two boys locked in very physical violence. Freddy fought well, managing to roll out of Gordon's headlock and turning the tables to apply his own with sufficient force to hold his aggressor down, where they had rolled the long grass flat in their struggle. After a moment, Gordon capitulated.

Freddy released his hold and Gordon scrambled to his feet in tears, mouthing obscenities and swearing like a trooper. That was the main reason why he wasn't part of the gang. His profanities were abominable for one so young. Gordon rushed away shouting that he'd get even, and the rest of the kids jeered, glad to be rid of him.

Later, as they made their way home, Freddy's fight long forgotten, the group of tired kids took the pathway leading to the road where they all lived. Edward was in the lead as they marched in single file, heading for a stile that prevented the farmer's cattle from roaming out of the field. Beyond the stile lay a brick and concrete air-raid shelter against which the district council had left several heaps of grit and sharp, flint chippings for use in road repairs. The path passed next to the shelter and the grit-piles, opening out to meet the side walk of the main road beyond.

As leader, Edward was first over the stile. Suddenly, he faced Gordon, who had leaped from behind the shelter, grinning wickedly as he prepared to hurl handfuls of chippings at the oncoming group. The first barrage took Edward full in the face as he stepped astride the top timber of the stile. Momentarily blinded by the dusty grit and hurt by the sharp-edged chippings, Edward stumbled and sat heavily upon the wooden plank to stop himself falling. He cried out to the others as Gordon stooped to gather more stones and grit. The rest of the group retreated in disarray, but Freddy, bringing up the rear, came forward like a tornado, clearing the stile from which Edward had by then speedily removed himself, in one gigantic leap. He landed on the run just as Gordon let fly with another shower of stones. Gordon took-off in haste, pelting away with Freddy hot on his heels. It was just as well that he ran, for Freddy was as angry as a hornet.

Edward had no stomach for such behaviour and hated it when his brother had to resort to violence when bullies tried to intimidate either of them. He'd never forgotten the occasion when another neighbour's son attacked Freddy outside the school one evening, as the kids made their way on a waste-ground track to reach the Post Office at the edge of the village. It wasn't clear to Edward just why the fight started but he remembered his fear, and his own tears, as the two older lads fell amongst the brambles and grass, locked in a deadly combat. The other boy had a reputation for scrapping; he revelled in it. He was one of those short, stockily built lads, with a low centre of gravity and muscular limbs, but he hadn't reckoned with Freddy's strength and able competence in a wrestling match.

At Edward's tears and fearful shouts the two stopped fighting and scrambled to their feet, to dust themselves off, pick up their satchels and caps, and march off: match drawn and honour on both sides apparently satisfied. Freddy told Edward that he could have beaten the other fellow easily and that tears were unnecessary but at least they stopped the fight.

6

On a piece of overgrown wasteland, at the farthest end of Boundary Drive beyond the farm, the sound of clattering, iron wheels on steel rails announced the birth of the Rhodes-Weinel-Railway. A builder's narrow-gauge railway trolley rolled heavily, attended by a group of grinning, puffing children. The gang of eight or nine potential railroad engineers had cleared under-growth and thickets of vegetation after discovering a construction-site narrow-gauge railway. They had oiled the axle bearings and bodily lifted one of several disused trolleys onto long-dormant and rusting rails.

Building had ceased on this site when war broke out and the authorities must have overlooked the small railway when they scoured the district to gather metal for the war effort. The site became overgrown with disuse and the rails were practically invisible until the youthful gang stumbled upon them whilst exploring amongst the tall grasses and undergrowth.

Their own railway – what fantastic wheeze, but what dangers: a quarter-ton of rolling stock and no brakes! Rusting, upturned trolley-buckets balanced on bricks to make huts! What ingenuity! They could hardly wait for the weekends to come. Plans for their little railway grew as they developed new ideas for its improvement. There were plenty of loose rails in the undergrowth and they worked out how to bolt them together. Extension of the line came high on their priority list.

The builders probably used an engine of some kind, to pull the trucks around the site when the rails were in proper use. This, of course, was absent but there was a siding with proper switch-points, complete with its heavy lever bearing a big, round counter-weight. The discovery of several more abandoned trolleys, lying in the deep undergrowth nearby, excited their thoughts towards further extension. The question was, however, could the gang get two or three of them on to the rails to make a real train of coupled-up trucks?

What days of fun they had, but what trouble they got into when a stern man, not unkindly, advised the children to desist from balancing rails on pillars of loose bricks, to traverse a dip in the ground and onto the unmade roadway. He was right of course. There would certainly have been a few broken bones or lost limbs if they had completed the ambitious task.

Eventually, enthusiasm waned, as with all things involving imaginative enterprise, and the RWR fell into disuse – submitting again to the wild grasses, bracken and undergrowth, as the gang members dispersed and their interests changed with winter's approach. During the autumn months, the boys gathered horse chestnuts from under the great trees at The Grange, a bleak old house, set in its own grounds up the road from the Weinel home. After piercing the nuts, they treated them with vinegar and baked them in the oven to make them hard. Selecting the best and threaded them with good stout string, they took the hardened nuts to school for the playtime sport of 'Conkers'; a strange recreation, played with the seriousness of older, wiser people, whereby technique and style often determined the outcome. The hardest conker won, its numeric value increasing with each winning stroke that broke an opponent's best. Turn for turn, swing for swing, until the weaker nut split or shattered.

Had it not been for the bus stop beneath the branches of its trees and the Anti-Aircraft gun emplacement across the road on Lightfoot's farm, the Grange was surely a Victorian mystery-writer's dream come true. Surrounded as it was, by tall trees in an overgrown garden that hid its secrets from the outside world, the gloomy house stood in constant shadow. Perhaps its history did have dark secrets and stories of dastardly deeds to recount. Built from aging sandstone and with its steep gables, slated roofs and church-like windows, the house was, in essence, every boy's fantasy of ghostly haunting.

Edward never saw the old, iron-studded, wooden gates open, but the house was visible, vaguely, through the stave and wire fence and Larch-trees that lined its perimeter against the road. Among the lads there was much talk of a wicked witch, who lived there, wielding a knife on the end of a stick. The brothers never saw her, but the conker-nuts they gathered from under the Grange's trees, however, were of the finest quality.

Football was typically the standard winter sport at school. It took the boys' fancy in the fields at weekends too, when the season came. Shorts, shirts, and heavy woollen socks saw the light of day again, after their summer hibernation with moth-balls in dressing-chest drawers. Boots and the freshly inflated and re-laced ball received liberal applications of 'Dubbin', to make the leather supple and new boot-studs replaced the old, worn-out. Before long, the ping and thump of boot on ball filled the winter, weekend air, as the boys practiced for the Cup Final at Wembley: one day, perhaps.

Invented games were also among their sporting pastimes. None-Stop-Hand-Ball was a special one that certainly consumed much of their energy. The game only needed a tennis ball and a large, enclosed space upon which to play. Once in motion, the ball had not to stop. Back and forth, scrambling and jumping – the imaginary crowds loved it. They roared with delight at the agility with which the players kept the ball moving. The game never stopped until either exhaustion forced a halt or the ball went over the fence.

Breathless youth on bounding feet, hands scooping and whacking to keep the ball in motion; what a sight they were, capering, leaping and diving, shouting and cavorting on the lawn of their back garden. Bushes and plants sometimes suffered at the intensity of the game. They managed to destroy half a flowerbed of Lupins one evening, much to the annoyance of their parents, but the game needed an enclosed area with vertical surfaces from which to rebound the ball; the garden was ideal, with its large expanse of lawn, plank fences and out-house walls.

Windows, of course, were very vulnerable to the missiles of such play. Edward remembered in later years, how his father remonstrated and put a stop to the wild game, when windows suffered the consequences of impact. The large panes shattered easily, splintering in all directions, without the protection of the war-time, brown-paper tape; long since removed after the bombing stopped.

The misuse of a wooden spinning-top with a metal point was another example. Once, when Edward hurled the top down the lawn in a rolling fashion, the cricket ball sized missile hit the edge of a concrete path that passed beneath the bay windows. The spinning-top leapt upward in a curving trajectory and much to the child's dismay, propelled itself straight through a newly repaired glazing with a terrible crash. His Dad was furious, and Edward learned a hard lesson in propriety.

A simplified version of cricket helped to develop the boys' skills in that game too until several more shattered windows put an end also to its playing in the garden. Their Dad managed to break one himself when he tried to show his sons how to keep the ball low, off the bat. Scooping up one shot, when Freddy bowled a ball down the offside, he glanced the ball very neatly into the living room, by way of a glass pane in the bay window. In other circumstances, a good wicket keeper or a slip fielder might have taken a splendid catch from it; so much for keeping one's elbow up and stepping out to the pitch of the ball! The episode upset their Dad considerably – having scolded his sons often, over the use of hard objects near windows. His efforts to guide them in the techniques of keeping the ball low had ended in disaster; at least the demonstration of how NOT to do it proved his point. Thereafter they continued to play their cricket only in Lunty's field, on the far side of the garden's thick, privet hedge.

Occasionally, the boys' parents liked to go out to the cinema at Woolton. When they did, they left Freddy in charge, at home. It was on these occasions, during the absence of watchful parents, that garage and out-house Roof-Hopping became an exciting if dangerous pastime. No parent would have permitted their children to play this game, under any circumstances, had they been home to see.

The objective was to climb out of a bedroom window onto the leaded roof of the living room bay windows, then, to make a leap across a yard-and-half gap onto the corrugated roof of the garage. What folly at ten feet up! Therewith, a light-footed caper over the asbestos roof, a precarious scramble across the pitched tiles of the wash house, past the chimney and down onto the coal shed corrugations, before climbing down to the garden fence and ground level again at the far end of the garden. Football and non-stop-handball were safer but the challenge of roof hopping excited the brothers more.

On other occasions, they played in Tom Lunt's fields until quite late. At the appropriate hour before bedtime, their Dad would hail them from the hedgerow at the end of the garden. Since the garden backed onto the fields, the boys were usually well within earshot and they'd run home quickly, with a shouted response to their Dad's call.

One evening, they were playing a spying game on Mr.Lunt the farmer, thinking at the time that he was not aware of their presence. The sky was getting dark and they soon realised that he had indeed spotted them scrambling about in the hedgerow, near the farmyard. Edward wondered if Lunty thought they were poachers after his chickens or something. Anyway, the farmer chased them silently down the pathway behind a hedgerow. The gleeful brothers dodged through a gap in the hedge and doubled back on the other side, hoping to dive through a much smaller hole farther along, taking them onto the first side again to give Lunty the slip. Unfortunately, for the boys, he foiled their plan.

Their surprise was unimaginable when, ducking down to scramble through the get-away route, they fetched up with Lunty's great bulk, on all fours, crawling through the same hole in the hedge. His cap got caught up on a twig, exposing his grizzled, grey hair, as he lunged through the briers and hawthorn.

"Got you!" Lunty's voice rasped gruffly through his enormous grey moustache, as he lumbered to his feet and retrieved his cap. "What are you two up to, Freddy and Teddy?" The stern enquiry stopped the boys in their tracks, as Mr. Lunt reset the cap upon his sweating brow, but Edward noted that there was a merry twinkle in the farmer's eye. He was trying hard not to smile at the success of his table-turning trick to catch the pair of them.

"Oh, we're just playing," announced Freddy, tugging at his brother's sleeve in a bid to make off.

"Is that so? Well now you can help me get the ducks in," rejoined Lunty, with mock severity. "Come on, make yourselves useful then." He probably knew that this would make them late getting home and that their Dad would be shouting for them soon. Reluctantly, they marched off with the farmer to fetch his ducks from the pond in the next field.

The sky grew darker with the last receding rays of the sun, as they scurried and fussed about behind the ducks, to keep them on the right track to the farm. It was labour in vain really. The ducks clearly knew the way home, waddling in line astern in the right direction, without energetic ministrations of two inept kids. Old Lunty must surely have grinned to himself in the twilight, enjoying the discomfort that his mild punishment caused!

Their Dad's voice, raised to call the boys in, echoed across the darkening fields. They made to rush off in response but Lunty gruffly told them to stay. Lifting his own voice, he shouted to their Dad, "It's all right Mister Weinel, they're with me – helping with the ducks. They'll be home soon enough!"

That took the pressure off a little, though both Freddy and Edward knew they'd be for it when they got home. Darkness had fallen by then. They were for it, too! Edith, quite rightly, always worried about the boys if they were out after dark, and both she and their Dad gave them a hard time when the lads finally arrived on the doorstep.

Although Mr. Lunt did not encourage it, Freddy and Edward sometimes played in the farmyard area, among the stables, sheds and around the big barn. One afternoon, Edward managed to fall from a stack of baled straw in Lunty's barn, bearing out the farmer's contention that farmyards can be dangerous. The gang, for this was one of those days when they played together as a group with neighbours' children, had been tunnelling inside the stack. They discovered that by removing loosely set bales, on the outside, it was possible to move others, farther inside, to form narrow passages. In this way, they developed a small labyrinth of tunnels and spaces within the top part of the stack: really, quite a risky business!

Edward's accident happened when he came out of a tunnel on his belly, but backwards, to where we had placed a length of broken ladder to climb down to ground level. He got his feet on the top rung and started his descent. One of the broken ladder shafts was shorter than the other end and it caught in a leg of his short trousers as he moved hastily downward. The ladder leaned at a steep angle, its foot much too close to the stack. Edward's impetus, transferred through the impaled garment, dislodged it. Then, ever so slowly, the ladder began to tilt away from the stack of baled straw. Arcing through the upright, the ladder began to fall the other way. Still impaled by his shorts, Edward clung to the rungs for dear life: terrified. It was like being in slow motion until he hit the ground, belly up. He might have broken his back because he fell across an old wooden beam, half buried in the farmyard dirt. As it was, the fall certainly knocked all the wind out of him. He thought he would never get his breath back. So much for that game!

It wasn't all play and childish adventure on the farmer's property, however. There were other times, when the boys helped on the farm too. They fed the chickens and ducks, collected eggs and picked up windfalls in the apple orchard. Tasks of that nature gave them interesting outdoor activity during school holidays and at weekends. Mr. Lunt could not pay them in cash. He often gave them eggs and sometimes apples or pears for their time, although they didn't go to the farm just for that reason.

For all his stern demeanour, the farmer had a dry humour that often made the boys smile. For example, after weighing out seven pounds of potatoes, he would ask for their school caps to carry home the potatoes. Old Tom Lunt really was a kindly and generous man, in spite of his gruff way and rough appearance. Edward never knew if the farmer had children of his own. If he had, then Edward supposed that they were grown-up, with families of their own or perhaps lost at war, killed or captured in some far off land: for King and Country. Lunty did look so sad sometimes.

Occasionally, Edward accompanied his mother on her shopping trips to the village. Hunt's Cross had a number of shops grouped on two sides of the circle, a branch of '*Lloyds*' bank, a '*Greenall*' pub and a petrol station. A branch of the *Co-op* occupied one of several shops in the parades that surrounded the big roundabout. Next to the bank, a small plot with railings contained the stone cross to which the village attributed its name. The 'Hunts', from nearby villages, apparently crossed at this point and the riders probably took a cup or two at the pub before going on with the chase. However, this local activity was of earlier times, before the village became absorbed into the regional development plans of the late thirties, changing the character of the village as Speke's industrialisation expanded.

Edith was a member of the *Co-op*, or CWS, as it was then known. She did much of her shopping there, since the shop provided both butchery and grocery. Clean sawdust always covered the tiled floors and the assistants wore little white hats and white smocks or aprons. When money changed hands, the pull of a hanging cord sent it hurtling away on a taut wire, by means of wheeled rollers with a small cup-shaped canister clipped underneath. The rollers zoomed from all directions, up the wires to the cashier's glass paned office, propelled by mechanical catapults fixed at each end of the wires.

Change came back to the counter in the same way, the little cup coasting down the wire again, to be retained by a caliper as it crashed into the mechanism above the customers' heads. It was all quite fascinating to Edward, as a small child. Edith always kept the receipt tickets so that she could check the tally, when the payment of dividends fell due. Her society number was 5756, a number that always remained in Edward's memory, when most other numbers eluded long term retention.

Ration books were a common sight. Everyone had one. Rationing ensured the fairest and best economic division of essential food among the population of Britain's war-isolated islands. Despite the frequent shortages of almost everything, everyone seemed to accept the war and its inconveniences with grit and good humour. The art of queue etiquette became highly developed in those days. Anyhow, Edward supposed that the need to line up gave people an opportunity to catch up with gossip and latest in local news. Bearing witness to this, the wartime cartoonist, Lee, put queue-gossip in a nutshell. In one of his humorous depictions, two ladies stood in line discussing Hitler's latest secret weapon. One lady fearfully confides to a shocked and horrified companion, "....and d'you know, you can't hear it, you can't see it, you can't smell it and they say it does you absolutely no harm at all."

For all the civilian hardship that conflict brought during those war years of the nineteen-forties, people were healthy, if lean. Rations were sparse but nutritious and sustaining. Dental problems were almost unheard of among children, for there was precious little in the way of sweets and sugar-saturated goodies in the shops. Oatmeal porridge for breakfast, with milk and sweetened with black treacle, found healthy appetites amongst the kids, but bananas and coconuts remained unseen until the war was over.

Edward's clothes were often hand-me-downs, darned and patched by a caring mother, when rips and holes appeared. A very shallow bath, once a week and often shared, kept the boys clean enough, whilst cod-liver oil followed by a spoonful of malt each morning, kept the colds and flu at bay in the winter months. The malt came in a large jar, with its label bearing the picture of Atlas or Hercules or some-such heroic champion, supporting the world upon his shoulders. Edward asked his mother, one day, if he would grow up to be like the man on the label.

She replied, thinking that her son referred to the hero's muscular build, that surely he would, and he'd grow up to be very strong if he took his malt each day. Edward told her that he didn't want any more. Surprised, she asked him why ever not.

“I don't want to walk about in front of everyone, without any clothes on, wearing only a leaf,” he confided, in misery at the thought and close to tears. His mother gathered him up in her arms, hiding a smile at Edward's vivid fantasy. As luck would have it, no such imaginary experience ever occurred and his personal modesty remained intact. A spoonful of malt remained a daily routine.

Part 2

Growing Pains

7

The war against the Third-Reich was over. Hitler had taken his own life as the Allied spearheads and the Russian hoards bore down upon the last of his Wehrmacht defences in Berlin. Italy's defunct Fascist leader, Mussolini, was long-since dead; taken and shot by partisans in April 1945, then hanged along with his mistress, the bodies left to dangle from a lamp-post in a Milan square, to be gladly forgotten by the war-weary, disillusioned Italian people.

Revulsion filled the free-world as advancing Allied forces discovered the horrors of what we now call ethnic cleansing; the camps of death and concentration where the Semitic peoples of Europe had faced their Holocaust. Stalin's armies had met the Western Allies in Germany as they released the subjugated nations from the Balkans to the Baltic, but deliverance came at a price. To be rid of the Nazis brought relief to the liberated nations.

In the West, in France, in Holland and in Belgium it brought jubilation, but in the East, such euphoria was short-lived as Russian armies imposed an occupation of a different kind. Stalin's forces scrambled to grab as much territory as they could, encompassing half of Germany and absorbing Poland and the Slavic, Eastern states into their own growing, totalitarian, communist empire; in time, to be locked away behind frontiers that were to become known as the 'Iron Curtain'.

In May 1945, another uneasy peace finally descended upon Western Europe. Later that year, fighting in the Far East ceased. Atomic bombs devastated Hiroshima and Nagasaki, bringing the war with Japan abruptly to an end. The world had entered the thermo-nuclear age. Of course, at only seven years old by then, Edward had little knowledge or real understanding of these things. To him, peace meant no war, but in Britain, austerity, rationing and shortages remained firmly in place. Life remained no less and no more than perfectly normal to a boy who knew of nothing better than that which he already had. However, changes at home did make him feel uncomfortable. To him, that was more important than the ending of a silly war.

Edith was expecting a baby in November. Freddy had passed the eleven-plus scholarship examinations well enough to go to Grammar school at the Wade Deacon, in Widnes and it was, therefore, with some uneasiness in September of 1945, that Edward started a new term at Primary school without the companionship of his elder brother. It wasn't that he'd lost sight of his brother completely, but Edward missed Freddy's presence during the daytime, and his company walking to and from Kingsthorne.

The Wade Deacon Grammar school was some distance away. To accommodate travelling arrangements, the boys' father refurbished the old *Rudge* bike and gave it to his eldest son so that he could cycle the couple of miles to Halewood railway-station, to catch the train for Widnes. Each weekday morning Freddy pedalled off, leaving the bike locked up in the station's waiting room, ready for his retrieval on the return journey in the afternoon. He took lunch at the school because of the distance, and so it was well after five in the evening before he came home again.

Edward too, stayed for lunch at school but, in his case, it was simply because of their expectant mother's condition and her confinement in preparation for the birth of the baby. The boys' father worked away from home during that time, temporarily posted to an aircraft factory in another part of England. Perhaps it was because of these deviations from the norms of home and school life he'd known until then, that Edward experienced a feeling of loneliness. He only saw and talked with his brother first thing in the mornings, and again in the evenings. They saw their father only at weekends: two joyful days when the family was together, but Edward certainly felt lonely, during the long weekdays between.

Victory celebrations at the ending of war passed without sufficient import for Edward to recollect with clarity in later life. Disturbances to the normally well ordered family and household routines overshadowed the sudden advent of peacetime. Apart from the victory announcement, a little flag-waving and some cheering at school, the day seemed to pass without drama. To Edward, the skies were not any bluer, nor were the trees any greener. Rationing remained and shortages of everything persisted.

Just after mid-November, a new baby brother was born into the household. He arrived with little commotion. Edith wanted a girl but, by nature's way, she proudly presented another boy to her family. Fred was home on leave for the event: a proud father for a third time. He woke the sleeping, older boys in the middle of the night to announce the happy event. Delight and excitement reigned.

It was not unusual for mothers to have babies at home. Hospitals, still overloaded with war casualties, were unable to cope with natural phenomena like straightforward childbearing. The family doctor arranged for 'home-help' during Edith's confinement and convalescence, which was quite normal in those times.

Always in uniform with starched white cuffs and pinafore, smelling of lavender and having a brisk manner, the Weinel household received the daily ministrations of their Home Help. She was a dear lady, but quite stern – rebuking the older boys frequently when their boisterous antics got out of hand.

The baby's arrival was a joyful, new and interesting experience for all of them and, clearly, the event delighted everyone. He was Christened David Rex; but Edith was adamant about the use of this name, making the older boys promise not to shorten it to Dave. The Latin, Rex, sprang from Edith's maiden name, King. Unlike Freddy and Edward, their new baby brother had blue eyes like Edith's and, eventually, fair hair. The two elder boys had dark hair and hazel-brown eyes, like their Dad.

Freddy's Grammar-school studies produced reams of homework, taking up much of his time during the evenings, reducing his recreational activities and distancing Edward a little more. Nevertheless, the joys of brotherly companionship continued to fill their spare time on Saturdays and Sundays. A fresh posting back to Liverpool allowed their Dad to be at home every night again, instead of only at weekends. This gave him the opportunity to introduce the boys to the making of model-aircraft, an interesting diversion aimed at capturing the boys' enthusiasm. He started with the construction of an elastic-band driven model that he'd bought at the Hobbies shop, during one of those family excursions into the city.

The making took several weeks of painstaking work. The boys watched with anticipation and mounting excitement as the balsa wood structure took shape on their father's board, with glue and pins to hold things in place. A fine tissue-paper covering, doped to make it taut, gave each component shape and form, and the moment for final assembly approached. They fitted wheels to the piano-wire undercarriage, mounted the wings, held in place with strong elastic bands and, finally, threaded the thick elastic cords that would drive the propeller through the nose of the fuselage, for attachment to a strong point in the tail. At last, the long awaited moment came. The plane was ready to fly.

Out in the garden, that weekend, Fred Weinel wound up the elastic motor with care, counting as he went and watched intently by his two eager sons as he turned the propeller backwards with his index finger. The required number of turns escapes memory but the instructions were very specific about not over-winding the elastic on the first occasion. Yet, to their utter dismay, and before the final, counted turn of the propeller, the model fuselage collapsed like a concertina, with the sound of crunching balsa and tearing tissue paper. Fred's face dropped. To him, as an aircraft engineer of some considerable experience, the episode was a major disaster.

Ultimately, he rebuilt and strengthened the model, after which it flew very well, giving them all great satisfaction. Edward's own interest in aviation took another leap forward. Not that he yet considered a career with aeroplanes, as such, but model making appealed to him.

At school, he moved upward each term, having passed into the junior grades, enjoying the privilege of classes on the upstairs floor. New teachers appeared as older ones retired and as extra classes formed to accommodate the increase in attendance numbers. Most of the new teachers were men, released from the armed forces at the ending of war and returning to the careers they had trained for and practiced, before call-up. There were others too, who, having given (and survived) their military service to their country almost straight from school or university, had entered the teaching profession after demobilisation. Their first civilian employment followed courses at post-war, teacher-training college.

They brought a new form of discipline to the school, along with new ideas and fresh attitudes towards studies in Geography, History, Craftwork and more especially in sporting activities. Until then, for the boys anyway, football had been more of a general kick-around on the playing fields of Kingsthorne, using piles of blazers for goal posts, and under the direction of lady-teachers who knew little about the rules of the game, bless them. Real Cricket was unheard of under the earlier school-regime. Instead, they played Rounders, the nearest equivalent and a game in which the girls could participate, too.

Edward wasn't too keen on Rounders. It was a game rather like American Baseball, played with a soft (tennis) ball and a silly bat not unlike a cricket-bat, but shorter. He usually played at longstop when fielding, but one day he took catcher's position, under protest, immediately behind the batter. During the game, a high ball caused the batsman to back off smartly, and Edward took heavy blow on the side of his head from the batter's back swing, followed swiftly by another, on other side, from the through swing as the batsman, swiping wildly at the ball, reversed into the hastily back-pedalling and unwilling backstop-catcher. He missed the ball but not Edward's head.

Mr. Meek was in charge that day and immediately stopped the

game as Edward screamed and clutched at his well-thwacked temples. A sponge of cold water from the drinking fountain in the yard took care of the swelling, but bruises soon appeared when the initial shock had subsided. After the tears, Edward declined to rejoin the game.

The new teachers, Mr. Lister, Mr. Exell and Mr. Meek, redirected the school's sports-funding for the acquisition of proper goal posts, corner flags and a line-marking machine to provide the white touch-lines, goal areas and centre-circles for a full-blown football pitch. Stumps, bats, wicket keepers' gloves, leg-pads and real cork-balls appeared for the summer-sport of Cricket. Netball and Rounders remained in the domain of the girls, much to Edward's satisfaction. Football and cricket were much more to his liking.

Edward and his class companions studied under Mr. Meek for a while, although in spite of his name the teacher was far from mild in his personality. No doubt, he was a good man but his weakness was his rage.

Several of his pupils found great amusement in provoking the man into fits of anger that sometimes produced almost violent reaction, sending his already high-pitched voice even higher, in his futile attempts to subdue rebellious boys. One mischievous fellow, named Uttley, narrowly escaped a whacking after an insolent remark aimed at provocation. Mr. Meek lunged rapidly, with serious intent, to where Uttley stood beside his desk next to the classroom wall. Uttley scampered fearfully away, down the aisle towards the back of the class, and Mr. Meek skidded to a jolting halt as he hit the wall with his own shoulder, close to where Uttley had been standing. That seemed to bring him to his senses and his anger quickly dissipated. He smiled at his own clownish demise as the class giggled with subdued mirth. He quietly told his quarry to return to his seat and keep his silence. Uttley returned sheepishly, red-faced and suitably re-adjusting to a non-provocative status.

It was at about this time in Edward's young life that he discovered the consequences of telling lies. Normally, he was not an untruthful boy in the serious sense of the word. However, he fell into a self-made trap one day and it appeared to him that the only way to avoid trouble was to be evasive or simply lie.

A family with two daughters and a son lived next door and an accident in their household lead up to the problem. Edward often played with the boy, Trevor, at weekends, in the neighbours' garden where there was a swing that the boy's father had built: an interesting plaything. The daughters were in their teens and studied in the girls' department of the Wade Deacon Grammar. They were old enough to look after their younger brother when their parents were out. One evening, before putting Trevor to bed, the middle daughter had knocked loudly on the kitchen wall to call Fred Weinel for help. Trevor had fallen through the bathroom washbasin whilst she was giving him an all-over wash-down, standing in the sink.

He had cut his leg severely and Edward's father dashed next door to assist. After stemming the flow of blood, he called the doctor. The wound needed stitches and after the doctor had finished, it was clear that Trevor would be off school for a while.

Edward's part in all this was to take the doctor's note to the school next day, with instructions to hand it to Miss Kirkman, Trevor's teacher. In the morning, he pocketed the note in his blazer before leaving for school. He promptly forgot all about it. Edward re-discovered the undelivered envelope upon his return home that evening. His father wouldn't be home until after eight and Edward was frantic over the significance of not handing in the note. What should he do? His father would likely be very angry and he dared not tell his mother. The more he thought about it, the worse became his worry. He knew quite clearly that he should make a clean breast of it and own-up to his failure and forgetfulness but somehow he couldn't face the consequences of his error. Instead, he decided to flush the note down the toilet and bluff it out with his father if asked about it. He hoped he would not.

His father expressed surprise when he got home. Edward had not been at the bus stop to meet him, as was the usual routine when the weather was fine, but Edward didn't feel good about what he'd done and his conscience pricked him so much that he couldn't bring himself to meet his Dad off the bus. After his evening meal, Fred asked his son, casually, if he had delivered the doctor's note. Edward said that he had. The lie came out pat: well rehearsed many times in the mind of a distraught boy. What else could he have said? He couldn't say, “No; I flushed it down the toilet.” Fred pressed the point. “And what did the teacher say about it?” Edward tried to be just as casual as he answered. “Nothing; she just took it.”

Unfortunately for him, or perhaps it was fortunate in reality, his father found the note later, still floating in the lavatory. It hadn't gone down with the flush-water. After dropping the seat-lid, Edward hadn't even checked. Fred was furious. He took his son aside, upstairs, to show the boy what he'd found and for Edward's blatant lie, his father spanked him soundly, over his knee. It wasn't the note that mattered, it was the downright lie and the deceitful way in which Edward had tried to dispose of the evidence.

The incident left Edward tearful and shamed, to sit alone in his room, nursing a very sore posterior. He realised that his father had lost face with their neighbours, and his own integrity was also at stake. He had broken his father's trust, and that hurt: probably more than the spanking. Fred Weinel had a way with words too, a capacity to draw repentant tears from his sons when correction of their errors deemed it necessary.

After the heartache of punishment and a suitable time-lapse over several days, Fred could also express a father's love. That probably did more to keep his boys on the right track than any chastisement might have done. Edward came to understand that honesty most certainly was the best policy, in all things. Before long, the incident consigned to history, he sorrowfully returned to the fold after it became clear that he had learned his lesson well, albeit by the hard route.

8

Things were changing at school, too. For one whole school year, before reaching the top class, Edward and his classmates found themselves taking classes under Mr. Exell in the school canteen. This was due to a desperate shortage of classroom space with which to accommodate a growing pupil-population.

Without proper desks, they had to carry their books and materials in large boxes of thick cardboard, to and from a storeroom, several times each day. It was an uncomfortable way to study, seated upon benches at the long, refectory tables but they had little option. The learning processes suffered as a result.

At last, each pupil having reaching the age of ten, the class returned to the main building again – upstairs on the coveted top floor. They had reached their final year at Primary School, to struggle there with preparation for the 'eleven-plus' examinations under Miss Armor, who still taught the top-class at Kingsthorne. She remembered the Weinel name and how well Freddy had done in her class and took every opportunity to remind Edward.

However, during that last term, an interesting project came up – an unusual opportunity, redirecting and stimulating Edward's artistic abilities. It came about because of the rationing and post-war austerity that continued to press everyone into economies. Though some people still kept their gardens or allotments active for the husbandry of home-grown vegetables and fruit, many others had given it up after victory, as an unnecessary chore. Kingsthorne still had an excellent garden in which the pupils cultivated a variety of vegetables under teaching-staff supervision, as did many other schools of the nation.

To stimulate flagging, post-war, public interest in this form of economic activity, the local authorities proposed a horticultural show, to be staged at St. George's Hall in the centre of the city.

With cash prizes for the best garden products, clearly, the hope was that there would be many exhibitors and a revival of enthusiasm for home-grown produce. In parallel with this, as a complementary activity, schoolchildren in all the districts of Liverpool received invitations to prepare posters for exhibition at the show.

The idea aimed at graphic publicity to help regeneration of interest in the production of home-grown food from gardens and allotments: cultivating the cultivators, so to speak.

Edward's teacher, Miss Armor, asked several children in her class to create suitable designs for the competition. It surprised Edward to find himself among the selected few, for Miss Armor's request came in spite of her often critical remarks about his poor progress in the three R's. Miss Armor frequently made a point of telling him that his class work was 'not a patch on his brother's'. How he hated being placed in the position of competing with Freddy's notable success at school.

Nevertheless, since Edward's best subjects included drawing and painting, he presumed that she'd given him the benefit of her doubts, perhaps recognising some potential in his artistic leaning, contrary to his otherwise inadequate academic abilities. Edward worked diligently and vigorously at the poster task, along with the others of his class whom their teacher had appointed to participate in submissions for the competition. They worked for a week as a special group under the supervision of Miss Kirkman, co-ordinator for the school's efforts towards the show. On completion of the project she sent the work off for viewing by the selectors.

Only a short time elapsed and then came wonderful news! The selectors had accepted Edward's poster! His poster would be on display with its compelling message, "Hoe More To Grow More." The picture of a man in his garden, surrounded by all kinds of vegetables and hoeing with pride between cabbage rows, would go on show to the public. When Edward proudly told his parents, they were ecstatic. They had to go and see it!

St. George's Hall was an impressive building to the eyes of a ten-year-old. The tall, Doric pillars fascinated Edward. Blackened by the passage of time, soot from the fires of wartime bombing and industry's chimneys, they stood in geometric rows, supporting the triangular porticos that towered above the city roadway. The Weinels, along with what seemed like most of Liverpool's population, gained entrance by climbing the many broad, stone steps that reached up to, and between the noble columns. Thousands of people streamed towards the tall doors, beyond. What a crush it was! Edward and his parents reached the exhibition of posters, mounted in a side gallery, next to the main hall. They moved through the throng in search of his painting. Edward could not hide his astonishment at the number of posters on show, for there were hundreds. His pride was not a little hurt as he experienced his first encounter with serious competition against what he thought was his best attribute. It was a pang of childhood jealousy, perhaps.

Their progress through the gallery was slow because of the jostling crowds but the reward for this trivial discomfort was immense when, to Edward's complete satisfaction, they found his exhibit. He pointed it out to his parents. There it hung along with all the others, in no special place, but splendidly bedecked with a rosette for First Prize. FIRST PRIZE! How marvellous! How on Earth? Were they seeing things? Could this be true? Suddenly, Edward was famous! Was he? No, he thought not, but what happiness he felt. His parents were so proud. His Dad, wearing a broad smile, patted him on the back as his mother stooped to kiss him, a tear of her joyful emotion wetting Edward's cheek. Until that moment neither of them had seen the painting and they gazed upon it with uninhibited pride.

Prize giving came a week or so later. The Head Mistress called Edward to the stage at morning assembly, to receive his award. Ten Shillings amounted to quite a sum of money in 1948, and he put it straight into his Post Office Savings account.

Stimulated by this unexpected success, Edward increased his activities at drawing and painting. A new paint-box, given to him at Christmas-time, added to his enthusiasm; and his self-expression in painting supplanted his poorer abilities at other subjects. Edith encouraged him ardently, for she too, enjoyed sketching and drawing in her spare time as an outlet for her own artistic talent; its development frustrated by the loss of her parents all those years before. She had often told Edward of her own father's ability for painting and in the knowledge of this hereditary, family talent, Edward's interest in visual, graphic arts grew in a fertile environment. Thereafter, art became an important ingredient in his future life.

By then, he was a prefect at school, as was his brother before him. Divided into groups, the prefects at Kingsthorne assumed responsibilities to monitor corridors, entrance doors, playgrounds and milk distribution. Like Freddy, Edward volunteered to wear the blue and white, pin-button badge of a milk-monitor prefect. As such, along with the others, he had to be at school half an hour before assembly each day to distribute the many crates of milk to the classrooms.

Though the metal crates filled with third-of-a-pint milk bottles were heavy, the lads soon learned that they could better-manage those crates having round, tubular handles, than those with sharp-edged square type. They became stronger in wrist, arm and back. Carrying a full crate, single-handed up the stairways, earned admiration. Several boys in class had scoffed and joked at the Edward's appointment to milk-prefect. They thought he was too puny, but as it happened, Edward was quite strong, in spite of his slight build. Frankly, he found little difficulty in carrying a full crate, on his own, through two flights of stairway and to the farthest classrooms on the top floor corridor.

He couldn't understand why the other chaps made such a fuss about it. The scoffs and jokes soon disappeared, and he modestly accepted not a little unsought admiration and respect from his fellow monitors. Prendergast, Thomas and Harling, hitherto fervent critics, quickly became more amiable, adding to Edward's self-esteem.

The prefect's responsibilities didn't end at the morning delivery and distribution. It remained for them to re-assemble the crates of empty bottles, retrieved from the classrooms during afternoon recess, and deposit them outside the school gates, in readiness for collection at following day's milk delivery. As privileged badge-wearers, they carried a mild authority with their prefecture, such as to be obeyed by the other pupils wherever in the school precincts. They saw to it that there was no 'running' in the corridors. They secured doors at playtimes and silenced noisy infants and juniors. They tried to keep the peace at childish skirmishes in the playgrounds, preventing many fights by separating the antagonists for dialogue and negotiation: Edward's first brush with diplomacy and judicious peacekeeping.

One fellow they didn't tangle with however, was ‘Bonzo’. He was in Edward's class and sat at the next desk. Bonzo's real name was Reggie and he was big. He was very tall for his age, large boned and heavily built. He came from a family of physically big people. Bonzo's enormous father operated a fuel-merchant business, delivering coal and coke all around the district for the peoples' fires in winter. Although Bonzo didn't indulge in bullying, fortunately for the rest of the children, he did turn out to be a negative friend. This unhappy discovery came about when Edward found himself pressured into anger one day, during afternoon recess.

A younger fellow from a lower class, running up from behind, pinioned Edward's arms at the elbows. He almost bowled Edward over with the ferocity of the action. Bonzo was on the boy's tail in some wild chase. To effect his escape from the clutches of the colossal son of the coal-man, the younger boy had grabbed Edward with the idea of placing his captive between himself and the pounding approach of Bonzo. He swung Edward around to face the oncoming Titan but as luck would have it Bonzo managed to pull-up before what would have been a very nasty collision.

Edward's captor continued the torment by swinging his

human shield this way and that, in an attempt to impede Bonzo's flailing arms and grasping hands. Edward became angrier as his shouts for release went unheard. Finally, he began to see red and used some choice phrases aimed at his antagonist. At this, the boy released Edward's arms and spun him around, remarking pointedly. “Hey Bonzo, this kid wants a fight.”

It was a way of distracting and defusing his own plight. Bonzo grinned at this new development. He put aside his desire to get at his quarry, he stepped back, saying as he did so, “Go on then, give it to him.”

Pugilistic pursuits didn't appeal much to Edward. His glasses were vulnerable and without them, it was impossible to judge distance. His Dad had boxed for his squadron in the Air Force and had shown Freddy and Edward a few pointers so that they might at least defend themselves if need be. A fight Edward certainly did not want, but he was angry at the aggressive and rough treatment.

Like Bonzo, the boy was a member of the boxing fraternity and put up his fists, crouching in a boxer's stance. In defence of his innocent situation, Edward automatically did the same, though he was a Southpaw, leading with his right. For his trouble and failing to successfully parry the first blow aimed at him Edward took a very hard punch on the nose that bent his glasses and drew blood. He didn't want any help, nor did he get it but Bonzo just stood by, sniggering, as the other boy ran off, laughing. At the same instant, the yardmaster blew his whistle to end the recess.

Edward returned to classes, still mopping the blood from his nose. His seat was next to Bonzo's and until that day, they shared relatively mutual harmony. At least, there was no real friction between them but to Edward's astonishment, upon resumption of lessons after the playground incident, Bonzo took great delight in expounding, in a loud whisper to their class-mates, how comical it had been to see Weinel get bashed by a younger boy. Bonzo ceased to be a person with whom Edward fraternised.

This incident wasn't the last time that Edward got a bloodied nose but on the next occasion, he had the satisfaction of response, albeit not in quite the way he would have otherwise wished. Another very large boy, almost the size of Bonzo, had a smaller fellow in tears during a playground team game. It was a clear case of bullying. The youngster was playing Relieve-oh as a member of the team charged with avoiding capture by the opposing team whilst at the same time, trying to tag the Den to release those of his own group, already apprehended. The bullying-boy was with the opposition, chasing, tagging and marching captors to the Den but his game methods of intimidation by physical hurt took advantage of the smaller boys. His arm-twisting and dead-legging of the younger fellow appalled Edward. He tackled the bully verbally about it, in return for which he got an aggressive response that produced another punch on the nose, knocking off Edward's vulnerable glasses and drawing blood.

Yet again, in spite of his otherwise tranquil demeanour, Edward's anger rose and he lunged at the bully with determination, leaving the spectacles where they fell. It was an unexpected response and an expression of disbelief appeared upon the larger boy's face. He backed off a few steps before turning to run. Edward's anger spilled over and he gave chase. He was a good runner and within a few fleet steps, he was close enough to tap one of his quarry's flying heels with his toe. The boy tripped and fell heavily on his hands and knees with a resounding slap. The effect was dramatic, as impetus carried him forward so that his spread-eagled body slid upon the concrete flagstones for several yards.

Leaping up immediately, the fallen Goliath turned in dismay to face a bloody-nosed and angry Edward. Tears steaming, he took another swing at his pursuer, but Edward was ready and dodged; landing his own fist square on the nose of his quarry.

Edward wasn't very proud of the way the bully got his just deserts, but at least he'd responded. Aggression gone, the boy sidled off, bawling, to the jeers of a gathering crowd. His own anger dissipated, Edward walked back to retrieve his fallen glasses, hoping that they were still wearable. Fortunately, someone had the sense to pick up the bent frames before the heavy footfalls of running kids could shatter the lenses. The episode left him shaking. He had little stomach for fighting and his pacific nature felt ravished by his own aggressive reaction to physical hurt.

Later, he was in trouble with the new Headmistress, for his part in the affray with the 'younger' boy. The little snitch had tearfully reported Edward to Miss Evans. The telltale may have been younger, but standing head and shoulders taller than Edward, little he most certainly wasn't. He had struck first and Edward had responded in self-defence. Miss Evans would hear no explanations and delivered a stiff reprimand to both, bidding them to show restraint in future and to desist from common brawling in the yard.

Edward was in trouble again a week or so later when, in the face of new rules about keeping off the newly seeded playing field, someone gave him a tremendous push from behind. The force of the shove threw him headlong among the tufts of young, sprouting grass, just as Miss Evans blew her whistle for the end of afternoon break. The culprit was that snitching bullyboy again.

It was unheard-of for the Head Mistress to perform yard monitor duty, but there she was – and there was Edward, on his hands and knees among the forbidden, young, green, seedling grass at the edge of the field. Everyone had come to an instant standstill in compliance with the whistle.

Miss Evans cast her eyes around the playground, searching for the slightest movement among the motionless children, halted in their play, captured in stillness as if snapped by a camera. She spotted Edward (on the grass) and angrily called across the yard for him to go directly to her office for punishment. Her rules were explicit. The deterrent was plain for all to understand. She would cane anyone caught playing on the new, seedling turf. The snitching bullyboy openly sniggered as Edward marched off towards the foyer to wait in front of Miss Evans' office. It would be the cane for sure this time. Edward knew it but he wouldn't make any excuses or explanations. She would not listen anyway. He removed his prefect's badge and popped it into a blazer pocket, out of sight, out of mind. He really didn't want to forfeit that, too.

He'd been well and truly set-up by the snitch and for the first time in his young life Edward felt the pangs of vengeful hatred. He'd get even, just wait and see. For the first time in his life, too, he was in for a caning.

The juniors marched past him as he stood in the foyer, on their way to their final classes of the day. Some eyed Edward with vicious grins, others with sympathy. Edward felt very uncomfortable. The angry Head Mistress arrived in their wake. She stormed into her office, quickly returning with her cane, bidding Edward to hold out his hand. Her utterances, declaring him a naughty boy who had broken the rules, cut him to the quick. He felt shamed, but he could not, or rather he wouldn't, respond to her question as to his reasons for being on the forbidden grass. He was no snitch.

The single stroke of the cane stung but it didn't hurt that much; his pain stemmed more from the demeaning nature of the castigation. He'd always tried to keep on the right side of the rules, although boyish pranks had sometimes drawn a teacher's verbal chastisement or half-an-hour's detention after school, to write a-hundred-lines. No one was perfect but he'd never had accusations of outright disobedience levelled at him. It was another new and distasteful experience.

Edward heard Miss Evans's voice only distantly as she warned him vigorously about behaviour and disobedience before dismissing him. He marched away to his class without a word, fingering the pocketed badge. At least he'd saved what little remained of his honour. Miss Evans most certainly would have confiscated the badge and reduced Edward to the ranks, had he worn it at the caning.

9

The eleven-plus examinations loomed ahead. Studies intensified as the appointed date approached. Two things were paramount in the thoughts of all in the top-class. First, a change of school was imminent as their primary education ended. Second, their futures rested squarely upon the result of this first important and major test of abilities and learning. Whatever the outcome of the exams, each of them in that year's top-class at Kingsthorne now faced changes. Those who passed would go to Grammar school: those who failed, to Secondary Modern. Few considered failure, though for his part Edward found the going hard, in the run-up. Miss Armor regularly compared his poor marks in most subjects with those of his elder brother's. This did not help his morale and he often felt despondent.

When Freddy passed the Eleven-Plus, people who lived outside the boundaries of Liverpool, like the Weinels, could not apply for entry into any of Liverpool's several grammar schools. Four years later, this situation had changed and scholarships to schools in the city were open to all. Edward's choice would be The Holt.

It had an excellent reputation and was within easy reach. Other possible schools included The Collegiate, The Institute or The Blue-coat Grammar schools but the Holt seemed the best for him.

Examination day came. Edward was nervous and couldn't concentrate. The words on the question paper whirled before his eyes incoherently. He struggled on, but he knew that his heart was not really in it, nor was his learning up to it. He came away disillusioned and unhappy and when the published results came through, it didn't surprise him to learn that he had failed. A Scholarship to Grammar school would not be his.

He'd gazed more than once at his elder brother's name upon

the school's Honours-board in the lobby, with a vain hope that he might see his own added to the list of scholarship winners. Alas, it was not to be.

Edward's failure surely disappointed his parents, but they didn't show it, and they never rebuked him over it. Indeed, their supportive attitude sustained him in the face of his own dejection, for although he had no great liking for school, he really did want to please his family. Together they selected a suitable Secondary Modern, in readiness for a new start in the following September.

David had grown up fast. Babyhood gone, he plummeted about with vigour and zest. The seven-year age difference between Edward and his younger brother was not hard to bridge, although David was quite self-willed and not a little aggressive. In another year, he would follow his elder brothers' footsteps through the rigours of Primary schooling at Kingsthorne. He seemed eager to go.

Meanwhile, during that summer break, a neighbour's grown-up daughter gave Edward her old bicycle. She had worked in the village at Miss Duncan's post office for some time but now, with marriage in sight, she no longer needed the bike. This was Edward's first bicycle. It was immaculate, with a Lucas dynamo powering lights at front and rear, and Sturmey Archer three-speed gears controlled by a little quadrant lever mounted on the down tube of the frame. It was a lady's bike and didn't have a cross-bar. A large saddlebag hung from the saddle above the back wheel and the tyres were in near-new condition.

With the seat suitably lowered and the tyres pumped up, Edward prepared himself for his first encounter with the laws of gyroscopic balance. His Dad went with him to show him the ropes and to provide moral support. Selecting the lowest of three gears, taking a good grip on the handlebars, and with his father's hand firmly grasping the saddle behind his rump, Edward set forth with a tremendous push-off, tentatively pressing the pedals around energetically, if not without a little apprehension. The handlebars twitched this way and that as his unaccustomed senses came to terms with a new and wonderful experience.

His Dad encouraged him whilst keeping a firm, stabilising hand on the machine for his excited son's exuberance. “Keep going. Don't slow-down!” He wobbled on, thinking the kerbstones were passing quickly by then, and wondering if he could reach them with his foot if need be. “Look up, boy, don't look down!” His speed gathered and with it more stability. “Keep your eyes on the road ahead!” The wobbling decreased as he leaned harder into the pedals. “Wait for me, not too fast!” Round the bend, he peddled, off the main highway, into Wood Road and towards the country lanes of Halewood.

He felt elated; but not yet completely under control. His Dad was still with him, trotting alongside but only just, and puffing somewhat by that time. Then, dropping behind as Edward wobbled on without impediment, nor aid, he called out. "That's the ticket lad, keep going and keep your head up."

Moments later, Edward sensed traffic behind him and heard the sound of a motorcar following. The road narrowed between hedgerows. The car came closer. Panic gripped him, for his plan was to turn at this point and go back to where his Father stood, some hundred yards behind. Instead, he kept straight on. He was unsure of the best way to stop. A toot on the car's horn warned him that the driver was about to pass. Edward applied a touch of brake. Too much! The car drove by, giving him a wide berth. Then the jutting hedge came at him! He applied more braking. Foot down! Rump off the saddle! Nasty jolt in the groin! Ouch! He grazed his shin on a pedal as motion ceased. Oh well, at least he managed to stop without ending up in a heap on the road, or worse, tangled up in the Hawthorn hedge.

Still astride the heavy machine, Edward turned his bike, walking it through a hundred and eighty degrees to face the other way. His Dad came trotting up, puffing a little but grinning all over his face. "Well-done lad," was all he said, but it was enough, and Edward felt very proud of himself and very content with his first two-wheeled excursion into the realms of cycling. He began to feel grown up.

After that first experience, he practiced more and quite soon, he mastered the techniques of two-wheeled balance. It was as easy as falling off a bike: a phrase Fred Weinel often used to describe simplicity. The bicycle became Edward's pride and joy. It was in excellent condition, and the paintwork and chromium shone brilliantly when he cleaned it, which was often. Before the end of summer, he was quite proficient, practicing with the three gears and pedalling about the locality at every opportunity. When Freddy had time at weekends, the two boys cycled together, riding to Huyton, Ditton, Woolton, Garston – even to the airport once, and all around the district, exploring places they'd heard of but never seen before.

Earlier that year, their Dad had left his government job with the Air Ministry, to join Martin Hearn's company based at Hooton Park airfield on the Wirral peninsular. Along with other aircraft work, Martin Hearn built and assembled Slingsby sailplanes. Fred filled the post of Manager.

One Saturday, he took the boys to see the aeroplanes and gliders. It was an unusual excursion, an exciting diversion: travelling by bus into Liverpool, then by underground train in a tunnel beneath the river, and on by bus again to the airfield at Hooton. The trio arrived in time to watch the launch of a Slingsby glider, scheduled for air-test.

The launch-team was about to use an estate car, with its rear doors removed and a cable winch mounted in the back, to pull the aircraft into the air. A member of the team had already attached the free end of the cable, strung out behind the car, to a towing point mounted beneath the nose of the glider. The car's engine roared as the vehicle drove away across the grass, into the wind. With the slack taken up and the car's speed gathering, the glider trundled off on its single main wheel under the fuselage, its tail-wheel bouncing in the grass-tufts of the field. A couple of chaps ran along with it, hanging on to the wing tips to give stability.

Eventually, they let go as the speed gathered and the pilot (Lance Rimmer) gained aerodynamic control. More wire ran out as the winch-car sped into the distance and the glider soared majestically on the end of the cable, rocking gently in the gusty airs aloft. After a few moments, when the pilot had gained sufficient height, he released the tow-wire, letting the cable fall to the ground, like an enormously long snake. The winch-men wound it in before bringing the car back to the hard-standing in front of the hangar whilst the glider rode the windy airs, turning and soaring as the pilot put his craft through its paces over the airfield. To Edward it was a fantastic demonstration: impressive in its silence: beautiful in its smoothness of motion.

A member of the winch team came to speak with the boys' father. Apparently, there was a problem with the winch-mechanism. After some discussion with the team, Fred took his sons into a workshop area, where they watched with interest as he cut a piece of two-by-four to produce a short length of about fifteen inches, using a circular saw. He cut a deep V into one end, this time using what he called a band saw, before taking the finished piece back to the winch-men. The idea of the wooden tool was to provide an additional braking load on the winch-drum reel-flange to eliminate snatch, Fred told his sons. However, the technicalities of this meant little to Edward at the time.

Later, they returned to the hangar to look at the other aircraft, parked inside. The boys climbed into several in turn, to sit in the pilot seats whilst their Dad explained the controls. Finally, they clambered into an *Anson*, making their way up the inclined floor of the cabin and scrambling over the centre-section's main-plane spars to reach the cockpit. Once the boys had seated themselves in the pilots' seats, their Dad leaned over their shoulders to point things out and to answer the many questions that the two lads fired at him.

The restful, pale green paintwork and the smell of aeroplanes appealed to Edward. He gazed about him, drinking in the complexities of the cockpit, the instruments, the levers and the control yokes. He noted several attachments to the seats, not least of which, was a cone-shaped funnel fixed in a clip and fitted with a rubber tube that disappeared into the floor. Edward was curious.

"What's this for Dad?" He asked.

"Well son, that's for the pilot to use if he wants to spend a penny."

"Oh," said Edward, but he could not, for the life of him, see how one could get a penny down such a small hole at the bottom of the funnel. Then other questions overtook the boys' interest and he put aside the penny-in-the-funnel puzzle. It was something he didn't quite understand.

The day ended and on the way home they all agreed that the trip to Hooton had been great success. The boys had never been so close to aeroplanes before and their interest in aviation took on new dimensions. Perhaps that was just what their Dad wanted, after all. Later, Freddy put Edward right over the penny-puzzle. "It's for when the pilot wants to do a pee after he's been flying for a long time," explained Freddy and for Edward, the penny dropped! Until then, the phrase 'to spend a penny' as applied to having a pee, was foreign to Edward; he'd never heard the expression before. So, with the question over Dad's phraseology explained, the two boys chuckled over the simplicity of such a device but Edward secretly wondered if any pee got sprayed over Speke and Hunt's Cross during war-time raids. After all, he supposed that they used similar funnels and tubes, through which to spend a Pfennig, flying so far away from home.

The Summer holidays ended and it was time to start at the new school. Gilmour, the Secondary Modern at Allerton was one of those new-concept, gender-separated, twin-schools in which boys and girls attended classes at different locations. At Heath Road, the school took boys only, and at Duncan Road, not far away, girls only. The notion of gender separation, boys from girls, brought the Secondary Modern theme into line with the Grammar schools of those times.

Bonzo had passed the eleven-plus and won a place at the Holt Grammar. During the last days at Kingsthorne, he took great delight in tormenting Edward with stories he'd heard about the formidable Gilmour, Heath Road.

"You'll get bashed-up every day there," he'd said; a cruel reminder of the incident with the boxing-team boy. "It's full of ruffians and tough kids from Garston. No one in their right minds would want to go to that school."

This thought nourished Edward's anxiety as he took off to Allerton on the first day of his secondary education, wearing the uniform of his new school. His worries were needless, for the induction was easy enough as the new intake of first-year boys crowded into the recreation yard at the back of the school, albeit, under the scrutinising gaze of the school's older student population.

Second, third and fourth year veterans eyed the new lot with open curiosity but they kept their distance until the whistle blew, and the masters came out to line the boys up for assembly.

Having moved up into their new grades, the experienced veterans marched off in perfect step, with disciplined order: wheeling left on command, like well-trained soldiers. Those of the new first-year entry lined up by roll call into four, ragged sections. All of them had failed the dreaded eleven-plus exams but they were 'streamed' according to the marked percentage of that test, along with their final year's progress at Primary. As a result, Edward was lucky enough to find himself in the 'A' stream along with a number of his old friends from Kingsthorne. This made it easier to settle down in surroundings that had a much stricter order to discipline than their previous school. Along with this, Edward made new friends amongst the 'rough kids' of Bonzo's fantasies.

In truth, many of the boys at Gilmour were clearly less fortunate than Edward. They came from poorer circumstances, children of working people who had given equally to the war effort: been bombed-out: lost a loved-one in the fighting or simply orphaned by the war. This was obvious from their attire, and from their manner and speech. Edward concluded that if these were the 'ruffians' and 'tough kids' that Bonzo warned him against, then there was little for him to worry about. Rough around the edges perhaps they were, but he made many friends among their deprived ranks.

Although the school was large and well established, it was clearly hard to accommodate the growing child-population that then pushed class numbers up to more than forty. Space was at a premium. The 'A' and 'B' streams of the First Year found themselves housed in new classrooms, built at one end of the main building, as annexes. These were purpose-built, however, not at all like that awful year of schooling in the canteen at Kingsthorne.

All the teachers were men, at Gilmour. Miller the Killer was a master to beware of, some said. As an ex-regimental sergeant major, clipped of moustache, clipped of voice, and his back straight as a ramrod, he brought discipline into youthful lives in true army style. He wasn't averse to applying the cane to the posteriors of those who transgressed. The lads marched well for him and, indeed, after the initial shock of their encounter with his strict ways, the boys eventually grew to like him. He was a good teacher.

Breezy Bill was another cane-wielder; a specialist in mathematics and music. Mr. Blease was older than most of the other teaching-staff at the school, and he suffered breathing impairment through being gassed in the trenches of the first war. His droning, tired, undulating voice was so monotonous that he frequently found it necessary to whack the desktop of a sleep-nodding boy with his cane, to regain lost attention.

Old Tom 'Gibbo' Gibson taught English Language and Literature. He also was a veteran of the first war and some boys often prompted him to recount tales from his wartime experiences: hardly fair really. Their prompting questions smacked of red herrings, rather than of real interest. However, Gibbo's cleverness matched theirs for he was still able to impart the subject matter of his lessons, woven into the fabric of his stories. Almost unwittingly, the boys learned much from him and Edward enjoyed the elderly master's classes.

Mr. Calcut was their form master, proper. His specialty was mathematics. He was a tall, dashing man with a loud voice, keen on sports and physical fitness. Twice a week, he took his class for PT in the assembly hall that also served as the school's gymnasium; insisting that everyone stripped to shorts and plimsolls for participation in mat and apparatus activities.

Under his ministrations, pale, bony torsos and spindle-like limbs soon developed sinews and muscles that would have made any lad proud. Press-ups, forward-rolls, walking the narrow timbers to develop balance, tossing and catching the heavy medicine-balls, and general exercises to improve articulation brought team spirit into the lives of spotty-faced, youthful athletes. Vaulting at horse and box were favourites, although many of the boys ended up with bruises and grazes from the course coconut matting they had piled on the reverse side to break their landings.

Mr. Calcut took them for field sports too, rushing about energetically with a whistle in his mouth; blowing loudly upon it at even the smallest infringement, as the youngsters kicked and jostled at soccer. In the summer months, he taught them more of the finer points of cricket, resplendent in his 'whites.' The boys didn't have a nickname for their class-master except for calling him Cal in their schoolboy conversations.

Of two craft masters, Mr. Brown and Mr. Butcher, the latter taught 'A' stream woodwork. He was another disciplinarian like Miller-the-killer. Short in stature, smaller than most of the boys, even shorter than many in first year, rigid and upright, he demanded proper conduct. He was a good master though, and clearly demonstrated the care of tools and the simpler arts of carpentry.

At first, in the artistic field, it was Mr. Munday who conducted their classes, until he moved on to greener pastures. The encounter was brief, but Edward found him to be a sympathetic teacher whom he was to meet again in later times. Mr. Munday's abilities and teaching methods were to influence and enhance Edward's talent in the years to come although little did the youngster know it, then. The replacement arts master, Mr. Taylor, was equally likable and Edward's enjoyment of art continued to flourish under fresh guidance. He joined the new teacher's after-school painting club, along with several other classmates, held for an hour, once a week in the art room.

The school building probably had Victorian origins, with its steeply pitched, slate roofs, redbrick walls and ornate, wooden-framed windows. Standing in its own grounds, it was symmetrical in design and layout, long and low with a corridor running its whole length. The classrooms occupied the front side, with windows facing the road beyond a narrow, well kept, garden. A large assembly hall, serving as a well-equipped gymnasium lay centrally at the farther side, flanked by the headmaster's office, staff rooms and toilets.

The entrances from both front and rear lay at each end, in adjuncts containing upper floors and staircases. Two woodwork classrooms occupied the northern end, one above the other, whilst the large Arts room occupied the upper-floor, above the Science lab in the southern end.

10

At the beginning of his first term, Edward used the bus for school, but in due course, he won his parents' agreement to make the journey on his bike. It took no more than twenty minutes of steady pedalling to reach Allerton, a four-mile trip through the greenery of Woolton's lower end, but because of the distance he stayed for lunch in the school's dining-hall, a separate building built at the south end of the yard.

School dinners did not appeal very much to Edward. They came from a central kitchen somewhere in the City, delivered by van in canisters and covered metal trays. That which passed for food, found its way into the ovens in the canteen's kitchen, to be kept hot until the mid-day bell rang. Each day at twelve o'clock, Edward joined the line of hungry boys at the entrance, until the kitchen staff, wearing aprons and turbans, threw open the doors, allowing the hungry hoards to file in for serving at the counter-hatches. Although the meals were wholesome enough in content, presentation lacked any kind of appeal.

There were some good days, however, when Cornish pasty appeared on the fixed menu, served with mashed potatoes and bullet-like peas. Those were days for 'seconds.' Not so with dessert however, when milk puddings in the form of Rice, Tapioca, Semolina or Sago filled the dishes: ghastly stuff, topped with a dollop of red jam. Edward hated them then, as still he does.

It was outside the dining hall one lunchtime, when Edward got the worst of another physical confrontation with a bigger boy. The whole thing was grotesque, but in the end, his singularly defensive conduct won him newfound respect among his peers.

Edward had lined up with the rest, waiting for the dining-room doors to open. He was about mid-way in the long queue chatting with companions, when, without warning, he took an unexpected and violent shoulder-charge from his right. The force of the fierce shove threw him against a wooden, paling-fence. His head hit the fence with a resounding thump and splinters from the old wood tore through his blazer jacket and imbedded themselves in his left shoulder and upper arm.

A larger boy, a first year 'D' streamer, had arrived late from class, apparently having been kept back for misbehaviour. Edward knew of him as a bully, a big-framed lad, rather like Bonzo in build but as thick as two short planks in spite of his clearly more affluent family background. Seeing the long queue, he decided that he wasn't going to wait at the back, for anyone. He'd tried barging-in farther along, near the front but the senior boys in that part of the line sent him packing. Still determined, he selected Edward as his next target, clearly expecting no real opposition.

When Edward picked himself up from the brutal shove, he was very angry. He returned the shoulder-charge, displacing the aggressor, by then standing in Edward's place looking as if butter wouldn't melt in his mouth, and told the surprised bully to push-off to the back of the line. Others around the scene raised their voices in support of Edward's demand. The big fellow stepped back grinning wickedly, and asking who would make him. To Edward, this was like a red-rag to a bull. He stepped forward and threw a well-aimed punch at the aggressor. Much to Edward's satisfaction, his fist found its mark. The bigger lad reeled backwards clutching his mouth but he came back like train. He didn't intend to enter into fisticuffs though. Instead, flailing wildly with his arms, he warded off several of many more punches and dived, in a flying leap, to encircle Edward's neck with a thick arm.

His velocity, height and weight did the rest, forcing his opponent to the ground. Edward heard cheers of encouragement from the other boys, but under the pressure to the side of his head and neck, and with the weight of the bigger boy pressing him down, he was helpless. He managed to squirm into a prone position, lying on his stomach, but that was a bad mistake. His aggressor changed position rapidly, to sit heavily upon the small of Edward's back pulling mightily with cupped hands beneath his captives chin. The effect was lethal. Edward thought his back would break and he cried out in pain as he felt something click, or rather, displace, in the lower part of his spinal column, where the boy now sat with all his weight.

Suddenly, the pressure ceased; the weight lifted from Edward's back and the hands at his chin released their grip. Edward fell forwards, gasping, only vaguely aware that Kenneth Atherly, one of the hefty seniors, a prefect supervising the line of boys, had hurried to the scene and dragged the bully away. Atherly gave the bully a good whacking whilst the other lads helped Edward to his feet.

Painfully re-joining the dinner-line, Edward felt his hand pumped many times as congratulatory comment passed among the gathered throng. His tall rescuer came back to see if Edward was all right and to tell him he'd put up a good fight. Moreover, the bully wouldn't be eating that day, for he had several loosened teeth and broken, swollen lips – not from the senior's beating, but from the few resounding shots Edward had landed, before being wrestled to the ground.

When Edward had recovered his composure, he found himself in the limelight with his peers, but it wasn't something he enjoyed at the time. He was breathless and his shoulder, back and neck hurt abominably; though his injury might have been worse had it not been for the senior fellow's intervention. Thereafter, however, the bully kept his distance, giving Edward a wide berth whenever their paths crossed.

Eventually, and not because of the fight over line-rights, Edward made other, more agreeable, arrangements for lunch. School dinners remained quite unpalatable and not at all to his liking. By keeping the weekly school-dinner money that his mother gave him every Monday and by supplementing a shilling from his own weekly pocket money, he could eat like a king from the parade of shops, not far away from the school.

Quite without realising it, Edward had started a popular trend that led to an increasingly large group of his companions dining on the culinary delights of the Chippy, delicacies from the baker's shop and fruit from the green-grocer's, consuming their purchases beneath the trees that shaded the broad, flagged pavement in front of the shops. Four-penneth of potato chips and a bread roll for a penny went down a treat. Sometimes, with care, they could stretch their funding to include a yesterday's cream cake, at a ha'penny, or an apple or two as dessert to round out the picnic-like meal.

Not unlike the Grammar schools of those times, the regime at Gilmour divided the student population into a ‘house’ system. Designed to encourage team spirit and a sense of competition, each house carried the name of a famous personage, in Gilmour's case, Grenfell (after the medical missionary), Gladstone (after the politician), Masefield (after the poet) and Ross (after the explorer).

Mr. Calcut had randomly divided his new, first-form class into four almost equal groups on their first day, and issued each boy with a small, button badge to be worn on the lapel of their blazers. Edward wore the yellow house-badge of Ross. Inter-house competitive activities were intense, both at sporting events and in scholastic pursuit.

Soon after his induction into Gilmour, Edward stepped into his first long trousers. They were brand new, not hand-me-downs.

At last he'd joined the ranks of the 'dropped'. He had not worn them for more than a week when, one day in the village, he met Mr. Cave, the caretaker from his former school at Hunts Cross.

"Aye-Aye," said Mr. Cave, "I see you've been dropped then," referring his eyes to Edward's new, flannel drain-pipes, "watch out for the girls, sonny. They'll be after you now, you know."

Edward felt a bit self-conscious but managed to grin in response. He was as yet hardly accustomed to the feel of covered knees, let alone the imports of looking older than his years in long trousers and the possibility that he might now be a subject of attraction to the adolescent eyes of watchful girls. He hadn't yet thought too much about the latter; but secretly, encouraged by Mr. Cave's observation, the imports of his new, grown-up status pleased him.

The creases wouldn't stay for long in the thick flannel material, especially in the rain. The legs of the trousers always took up their natural drainpipe shape when they got wet. They had turn-ups too, that turned down in the rain. Anyway, for all that, gone were the days of short trousers and the long socks with elastic rings to hold them up. Edward felt all but half a man!

He continued to ride his bike to school each day (sporting cyclists trouser-clips to stop the new 'longs' catching in the chain) and he began to call at a friend's house, on the way. They cycled to school together and back again in the evenings. His new friend's name was Robert and he still wore short trousers when Edward first met him, a mid-term entry who looked lost and bewildered in the school yard, on his own first day at Gilmour. Veteran's scrutiny had disconcerted him and he clearly felt self-conscious at comments passed about his short trousers. Edward befriended him, showing him the ropes and shoving him into place with his class at assembly time. Robert was a 'B' streamer but only because he joined at midterm and there wasn't a place for him in the 'A' stream. His family had moved to Liverpool only recently, which accounted for his late entry into Gilmour.

They discovered that he and Edward shared similar interests especially in artistic pursuits, for which Edward persuaded Robert to join as a member of the after-school art club. Their companionship blossomed as they developed ideas and interests and they became good friends. It was a brotherly relationship, which clearly proved to be of mutual benefit – in Robert's case, because he was an only child without brothers or sisters of his own. In Edward's case, perhaps it was because his elder brother had moved on into a different schooling environment, whilst his younger brother was still too young to be on a par with Edward's activities. Robert (Bob) was 'dropped' too, and soon after, with long trousers, began to feel less conspicuous in school.

Whilst thus engaged in settling down at Gilmour, Edward's younger brother, David, started his education at Kingsthorne Primary but, being eleven years behind Freddy and seven behind Edward, he probably escaped the comparisons that teachers had sometimes drawn between the two elder brothers.

There was no doubt that David received recognition as one of the Weinels, flavoured by Freddy's scholastic achievement, recorded on the honours-list board for winning his scholarship. David didn't have to worry about comparisons, however, because he turned out to be quite a clever chap in his own right.

He and Edward had grown closer since Freddy started at the Grammar. It wasn't that Freddy and Edward had become so distant, it was just that the Grammar School homework-load was very heavy and free time a great deal less.

Before starting school, David had suffered with pneumonia; a complication brought on by the contraction of whooping cough and the measles, simultaneously. He was gravely ill and the whole family were beside themselves with worry. The emergence of new penicillin-based drugs developed during and after the war, narrowly saved David's life. The drug was very expensive, but it was available under the new National Health system. Clearly, had it not been for the system and more importantly, the existence of the drug, David most surely would have died.

When their family doctor prescribed the new medicine, he was not sure of its immediate availability. Fred and Edith dispatched Edward to the village on his bike, to find out if the local chemist's shop could fill the prescription. As luck would have it, the pharmacist had just two bottles. He wrapped one very carefully and charged Edward to be very, very careful in getting it home. He knew its value to David's well being. Edward thanked God for it, for David made a good recovery.

Freddy joined the Combined Cadet Force at the Grammar. He came home one evening, loaded down with parcels containing uniform and equipment, his bike laden with packets tied up with string, dangling from every conceivable protrusion. Edward helped him to unload and check the contents. They spread everything out in the garage as Freddy called out each item and Edward ticked them off on the list. There was one tunic, button-to-the-neck, Khaki, cadets for the use of: one pair, trousers, khaki, (with many pockets) also cadets for the use of: gaiters, boots, webbing belts, satchels, pouches and, to cap it all, an enormous, dish-like beret, cadet-heads for the wearing of.

His new kit needed much work to prepare it for use – ‘Khaki Blanco’ applied to the gaiters and webbing equipment, with *Brasso* metal polish and elbow-grease to get the dull buttons and clips shining. His father's long experience in the Royal Air Force helped, when he provided several pieces of essential equipment for cleaning the brass, including a brass two-pronged device to slide beneath the buttons and protect the garment material from the *Brasso*.

The heavy boots acquired proper shine only after many hours of rubbing and buffing but the breaking in of the stiff, heavy footwear must surely have been a painful exercise for Freddy, on the parade ground.

His sights by then, seemed firmly set upon the army as a career, perhaps influenced by the fact that the grandfathers, on both sides of the family, had served in the army, and died, at war, in the service of their country. Freddy's interest in the military prompted the boys' father to fetch out some old medals and ribbons in a box, along with buttons and badges that had once belonged to his father, a professional soldier who had served with the Royal Artillery.

Before the summer recess of 1950, notification for the annual School Camp appeared on the notice board at Gilmour. Edward put his name down along with the other chaps. This was to be his first excursion away from home and family. Two glorious weeks on the Isle of Man, under canvas, eight to a tent, porridge and kippers for breakfast and torrents of rain, too, it turned out. The advance party travelled the day before, to prepare the site and put up all the bell tents in regimental lines. They dug the latrines and laid out the components of the big marquee, ready for when the rest of the lads arrived, on the following day.

After a very rough, four-hour crossing from Liverpool to Douglas, and a delightful light-railway journey southward to Port Saint Mary, the main body, marched in good order to the campsite. No one complained about the effects of the sea crossing by the time they reached Rushen United football field, where the advance party had pitched the tents. The camp functioned on semi-military lines and most certainly the objectives were to enrich self-confidence, along with a spirit of team work. The Camp Captain welcomed the lads to the camp with a short address before allocating tasks.

"Right lads, your first job: Collect your ground sheet, palliasse and blankets from Mr. Inness. Second job: Take your kit and put it in your allotted tent." Edward's was No 8. "Third job: Assemble at the marquee with your coats off and your sleeves rolled up; there's work to be done if you're to eat tonight."

It was beginning to get dark and the boys were all very hungry. The marquee was to be their only source of food and sustenance, so they pulled with a will to get the great poles erect and the canvases up. With tent pegs hammered home and guy ropes tensioned, the canvas side-walls didn't take long to hang, as they completed the job. It didn't take long either, to prepare and arrange the folding tables and chairs in readiness for the ravenous hordes to descend upon the food, the succulent smell of which had already reached their nostrils.

The advance party had also organised the field kitchen as a primary task on arrival, producing the first, full-complement meal at that year's camp, from the wood-burning, smoke-belching stoves. What culinary delights there were: sausages and mash with a fried egg on top, two or three doorsteps of fresh bread from locally baked, yard-long loaves: dabs of Manx butter and a mug of steaming hot tea. Everyone rushed back to their tents to fetch tin plates, mugs and cutlery. It began to rain and most got wet... but who cared?

They were all hungry enough to eat a horse. In the twilight, a few unaccustomed fellows fell headlong through guy-ropes or stumbled over tent pegs but nobody hurt themselves enough to miss the food; the pangs of hunger saw to that, after a very long and strenuous day.

Following the meal they sang songs, ending with a well-known ditty, depicting life at camp.

There is a happy camp, far, far away
where we have bread and jam, four times a day.
Eggs and bacon we don't see, ruddy great beetles in our tea.
That's why we are gradually-yyyyyyyy. (Note held)
Fading away, fading away, fading away, fading away..........

There was another verse too – in essence, having something to do with the latrines, but recollection of the words remains obscure.

The anthem over, the boys raised three cheers for the masters who had prepared and served their first meal at camp in spite of the rain that, by then, fell heavily. It promised to be a very wet night indeed. Then it was time to arrange sleeping kit. The advance party had taken in a delivery of straw and prepared it in a spare tent next to the store.

It was fortunate that there was a spare tent available, otherwise the campers might have had to sleep on damp bedding. The order to fetch palliasses came, and again the boys belted off in all directions, in their Macintosh ground sheets, through the rain to their tents, to fetch the empty, rolled mattresses – then, over to the store tent to stuff them full with straw.

"Make sure you put plenty of straw in the bag, Son." Edward's Father had told him. "Otherwise you'll sleep hard and feel every lump in the ground under you."

When his turn came, Edward stuffed as much course straw into the bag as he could, and more! A teacher, watching him and with a knowing grin and a twinkle in his eye, said, "Done this before, lad, have you?"

Edward just grinned and thanked the Lord that his Dad had

given him the benefit of his experience before his son left the comforts of home.

Back in the tent, by torch light, they laid out their ground sheets around the tent pole, placing the palliasses so that all eight pairs of smelly feet would be in the middle. They placed kit bags at their head for use as pillows and spread out the thick army blankets in readiness for sleep. There wasn't much space to spare, but finally everyone settled down with some measure of comfort.

Someone did a quick circuit of the tent ropes, putting a little slack in them to allow for shrinkage in the rain; buttoning up the apron and flysheet upon his return. The boys lay thoroughly exhausted, secure in unaccustomed and somewhat damp surroundings, listening to the rain pounding on the canvas.

Earwigs were a major source of concern. They tended to lurk under the canvas floor sheet and beneath the slopes of the tent, right next to the sides, in the eaves-like space, sheltered from the rain.

"They gerr'in yer ears y'know," a broad Liverpool accent piped up from the darkness, referring to the earwigs, after someone squawked and swore roundly about creepy-crawlies scuttling about in the dark.

"They don't, do they?" A worried Garstonian voice asked. Di-don't-do-di, is probably a better transcription of the accent.

"Di-gerrup-y'nose too," another voice chipped in, "shrrup yous lot, will ya." Someone started to cry a little, just a sniff or two.

"What's up wid 'im den," asked the Scouser, a veteran of many camping excursions, "gorra-touch of 'ome-sickness?"

"Yes." The answer came, with a smothered sob.

The boy's malady surprised Edward. It never occurred to him that anyone might pine for home-comforts on an adventure like this. There had been so much going on around him that he didn't even think about home-sickness. His parents had seen him off at the Pier Head and his Mother had been a bit weepy but although this was his very first expedition away from home and family, he had felt no pangs about it. It was all too exciting. For those who did suffer the twinges of fretting, and there were quite a few, they soon forgot their feelings once they got into the swing of things and began to enjoy the many activities available for their participation.

Two weeks later, tanned brown, feeling very healthy and with a fantastic camping experience behind him, Edward returned to Hunt's Cross. He'd thoroughly enjoyed his first holiday away from home and family. It had been a grand adventure along with his friends from school, but he was glad to be home again.

After their camping expeditions, Freddy with the cadets and Edward with his school chums, the brothers spent time at the farm again during the remainder of the summer break from school. They asked Mr. Lunt if they might help, and he willingly agreed. Extra hands at harvest were always welcome.

Harvest time was particularly interesting for Edward because of the horse-drawn, mechanical machinery and the traction engine, later used for the threshing. Mr. Lunt had four Shires with which to work his farm: two mares and two draught horses. They were huge and beautiful animals, soon giving way to age and to tractors, sad to say. They worked hard pulling ploughs or great, wooden, two and four wheeled carts, potato diggers, drills, harrows, rollers and reapers, working sometimes singly or at other times as pairs.

It was George Young, mostly, who looked after the horses. He was one of Lunty's best farm workers and the father of two lads and a daughter, with whom the Weinel boys often fetched-up, both in school and at play in the field, at weekends. They lived in poorer circumstances than the Weinels however, their home being a farm-cottage, not far along the road on the edge of Lunty's land. They had no electricity, and used paraffin lamps to light their rooms after dark.

A huge kitchen-range provided the heat by which to cook and boil water and their toilet was at the far end of the long garden in a little wooden shed with a heart-shaped hole cut in the door. A battery of those old-fashioned, glass-encased accumulators that needed re-charging from time to time, powered their radio.

Having agreed that the boys could help, Lunty detailed the workers off to the two horse-drawn carts, waiting nearby. It was a joy to ride out to the fields in that manner, the carts swaying in time to the gentle plodding of the two horses, creaking and rolling with each step.

However, there was much work to do, once they reached the cornfields. Lunty's men showed the boys how to put up ‘stooks’, stacking four sheaves of reaped and bound barley, oats or wheat, two against two in an inverted ‘V’, with corn heads upward; then two more, one at each end of the ‘V’ to make a total of six in each stook. The men explained that this formation helped to prevented the corn from rotting, by allowing the air to circulate, as an aid to drying out before collection, a few (hopefully dry) days later.

The work was hard, for the horse drawn reaper-binder produced many sheaves in its ever-decreasing circuits of the field. Carrying two sheaves, one under each arm, was a feat on its own, since the sheaves were heavy and almost as tall as the youngsters who carried them but the task did have its rewards when Tom Lunt's wife and her sister, Mrs. Rigby, the foreman's wife, came to the fields with food. They brought Billycans of hot tea, farm-baked bread, cheeses and fruit to slake the men's thirst and fill their hungry bellies. The horses too, had their rest-break, with a bag of oats each and several buckets of water. For the boys, there was hot, sweet tea from a tin mug, a bite to eat and a short rest before the farmer's men set the horses in motion again, with the click of a farm-hand's tongue.

"Gidd'up Captain!"

"Hup-then Dolly!"

The boys stayed with it until the reaper cut the last swathe from the middle of the field, which by then contained scores of field creatures, rabbits, shrews and mice: the small animals driven slowly inward by the machine's circulation. Those that had not escaped beforehand remained until the last moment. Then, when their final cover began to disappear under the reaper's blade, it was hell for leather to the nearest hedgerow. A few men and their sons tried, in vain, to pick off rabbits as they broke cover to run the fearful, tail-bobbing, zigzag path to freedom. They used catapults for this final chase of the day but, exciting though it was, it pleased Edward that no one scored a hit. The escape was sweet, final and complete. The brothers slept well and deeply at night, after those long, hot days in the cornfields: dead tired, but very contented.

Eventually, when the day came to gather up the stooked corn, the boys participated in this too, helping to pitch up the sheaves with pitchforks into the horse-drawn carts and riding back to the barn to off-load the sheaves, in readiness for the threshing. It was hard work but they revelled in the fresh air and sunshine. They grew stronger and their hands became hardened to the manual labour.

Farmers all around the district used steam power to operate the threshing and bailing machines after harvesting. It was a cooperative arrangement by which they used one set of machinery to serve all the local farms, one by one. A coal-fired traction engine, with all its bustle and steam, spinning flywheel and colourful paintwork, towed the machinery to and from each of the farms in turn. Once set up in line, with the machines coupled to the traction engine's huge flywheel, using stout, endless belts of leather, the threshing work began. Under power from the engine, the bouncing straps slapping and hissing against the drive pulleys, the thresher greedily consumed the golden sheaves as the men pitched them up into the machine's throat.

Heavy sacks hanging in a row beneath caught the grain, while the straw emerged at the top, to fall from a moving belt of broad canvas into the throat of the bailer behind. A stream of chaff from the threshed corn spewed out in a dusty cloud, to spill upon an ever-growing pile to one side while the bailer's hammer-like arm swung rhythmically up and down, to press and squeeze the straw into rectangular bails, slowly emerging, wire-bound and ready for stacking, at the farther end. What marvels of modern technology! Edward absorbed every detail, enthralled in the mechanical precision that reduced sheaves of corn to three usable commodities, ending with grain-seed ready for grinding to flour at the mill.

With the harvest home, the long, hot summer drawing to its end and school holidays almost done, the brothers reluctantly made preparations for their return to studies. It had been a wonderful year for Edward: his first three terms at Gilmour under his belt, cycling on his very own bike, camping and working on the farm. Now he was ready for his second year, in Form 2A. Enjoying the thoughts, he pondered over the coming term. On his return to school, he would be among the veterans, watching with critical interest the arrival of pallid, pale-faced 'new-boys' in unblemished uniforms, as they re-enacted the timorous entry into a new environment, just as Edward had endured on his first day at Gilmour. He could only wonder what the future held for him.

Part 3

Successes

11

Whilst Edward's academic abilities prospered only at a painfully slow pace, his interest and talent for the graphic arts developed eminently during his second year at Gilmour. At the end of the summer term of 1951 he would have another opportunity to secure a place at Grammar school by way of the thirteen-plus scholarship plan; a kind of second-chance at educational advancement aimed at late or slower learners. This time, however, it seemed that the forthcoming scholarship possibilities also extended to other types of school, alternatives requiring less emphasis on academic ability. Not least was Liverpool's Secondary School of Art.

As an 'A-streamer' in his second year at Gilmour, Edward had several avenues open to him. He could apply for a second attempt at scholarship into Grammar school, or a first try for entry into Commercial school, specialising in office and commerce related topics or he could solicit entry into Technical school for training in the engineering or building trades. Art school appeared last on the list of alternatives for further secondary education, with specialisation in the graphic arts.

Not for the first time in Edward's thus-far short life, nor for the last, his father took him aside to talk frankly about the boy's future. Edward accepted the premise that whilst his artistic talent might be good enough to get him into the Art school, his capabilities in other subjects were less than sufficiently developed to get him into Grammar or Technical schools. With only a year to go, Edward bent himself more diligently to academic preparation for the Art School scholarship exams.

There were only two other pupils from his form (indeed from the whole school) who also selected this alternative route. Places, however, were not easy to come by. Limited by numbers rather than solely by artistic ability and strength in other subjects, few applicants could expect to pass the entrance exams. In that respect, perhaps gaining a place at Art School would be harder than getting into any of the city's Grammar or Technical schools. Only one applicant from each secondary modern school entry was likely to win such a scholarship.

It was to be a sink-or-swim effort on Edward's part because failure to achieve a place at the Art school would result in continued studies at Gilmour. Unlike Edward, those of his class who elected to try for Grammar would have a second option for Commercial school, should their pass level not meet the grades needed for their first choice. However, Edward concluded that he had nothing to lose and everything to gain. He enjoyed the Gilmour environment well enough to accept failure at the thirteen-plus exams happily, in the event. If he passed, well and good, that had to be a bonus.

With his sights thus set under the wise guidance of his Dad, he redoubled his efforts to maintain the best marks possible over the coming months. History and Geography captured more of his attention, while Maths and English remained a major source of concern and hardship for him. However, Science was another subject that he quite enjoyed, along with his main interests in Art and Woodwork.

The alumni of Gilmour were lucky to have a Science Master whose popularity amongst the boys enhanced his ability to teach with positive results. Mr. Hogg taught rudimentary science and chemistry in a well-equipped laboratory at the farthest end of the school. The lads looked forward to their twice-weekly sessions with him, initially encouraged by older boys with tales of grotesque stinks and bubbling liquids emerging from complex set-ups of glass tubing, bell-jars and test-tubes. At the beginning of their secondary education, many lads enthusiastically shared juvenile ambitions in chemistry. Such aspirations ranged from Merlin-like wizardry enabling them to turn their enemies into toads, or worse, flies that they might swat, to mad-scientist-experimentation by which they might produce miniature atomic bombs.

Oh, how disappointed the workings of those youthful minds were. Mr. Hogg quickly dispelled their wild notions with kindly words and smiles. That wasn't to say that he was a push-over in discipline, for he was of strict character, demanding absolute attention to his lectures and practical demonstrations but he did have that rare, fatherly ability to capture the interest of his students in class. An example of this relationship came through one morning when Mr. Hogg was duty yardmaster of the day.

He initiated a delay for Morning Assembly just as he was about to blow his whistle for parade, dispatching a boy to run with a message for the Head Master. Most of the school's youthful population had gathered in a corner of the yard, the noisy press of boys hard against the paling fence separating the school from pasture-land. They jostled and craned to see what was amiss, as a cow bellowed furiously in the adjacent field. Mr. Hogg realised that the bovine creature was about to give birth to a calf and he gathered the throng of curious lads to watch in a more orderly manner. It was an excellent opportunity providing for a live performance of this natural, biological phenomenon: a perfect supplement to his classroom teaching, by practical demonstration. He explained the process as the birth progressed and answered the many questions fired at him as the boys observed in wonderment.

Of course, it was not unexpected that the voice of some anonymous wag in the gathering would ask the obvious question:

"Sir, how did the calf get inside to start with?"

After the laughter of older, more knowledgeable fellows of the upper forms had died down, the smiling, unabashed Mr. Hogg told the voice to wait for biology classes on the subject yet to come, later in the year. More laughter from the seniors cut short any further comment as the rest, who still had much to learn, cheered loudly as the glistening, new-born calf scrambled to its feet in the long grass, on wobbly legs.

It was a birth of some conspicuous fame, and the boys pronounced the calf's name to be Gilmour because of a marking upon its back, roughly shaped like the letter G, black on white. With such a fitting end to a very unusual event, the crowd dispersed in noisy discussion as Mr. Hogg re-established normal routine with a blast on his whistle. The lads lined up and marched off by class to Assembly in the hall, half an hour late.

Mr. Hogg was almost fanatical about demonstration by laboratory-presentation, though one of his group-experiments ended up in near disaster, when his class prepared to make gas from coal. The boys had divided into small teams of five or six. Each group set up their apparatus and fired-up the Bunsen-burners on the long benches. Bubbles of coal gas had already begun to emerge from the bent glass tubes submerged in water-filled, glass beakers beneath bell-jars when, suddenly, a shout and some commotion erupted from a group at the far end of the room. The test tube in their experiment had become molten in the flame, causing a perforation in the elasticity of the white-hot glass and sending a flare of flaming gas halfway along the length of the bench.

The group scattered but one brave soul had the presence of mind to turn off the gas supply-tap to the Bunsen, before the master arrived on the scene with a bucket of sand. All was well but it turned out that the punctured test tube was the wrong type, not meant for the higher temperatures generated in this experiment. The incident drew an impromptu lecture on the fundamentals related to reaction of materials to heat and the processes of glass production.

On another occasion, the lads made simple steam turbines using Tate and Lyle syrup tins with bicycle-tyre, inner-tube valves, soldered into the lids. The valves (used without the rubber sleeve) provided the tiny hole through which a jet of steam was emitted against a multi-blade turbine wheel, which the boys also fabricated from tin – mounting it on a bracket soldered in place, next to the tyre-valve.

In a more interesting exercise, their master aimed at demonstrating the power of a vacuum, or rather the existence of atmospheric pressure. Again, Tate and Lyle provided the vessel for the experiment. Mr. Hogg first demonstrated the science-lab's small Magdeburg Sphere: a hollow, metal sphere in two halves, with a seal between the two mating flanges and each half fitted with a handle. After evacuating the vessel with a suction pump, through an adapter valve in one of the halves, he showed that two teams of fifteen boys each, in a comic tug-of-war, could not separate the two halves of the hollow, cricket-ball sized vessel. The boys discussed this seemingly miraculous event in disbelief, until their master explained that it was the ambient air pressure all around them, which held the two halves of the sphere tightly together, there being no air inside. It was a difficult hypothesis for the class of twelve-year-old lads to understand fully.

In a follow-up experiment, they heated a little water in their syrup tins, over Bunsen burners, until the water began to evaporate, producing steam as the water boiled. Lids, quickly snapped on, sealed the hot tins. Then, rapidly cooling the closed vessels, under a cold tap, condensed the steam inside and the class watched in amazement as their tins crumpled before their very eyes, in the hand of an invisible giant.

Edward's old science notes reflect the lesson value:

1. Evaporating water (steam) replaces air in vessel.
2. Tight lid fitted. No air gets back inside.
3. Steam condenses under cooling and leaves near vacuum.
4. Tin not strong enough to withstand outside air pressure and therefore collapses.

Conclusion A: Ambient air pressure greater than the very little remaining air in the tin.
Conclusion B: Tate and Lyle should make stronger tins!

The latter verdict drew a rebuke from Mr. Hogg, about humorous remarks and sarcasm written in exercise books but it's likely that he chuckled later, over the logical origins of Edward's observation.

Furthermore, the explanation relating to their terrestrial atmospheric pressure also suggested that the human population of the planet walk about with a column of air balanced on their heads. The column reaching some two hundred miles high, with a pressure of 29 pounds per square inch at sea level, probably weighed-in at about a quarter of a ton. Amazing! If the masters had looked from their staff-room window as they took tea at morning recess, they might have observed a strange sight indeed.

They would most certainly have nodded approvingly through their smiles, to see a number of lads staggering about the yard with their eyes rolled skywards, their heads scrunched down into their collars, their legs bent at the knees and their arms held out like circus performers balancing a stack of buckets upon their heads. One fellow took the mime one step farther by pretending to stumble, then exclaiming: "Oh Crikey, now I've dropped the lot; anyone want to give us a hand to stack it up again."

Surely, the masters would have agreed that if such odd behaviour was a product of a lesson sinking in, then long live teaching as a profession.

Of many instructive out-of-school excursions devised by Mr. Hogg, two were of particular interest to Edward. The first took them to a milk processing and bottling plant in Speke, not far from the Primary School at Hunts Cross and the second to a water-works near Woolton. In spite of his young years, Edward had a deep interest in mechanical machinery. His capacity for detailed observation often surprised his family and his masters at school but to him it seemed like a fundamental necessity, a natural phenomenon that fed his need to understand how things worked, more especially if he contemplated drawing pictures of them.

The function of machinery together with aesthetic beauty frequently found in mechanical devices had enormous appeal to him and Edward's artistic pursuits often reflected this strong affection for technical and practical detail. Maybe he didn't realise it at the time, but his fascination with engineering was also to have a lasting effect upon his future life, along with his artistic ability.

He revelled in the practicality and organisation of the production line seen at the milk processing plant, later discovering that he could re-visualise every process accurately, distinguishing himself with good marks when Mr. Hogg asked for written summaries of the visit.

Empty bottles went in at one end for steam cleaning, disinfecting and drying whilst milk, fresh from the farms, arrived in large cans at the front, for decanting into the Pasteurising process and thence, after cooling, to the vats feeding the bottle-filling carousels. After capping, the bottles emerged at the farther end, crated and ready for distributing. His own conclusions on the subject observed that milk delivery, both to homes and schools, was something too often taken for granted as a daily occurrence, with little thought, perhaps, to the route by which it came to the consumers. Having regularly watched the farm hands milking the cows at Lunty's farm Edward was quite aware of its source. Now he knew what happened to it before it reached the doorstep.

Another essential commodity often taken for granted is water. At the turn of a tap, there it is, crystal-clear and ready to drink. Mr. Hogg's water-works excursion, to examine pumping machinery used in potable water production, filled another gap in his students' knowledge. Clearly, the origins of water weren't completely unknown to them. Starting with the oceans and seas, the lads were aware of evaporation by the heat of the sun, of wind-driven clouds releasing rain as a precipitation when cooled in the colder upper airs, to fall over land and mountains, forming streams and rivers, which would in turn find their way back to the sea: a near perfect and natural cycle.

They were also aware of underground water – that which had permeated the land to penetrate and trickle through many strata of rock to form subterranean lakes far below the surface. Their visit to the water-works showed them how pumps brought such water up again, by the sinking of wells. In this case too, Edward's fascination with machinery brought sheer joy to him, as Mr. Hogg's class passed along the catwalks surrounding an enormous, Victorian beam-engine, driven by steam.

Dark-green paintwork, polished brass and copper pipe-work, valves and fittings, gleamed in the immaculately clean, tiled interior of the engine-house. The gigantic beam, oscillating slowly under the power of pistons, cylinders and great, cast-iron flywheels drew hundreds of gallons of water up from the depths with each stroke of its lift-pump action. Edward marvelled at the elegance of the machine and its architectural surroundings and he won good marks again with his post-visit summary. Mr. Hogg certainly knew how to stimulate his knowledge-hungry lads. His scientific excursions were legendary. He was a much-loved master whose class attendance never faltered.

It was, therefore, a devastating shock to everyone who knew him, when, one winter morning, Mr. Hogg died in tragic circumstances just outside the school. The weather had turned very cold and ice had formed during the night. A thin layer of snow on top of this created dangerous conditions on the roads. The road surface in front of Gilmour was much higher than the flagged sidewalk, its camber curving steeply down to the gutter and deep kerbstones. A bus driver brought his charge to a precarious halt at the crown of the roadway, realising the dangers in trying to stop on the steeply curving camber. In spite of this, as passengers alighted, the bus began to slide sideways down the icy slope, until its wheels hit the kerbstones. At this point, because of the induced tilt caused by the declivity of the road, coupled with the momentum of lateral movement abruptly halted at the kerb, the double-decker fell on its side across the pavement. A number of passengers and pedestrians received injuries but only minor cuts and bruises.

Mr. Hogg was far less fortunate. He was one of the first to alight, before the bus began to slip sideways, and had stepped several yards forward alongside the bus, before picking his way across the icy road to reach the sidewalk. The bus's sideways slide down the camber must have swept him off his feet, carrying him to the kerbstones, where he died of the terrible injuries sustained under the weight of the toppled bus. The Head Master's tearful and sad announcement at morning assembly left the school's population stunned and distressed.

This was Edward's first encounter with the death of someone he knew. It disturbed him greatly. Obituaries and tributes appeared in the local newspapers and in the school magazine. Many of the students, at all grades, wrote something themselves, as the means by which to express their sorrow at the loss of such a popular master. The episode left a lasting scar upon Edward's thoughts and memories.

12

Freddy and Edward had long experimented with various aspects of the graphic and visual arts in their spare time. For a while, they produced their own monthly magazine, with articles, illustrations and cartoons. Initially, they laboriously hand-printed the text, with pen and ink, but later produced better copy with the aid of a John Bull printing outfit. This was even more laborious than the pen-work because of the accuracy needed in type-setting the tiny rubber letters in wooden blocks, but the finished product certainly looked more professional, and they could print more than just one copy for free circulation among the neighbours. They could not reproduce the drawings and illustrations in the same way, of course, so these were hand-drawn for each copy, but since they only produced three, the additional labour wasn't too difficult for them to absorb.

They created strip-cartoons along story-lines with military themes, based upon their own ideas about war heroes and persuaded their neighbour's father, an elderly Scotsman, to produce all kinds of word puzzles, Riddle-me-Rees and short rhymes. Mr. McGeorge was over ninety, but he enthusiastically used his still very sharp mind to meet the monthly deadlines. Such a man was he – tall, though bent with age, his white-haired pate always covered by a scull-cap, and in his hand a gnarled stick to aid his shuffling gait. His rheumy eyes never failed to twinkle, nor his kindly face fail to smile when he banged upon the Weinels' garage doors with his stick, calling the boys to the fence, to collect his latest offering for their magazine.

He was of such strong constitution that when he tripped at the head of a staircase one day and fell headlong down the whole flight, dislocating a shoulder he made not a murmur when Fred Weinel and the hastily called doctor relocated the joint without an anaesthetic, though he took a nip of whisky before-hand and another afterwards.

Along with the magazine, the boys took a serious interest in the performing arts. Hand-puppetry intrigued them and they got out the books to search for information. They were lucky because there were two sets of children's encyclopaedia in the bookcase. Their Dad was an avid reader and he thought the boys should be, too. It wasn't long before they secured the recipe for paper-mache and soon had prototype characters in the making. They built a theatre from boxwood and using some lamps and switches powered by two-cell batteries (the type used in the old bicycle front lamps), from an electrical construction kit Edward had received for Christmas, they arranged stage lighting. With a little ingenuity, they fitted curtains that Edith made for them, and they used bits of *Meccano*, small pulleys and cord, so that the curtains opened and closed at the pull of a string. Their mother also made the glove puppets and costumes, so the whole idea became a family project.

Eventually they put a show together and performed a trial run for their parents. This being entirely successful, the boys decided to go public, subject of course to some script and theme polishing. Their father suggested that the boys might use the empty garage to set up the theatre. The Weinels had no car so there was plenty of room. Deckchairs and stools, upturned buckets and wooden boxes made suitable seating for the auditorium. Old blackout curtains, draped over the garage windows ensured that they would obtain maximum effect from a darkened venue with their stage lighting, which by then they had modified to use power from the dynamo on Edward's upturned bike. However, this needed a third party to turn the pedals and the boys agreed to offer any volunteer a free show, exempt from admission charge.

They sent out invitations to friends with a charge of thre'pence-a-head. Strangely enough they all came with thre'pences ready and happy to pay. Entrance to the makeshift theatre was by way of the big gateway at the side of the house. An old fireguard and a plank provided a booking office counter, from which to issue the tickets. Yes, the boys made tickets too! Admit one. Price, 3d. At the prescribed hour, with their public seated and a power-generating volunteer suitably reimbursed with his thre'pence, they got underway with the show. It went like clockwork. Everyone enjoyed it, in spite of the whirring bike-wheel and whining dynamo. There were no technical hitches!

When Mr. McGeorge's daughter, Mrs. Freeman from next door, heard about their theatrical enterprise, she told the boys that she expected to see their names up in lights, one day. Her husband had been in the entertainment business between the wars and so the compliment filled the boys with ambition.

From puppet shows, the brothers graduated into image

projection. A second-hand epidiascope, from Lord knows where, had come into their hands. The device wasn't much bigger than a *Kodak Brownie* box-camera but it had a focusing lens and a lamp inside, powered by batteries. A slot at the box back provided the means through which to slide pictures or photographs, for projection. This gave the boys the facility to make up their own screen showings, using hand-drawn pictures on long strips of paper, cut to the right width for the slot, and rolled for easy handling. The idea worked well with verbal running commentaries adding to the interest. Even then, advertising seemed to be part of any theatre or cinema show and the boys projected images from their collections of cigarette packets or photographs of sportsmen, given away with snacks and sweets from the tuck shop in the village.

As a natural progression to their new hobby, their Dad bought a small, second hand, 9.5 mm, *Pathe* movie-projector for Edward's birthday. Now there was something! Moving pictures! A transformer plugged into the mains provided low voltage power for the light, whilst the film transport mechanism was hand-operated, using a handle geared to a heavy flywheel to ensure steady rotation. The machine came with several cartoon movies in tin canisters: *Felix the Cat* and some early *Mickey Mouse* epics. What fun they had! Home-movie nights had come to stay!

Eventually, Freddy sent off for another type of projector. This used ‘transparent stills’ or black and white photographic positives in strip form. Most of these were ‘clips’ from popular movies of the day. The boys really did have quite a collection of picture-throwing devices and thus, they could put together some quite respectable entertainment. They did not get their names up in lights, but in later years, the hobby was to have interesting significance in Freddy's adult life.

Cricket in the summer months took up much of the boys' spare time. Freddy already had a rather special bat. Made by the famous 'Gray Nicholls Company,' and having a ‘five-spring’ handle, the bat was Freddy's pride and joy. Fred Weinel brought home an enormous bag one day. Turning it out, the boys found several sets of stumps and bails, leg-pads for a couple of batsmen and a set for wicket keeping pads, gloves, balls and another bat. A colleague of their father had donated the bag and its contents on hearing about the boys' keen following of the game. The man had been a county cricketer in his youth, and he thought Freddy and Edward would make the most out of the equipment, well used and old though it was. Certainly, they did get many more years of use out of it. They could mount quite a game when they pooled all their newly acquired equipment with that of the other lads in the locality.

In one match, played on an unfamiliar pitch, the opposing

team elected to bat. Freddy, being the captain set the fielders of his team. He took his own position at midwicket, very close in, looking for a catch from a scooping-bat off a slow spin-bowler. However, a well-struck ball (travelling very fast, at head height, when it reached Freddy) broke through his catch-cupped hands. The ball hit him square on the forehead. Its force knocked him over backwards, and the ball went on to the boundary for four.

The game stopped abruptly because Freddy didn't get up from where he lay, knocked out by the tremendous blow.

An enormous lump appeared, as big as an egg. It turned red and blue very quickly. He was in an awful mess. Abandoning the game, somehow, the lads managed to get Freddy home. He bore a lump in that spot for the rest of his life. Call it a scar of youth if you will, but it might have killed him. He recovered well and it wasn't long before he was hitting fours and sixes like the best of them again.

By then, in those early days of the new media, sporting events, including Cricket, were televised frequently. British television in its infancy had arrived. For those who could afford the cost of a receiver, the Sutton Coldfield station beamed programs in black and white, to aerials in many parts of the land. The first "H" shaped aerial Edward ever saw belonged to Mr. Rhodes. He was the partner-owner of the only petrol station and electrical appliance vendor in the village. His shop was the first in the district to sell TV sets, and, of course, he had one in his house, too. He lived only a few doors down from the Weinels and his three children often played with the Weinel boys. Sometimes, Michael, the eldest of the three Rhodes children, invited Freddy and Edward to watch test-cricket on TV, instead of just listening to the radio-commentaries.

In those early days of television, cathode-ray tubes were very small and needed a magnifying lens to amplify the viewed image. Such lenses, fitted in front of the tiny screen tended to distort the picture; and if one was unlucky enough to sit at an oblique angle, as Edward frequently was, then images seen through the convex glass of the lens appeared only as warped likenesses, bent in optical parallax. It was another visual illusion, an optical phenomenon that fascinated Edward just like the illusion of foreshortening of objects he'd noticed from the upstairs of buses, all those years ago.

Television was an expensive luxury that few people could afford in those early days, just as with cars and telephones. Private cars were only just beginning to emerge onto the roads again, after lay-up during the war but, as post-war austerity eased and a little more affluence returned to ordinary folk, new telephone wires appeared, too. Busy workmen dashed about with their little green GPO vans and ladders, to connect those who felt they needed a private audio-link with the nation's telephonic communications system. At the same time TV aerials began to appear upon the chimneys of better-off homes.

The Weinels had none of these things but the boys often sat glued to their father's radio, listening to running commentaries on tennis at Wimbledon or when England played test-cricket for the Ashes. Sometimes they listened to a commentary for a county match, especially if one of their counties of birth played. The boys avidly followed their particular team's progress in the weekly county matches. Since Edward was born in Essex, this was his team. Freddy was born in Surrey like their Dad and so they followed that county. They kept scrapbooks, recording the scorecards of their respective teams and photos of their favourite players in action, cut from the weekend newspapers.

One sport that didn't appeal to Edward, however, was swimming. At Primary school, after moving up from the infant's to the junior grades, the classes marched once a week from Hunts Cross to Woolton's public baths. Edward's initial induction into this form of physical exercise was disastrous and he disliked the idea of swimming for many years after. He had retained an inherent fear of water, ever since the episode on the beach at New Brighton, years before. The sports teacher, Mr. Lister, was wont to target a suitable non-swimmer or beginner floundering about in the shallow end and, using an enormous ladle, he seemed to enjoy dumping some five gallons of water over the unsuspecting head. As if this wasn't enough, he would then use the empty ladle to dunk the child below the surface. As a non-swimmer himself, Edward considered such an act to be terrorism.

He found himself target-for-the-day on one occasion. As a wearer of glasses (removed for the swimming lessons), he was already at a distinct disadvantage. He did not see it coming, and half drowned under the torrent of water that Mr. Lister dumped over his head, to make things worse he then received a dunking. His ears roaring, Edward surfaced, choking, coughing and spluttering only to hear the sound of laughter and Mr. Lister's voice shouting words of derision. Edward did not like that at all.

It was altogether different at the Gilmour. They used the Public Baths at Garston, the weekly excursions being conducted with seriousness and in a more civilized manner. Under these more humane conditions, Edward did manage to learn the rudiments of floating, treading water and some swimming strokes but it was never really his kind of sport.

Soon enough, the second attempt at scholarship entrance exams came around. In their second year and having reached the age of thirteen, those who were in the ‘A’ stream were cramming like mad to prepare for the Thirteen-Plus.

The Art school exam was a two day affair. The first day for

the academic subjects and on the second day, candidates attended at the Art College, for the drawing and painting ability tests. This part seemed relatively easy to Edward and he came away quite pleased with his efforts. There was a waiting period of course and after several weeks, the results came through. Announcements for the Grammar, Technical and Commercial entries came at morning assembly, one Monday. Two days later, there was a special announcement after prayers.

The Head Master, Mr. Simpson, was quiet-spoken at the best of times, and half way back in the assembly hall it was often difficult to hear what he was saying. The hymn and prayers were over and Edward was not paying much attention as ‘Simbo’ made his usual announcements. His daydream came to an abrupt end with a sharp push from behind, accompanied by an insistent voice, fiercely whispering, “Go on Weinel, he's calling your name you idiot; you've passed.”

Edward made his way to the platform as if in a real dream. His heart pounded fit to burst. Passing the masters in the side aisle, he saw their smiling faces. He heard the applause loudly in his ears, and finally he felt the firm clasp of the Head Master's hand shaking his as ‘Simbo’ uttered congratulatory words that remained unheard for the clapping of many hands. Here was success and fame at last. On the platform, Edward turned to face his peers and his masters, buoyant with elation and proudly clutching the document proclaiming that the Scholarship to Art School was his.

Part 4

Adventures and Fresh Beginnings

13

The scholarship examination results delighted Edward. Now he had a real sense of purpose. The achievement gave him a sense of purpose, pleasing him beyond measure; and his parents didn't hide their pride either. It was 1951 and he would begin at his new secondary school after the summer holidays, following an interview with the Principal, scheduled for two weeks before the first term. Meanwhile, during his final days at Gilmour, what bliss! Feted by closest friends and by some of the more sympathetic masters; life just couldn't have been better.

Nevertheless, ever-lurking doubts lay beyond the euphoria of passing the entrance exams to Art School. Edward's success at the scholarship didn't alter some plain facts. Apart from his abilities in graphic arts and crafts, his school reports were not at all commensurate with his success. By no stretch of the imagination was he an academic. There were no accolades to his scholastic competence at other subjects. Edward's final report reflected this when some of his masters pointedly noted:

"There is no concentration at all in the important subjects. I hope, for his sake, that his ability in art will compensate for his other weaknesses."

"How a boy, who is so obviously good at Art, can produce such scruffy written work, is beyond my understanding!"

Another, adding an observation to highlight Edward's poor marks at mathematics, wrote:

"The gaining of his scholarship to Art School, though possibly well deserved, doesn't alter the fact that this boy should concentrate much more on other subjects if he is to make anything of himself!"

Edward wasn't too proud to concur that there was much truth in the these statements. But this didn't diminish his pride and pleasure at winning his place. After all, the Head Master's comment at the foot of the report seemed to contradict the other master's scathing remarks: "Very satisfactory. Success in art scholarship is very well deserved." Moreover, the prospect of specialisation in something he so enjoyed came second only to pleasing his parents.

Since the location of his new school was close to the city centre and because of its distance from Hunt's Cross, biking was out of the question; instead he would use the public bus services.

The newly opened Liverpool Corporation No.72 bus service that passed the Weinels' home hourly, along with the old Crossville schedules, would take him almost to the door of his new school and the city's Education Committee would issue a travel warrant or bus-contract to cover the fares.

Edward finished at Gilmour and summer holidays began. Freddy and he planned to follow their usual out-of-school activities but first, Freddy went off to his CCF camp, somewhere in the wilds of nowhere, while Edward went to Gilmour's annual camp on the Isle of Man for the second and last time. On this occasion he'd gathered his own friends to form a group of eight; enough for one tent. He was Tent Leader this time, carrying the appropriate responsibilities for the general conduct of his group, both on and off camp. They used the same football field as in the previous year, close to the little village between Port Saint Mary and Port Erin, at the southern end of the Island.

In those times, the town's main economies lay in fishing, mainly for herring to produce the famous Manx Kippers but tourism during the summer season also provided a major source of income to the local residents and business people. For those visitors to the island who enjoyed quiet holidays, away from the razzmatazz of fun fairs and bright lights, the locality was perfect. Wild shorelines, quiet seaside beaches, countryside and rocky moorland provided nature-lovers with everything they could wish for – not least of which was the view from Bradda Head, a prominent headland, with its ancient watch tower at the summit overlooking Port Erin.

The beaches were clean and relatively safe but the seawater was cold in spite of July's warm sunshine. Swimming was therefore not on Edward's own holiday-agenda of things to do. There were many small shops in the two towns, some selling all kinds of souvenirs and locally made craftwork. Edward spent time looking for a gift to take back to his mother and eventually he found a small, rose-tinted, glass decanter with a stopper and six matching glasses. The set wasn't real crystal but on his very limited budget it was all he could afford.

To some of the camp veterans, a fearful visiting place was the local abattoir. Many a first-year boy found himself taken to observe the slaughter of livestock, lured by gruesome stories of horror, blood and gore, or egged on by the 'dares' of older boys. They had dragged Edward to the place (under duress) during the previous year's camping holiday. It most certainly didn't appeal to him, and he decided to avoid a second visit. The gruesome spectacle of sheep and cattle under the knife or humane-killing machine was all too ugly.

Edward could not bear to watch the struggle for survival and the lifeblood pouring away, as a sheep died to provide meat for affluent dinner tables. He thought he could become a vegetarian quite easily, if it came down to it but, he supposed, there would be a few unwary first-timer lads taken to the place by the same older tricksters bent on the sight of blood and gore.

Brian Smart, known amongst the lads as Smarty, had joined Edward's tent-list at the last minute, as the eighth member of the group. A quiet fellow from an affluent background, Smarty had serious interests in both music and ornithology. At school, excused normal music lessons with the rest of Form 2A, he attended specialised classes for violin, twice a week in the assembly hall, whilst the others struggled without enthusiasm to acquire mastery of the dreaded recorder in Mr. Plumley's classroom. Whatever discordant scratching and squeaking emanated from the violin classes, no better wailing and howling could match that of 2A's terrible renderings of 'The Vicar of Bray". There were very few musically inclined fellows in the class, but Smarty never ever suffered either nasty jibes or derogatory remarks about his own serious interest in the violin. Even the roughest lads of his class appreciated that he had made a voluntary choice to learn a difficult musical instrument.

Fortunately, Smarty didn't take his violin with him to camp, but he'd filled his kit-bag with books and equipment relating to his other interest, bird watching. He explained to Edward that he wanted to make a trip to Castletown, where he could observe the seabird population on the rocky, estuary shores. It sounded like an interesting 'off-camp' spree, so three or four of the group decided to make a day of it, and accompany him. They made their intentions known to the camp captain, the Senior Master, and upon his agreement, they passed the evening planning for tomorrow's journey and studying maps.

The lads studied some of Brian's ornithology reference books too, in preparation to their excursion. In all honesty, Edward had little interest in bird watching, his thoughts lay more with the adventure of making a trip without the aid of grown-ups. Brian, for his part, was not a very practical chap and seemed pleased to have the trip organised for him, and to have the company of some uninitiated, if slightly disinterested, fellows.

Next day, after having an early breakfast and collecting packed lunches from the kitchen tent, they set off to the station with the airs of intrepid explorers bound for darkest Africa – or perhaps more like adventurers lured to the Spanish Main in search of treasure. The masters who saw the boys off must have grinned to themselves as the boys marched off with backpacks and stout sticks, plucked from the field's hedge-row; the sticks strongly recommended for their activities on the rocky shoreline at Castletown.

Brian had an enormous set of binoculars hanging by a strap from his neck and one of the other fellows carried a telescope. All carried backpacks, the weight of which caused them to bend forward as they marched off, lending an air of resolution to their stride. They only needed a one legged man with a crutch, a three-cornered hat and a parrot on his shoulder, to complete the band of intrepid searchers!

The island's narrow gauge railway system operated on a 'batten' system, and held a proud record for punctuality. Once underway, the train took the lads through picturesque scenery and quaint stations, making for an enjoyable and interesting journey, although the well-worn, wooden seats were hard and uncomfortable. The train puffed, hissed and rumbled through glades and dells, green then with the summer foliage of bushes and undergrowth that clung to the rocky terrain.

Spot-on-time they arrived at their destination and alighted in an unfamiliar town. It didn't take them long to find their way to the shore, and the bird sanctuary so dear to Brian's heart. The tide was out and indeed there was a vast expanse of wild estuary to explore. Unable to escape with the out-going tide, pools of captive seawater lay everywhere among the rocks. The troupe found their sticks very useful, as they clambered and scrambled over and around the tumbles of slippery boulders and sharp rocky outcrops. One thing they noticed, however, was the very strong smell of rotting seaweed. It pervaded everywhere, a stench that eventually became quite nauseating.

Birds there were, thousands of them! Probably millions! Brian was ecstatic. Excitedly, he swept his binoculars this way and that, calling out the names of birds he recognised and describing others that he did not. He had already seconded one of the group as his official scribe, and had him scribbling frantically in a notebook. Brian was in his element! This was his arena. Frequent reference to the small library of books, which he carried in his backpack, added even more sightings to his ever-growing list of observations. The day passed quickly as they scrambled about among the rocks, discovering fresh interest in fish, crustaceans and aquatic creatures, temporarily marooned in the rock-pools left behind by the sea.

Hunger eventually overtook them, and the boys stopped for lunch. They scaled the side of an island-like rock with a flat top, which provided them with an ideal and relatively dry spot to sit and eat. Opening their lunch-packs they found hard-boiled eggs, wedges of cheese and sandwiches, great 'doorsteps' of bread, spread with Manx butter, enclosing corned beef and salad. They ate their lunch, seated in a circle upon the rock's smooth surface high above the surrounding tidal estuary-floor.

The meal finished, and their bellies full, the lads spent some time gazing at distant coastal skylines through the telescope – each taking his turn as they discussed and argued over the topographical nature of their position, until one of their number drew attention to the sea.

"Tides turned." Their perceptive companion mused. Someone remarked on the nautical flavour of the statement. "There's a following breeze too," the young tar observed, lifting his eyes to the loom of gathering clouds, grey and ominous, drifting inland from the sea, while the others lifted their eye-brows in awe at the fellow's obviously knowledgeable, clearly superior, seafarer's comprehension of the elements.

"Is that good or bad?" Someone asked with concern.

"Dunno," the weather-wise sailor answered, dashing his companions' new-found respect for him. He seemed bemused at the question, but emphasised with sudden urgency that the sound of rushing water that slapped at the farthest end of their elevated island-rock would surround them and cut them off, if they did not move rapidly. "Hey, you lot – the tide's comin' in fast; have you seen all the water at the far end of our rock?"

"Well perhaps we should be getting back now anyway," Brian decided, "we've covered most of the shore and the town seems to be miles away. Let's hurry. I don't fancy swimming back."

They prepared their backpacks quickly, slithered down the rock's landward side and bent themselves to the task of 'getting back'. The rain clouds loomed ever closer and darker as the fine day began to give way to late afternoon and the probability of more rain. Ever-present was the sound of the surging tide, urging the boys on as it followed them towards the distant town. The rocks seemed steeper and more slippery on the way back. Someone fell into a shallow pool when he slipped on the slimy boulders. The others helped him climb out, chuckling sympathetically at his round utterances, as he wrung-out his shirt, but Edward promised him beans on toast and a cup of hot tea once they reached the town. The tide followed swiftly behind them, sloshing and swirling at their fleeing heels as they reached the flat, sandy shore and scrambled up stone steps to reach the coastal road, beyond the swirling water's reach.

None of the boys seemed to realise what a lucky escape they'd had. If they had remained upon the rock much longer they would have been marooned. The local lifeboat would most certainly have had to rescue them.

Later, still breathless from their exertions and not without a little hunger, they all tucked into the beans on toast and tea at a little side-street cafe not far from the station: Edward's treat. He really felt quite grown up!

The boys reported to the Camp Captain upon their return, grabbed some supper from the kitchen and retired to their tent to fall thankfully upon their beds, tired and ready for sleep. After the long day, they felt shattered. Amongst the other occupants of the tent, there seemed to be an air of indignation and it wasn't long before the someone complained about the stink. That awful smell of rotting seaweed pervaded strongly the confines of their canvas shelter. It clung, undiminished, to their clothing and shoes. It was bad enough to draw angry comments; rightly too, for it was a horrible odour.

To keep the peace, Edward suggested that the offending garments should hang outside to air, ready for washing next day. The heavy, over-night rain however, relieved them of that chore.

Among other activities, a supervised visit to the watchtower at the top of Bradda Head featured high on the agenda. This was an interesting interlude, since the tower was supposedly of early origin, possibly Viking, some said. The narrow winding staircase within the confines of the stone tower did not much appeal to Edward and he didn't venture to the top. A strange feeling struck him at the beginning of the ascent and he got no farther than two or three steps upward. He had discovered claustrophobia, a newly felt aversion to enclosed or over-crowded spaces. This was something he hadn't experienced before and he found it hard to understand. As a kid, he'd crawled through tunnels inside stacks of bailed hay or straw, in Lunty's barn. He'd never felt claustrophobic then, but at the tower, his senses rebelled and he elected to remain at ground level. Edward remained outside, sitting at the foot of the tower whilst the others went up to the tiny turret at the top.

Eventually he ventured to the edge of the cliff, to peer over and downward at the precipitous face. He remembered hearing reports of an accident from earlier years, involving a boy from the school. The boy had climbed upon the rocky escarpment and had taken a fall from a jutting ledge some distance lower, much nearer the sea. He had tumbled some thirty feet onto rocks and sand below, and was lucky to get away with only broken arms and collar bones. Subsequent visits to Bradda Head were thereafter under strict supervision. Edward sat for a while whilst the others called and hallooed from the turret, to hear their voices echoing within the stone confines of the tower.

He wished that he'd taken a sketch book with him, for the grassy cliff-top was several hundred feet above sea level, giving wonderful views of the port, the town with its small harbour and beaches. He realised that he had also found an affinity with the island that remained with him all his life. Alas, he never returned.

The Manx holiday ended and the Gilmour camping expedition landed again at Liverpool. Edward bade farewell to his school-friends before boarding the bus for Hunt's Cross. His time with them at Gilmour was over and he would make fresh friendships at his new school in September.

The remainder of the sunny, summer holidays drifted by at a leisurely pace. David was by then old enough at six to participate in some of his older brothers' activities. He and Edward had become closer. Chasing around the garden, as Fred and Edward had done, gave them hours of healthy exercise and fun. David had a few friends of his own, of course, and spent some of his time out of the house, playing with them but he had become very interested in agriculture. His heart was set on becoming a farmer.

As a prelude to this burning ambition, he had a fine collection of toy farm-animals, together with Dinky Toy tractors, machinery and implements. He played constantly with these and took every opportunity visit Mr. Lunt's farm with Freddy and Edward, when they were lucky enough to secure holiday or weekend jobs there.

When the two older brothers decided to do some farm work again during the remainder of the summer recess, David insisted on going with them. Lunty was more than pleased to have their help and they strove to please at whatever task he put their way. They even managed persuaded their Dad to go along, too, one day, although he had spent some of his youth unhappily, on an uncle's farm. No one could say that he found much pleasure in the notion that he should participate in his sons' proposal to help Lunty but he went along anyway, to please them.

Harvesting had begun and Lunty's men, hard pushed to complete the task before the weather turned, were glad of the extra hands. Mr. Lunt sent the Weinels up to his top barn, at the other end of the cow-field – the one with its narrow entrance where the 'Elephant' tree stood. The long meadow contained good grassy pasture, used mainly for grazing Lunty's dairy herd. The brothers often used it too, for cricket and sports activities because the grass was fine and well cropped. At the farthest end lay Old Hut Lane and the farmer's top barn.

Edward decided to cycle from home, up the well-worn cart tracks. Freddy, David and the boys' father went on the cart pulled by Lunty's shiny new, grey Ferguson tractor, the latest thing for small farms.

A new farmhand named Norman, recently demobbed from the army, drove the tractor, and Freddy had struck up a friendship with the fellow, because of his own interest in the Army. The day's task found the Weinel family, father and sons, on top of the sheaves already stacked in the barn, catching and stetting as fresh loads came in from the fields whilst the farm hands pitched-up the sheaves from the carts to the top of the stack, under the roof of the barn. Edward's job was to catch and throw the sheaves to the back, where Freddy and their Dad placed and packed.

Edward turned at one moment to catch the next sheaf but he was a little late, having lost the rhythm for a second. The sheaf caught him full in the face, plucking off his glasses as he toppled backward, knee deep in corn. Then another came, and another, and another, the sheaves burying him almost completely. His shouts for them to stop went unheard until the mounting pile of sheaves at the edge of the stack prevented the men below from throwing up any more.

"What's the matter then," came the shout from one of the hands below, "can't you keep up then?" There was laughter. Half buried as Edward was, and next to blind too, his glasses gone, he was lost beneath the pile of sheaves. His Dad scrambled over to where he lay and threw aside the heavy corn sheaves. His grin disappeared quickly when he saw that Edward's glasses were missing. They searched in vain but it really was like looking for a needle in a hay stack. The spectacles had those round lenses, held in frames of wire and horn, with springy, hook shaped arms to grip around the ears. The earpieces had not gripped very well though, and the lost glasses remained unfound, search as the family did. In the end, they gave up. Resignedly, Edward's Dad said, "Never mind Son, we have another pair at home, old ones true, but we'll fix 'em up for you when we get back."

The suggestion didn't help much then because Edward still had to get back home with his bike. He was as blind as a bat without his glasses. "Perhaps you'd be better off going back on the cart, Ted," suggested his Dad, "Freddy can take your bike if you like."

Determined to make a go of it, Edward said, "No, I'll be OK!" "After all," he thought, "what harm can there be in the fields. It's not as if there's any traffic to worry about."

The last load of the day was in and after a cup of tea, the workers began to move off towards the tractor and its trailer. Edward moved off on his bike while the men were loading up the cart.

Cycling with restricted eyesight is difficult. In Edward's case, it was a choice between crossing his eyes, seeing double but fairly clearly or uncrossing his eyes, seeing singly, but not in focus.

Another possibility, he discovered, was to close the worse of his two eyes and go for it with the better eye! This way there was no crossing and uncrossing to do but distance judging was almost impossible. Anyway, once out of the barnyard and onto the cart track in the fields, things weren't too bad. The track was wide, although in two well-worn ruts with tufts of broad leafed grass between. All went well until he got to the narrow part of the cow-field where the hedgerows closed in to form a funnel to the gate, close to the 'elephant' tree. There were ditches on both sides, well hidden by undergrowth, hawthorn and large clumps of nettles with their peculiar hairy, serrated leaves. He could not see them clearly, but Edward knew they were there.

At this point, things came unstuck. The deeply rutted ground was soft and wet near the gate, stomped by the cows that would wait there at milking time. Fortunately, there were no cows in the meadow that day. The bike's front wheel churned and slipped among the ruts and wet hoof holes, threatening to precipitate Edward into the horrid mixture of mud and cowpat.

He was traversing these impediments, and tossing up mentally whether to dismount or not, when the tractor and its trailer came up the track behind him. The trailer, loaded up with the men, their sons and the rest of Edward's family, legs dangling on either side, bumped and slithered through the mire. The closed gate was just a few yards ahead, but Edward could see that the tractor would reach it first, stopping for someone to open the gate and close it again afterwards.

Edward decided to ride on, pass the stationary trailer and tractor, go through the opened gate before the tractor, and press on into the next field. The plan came to grief when Edward tried to make his passage beside the hove-to trailer.

His progress, being impaired by the much deeper ruts and cow-hoof stomp-holes at that point, the front wheel of his bike slipped and churned ineffectively through the sticky, wet mud. As Edward came alongside the trailer, the space grew narrower, making it more difficult for his bike to pass easily. He was wary of the ditch to his left, and the beckoning nettles. Then the front wheel hit a deeper rut and Edward wobbled off into the waiting cluster of hairy-leafed stingers. It seemed like the biggest too! He went down in a heap and got nettle-stung for his trouble, cursing his bad luck. Meanwhile, someone had opened the gate and the tractor with its load drove through. From the cart, there were some ribald comments and laughter over Edward's predicament. That stung him too!

His people, his Dad and Freddy, seated on the other side of the trailer, did not see what happened until the guffaws of the others drew attention to Edward's plight. The gate opener was on the other side too, and he shut the blooming thing without noticing Edward, wallowing in the nettles on the edge of the ditch, trying not to roll down into the brook that trickled below.

Eventually he managed to retrieve his composure, close to tears though he was, getting his bike to the gate, opening it, wheeling the bike through and closing the gate again before remounting to follow in the wake of the fast disappearing trailer and its circus of amused farm-workers. Edward was angry and he itched something awful! So much for a disastrous day on the farm, he thought.

However, there was sympathy in abundance when he finally got home, and a bottle of calamine lotion for the stinging-nettle rash. If he'd had his glasses he could have seen well enough to pull some Dock-leaves, to rub on the stings but then, he probably wouldn't have had his spill into the nettles either.

There was definitely a humorous side to the incident. Edward couldn't think what at the time but he managed to raise a smile over it later.

Soon after the nettle incident and before the end of the summer holidays, Edith decided it was time to go into Liverpool, to buy Edward's new school uniform. They went to Manners, a specialist shop that stocked most school uniforms for the districts of Liverpool. They measured Edward up for grey trousers and black blazer, took head size for a black cap, and neck size for grey or white shirts. It was a sizable order, but Manners were specialists and provided everything needed, including the blazer and cap badges, in blue, yellow and red, sporting an image of Liverpool's famous Liver-bird.

At last, the day came for Edward's interview with the Head Master of the Art School. He and his parents took a sevenpenny bus ride into Liverpool, to keep the appointment. Mr. McFarquer welcomed them into his office. He was a big Scotsman with an Edinburgh accent and spoke at length about the themes and expectations of the school. He impressed upon them (Edward in particular) the importance of academic studies, even though the school emphasised the development of potential talent and capability in the arts. Furthermore, if, after two years, students showed promise, achieving the required academic progress, then they might expect a further year of secondary studies culminating in the GCE exams; the passing of which might then gain them entry into the Art College proper. At the College, successful students would specialise and polish their abilities in the chosen art.

With these goals fixed in Edward's mind, the Weinels returned home to Hunt's Cross to ponder the new beginnings that were inevitable and imminent. Edward was almost ready for the fresh-start promised by his success in the Art School Scholarship exams. That night he slept fitfully, beset with many questions coursing through his mind and with high hopes for his future in an artistic following.

Part 5

Failure and Disappointment

14

Standing on Gambia Terrace, directly opposite Liverpool's fine (though still unfinished) sandstone Anglican Cathedral, Edward's new school was within sight of the city's College of Art that lay just a short distance down the hill. Unlike the College, however, the Secondary School to which Edward had won his scholarship occupied two large terraced houses of the Victorian era, adapted and interconnected to provide the necessary access and space for classrooms. There were five storeys to each of what once were fine mansions, including the basements, which had accommodated kitchens, pantries and servants' working quarters. To reach the converted attics where third-year students studied for their GCE exams, there were therefore three flights of stairs to climb from ground level.

Of the two front doors, with their broad stone steps, ornate railings, Corinthian columns and imposing porticos, only one, number 12, served as the main entrance to the school. It bore a sign declaring this to be so. The door numbered 11, was permanently closed. Only the teaching staff had access through the front entrance. The students entered from the rear of the buildings, by way of Back Percy Street and a small gate opening into a large paved yard.

In times gone by, tradesmen with their horse-drawn vans or handcarts probably used the alley to deliver their wares to the gate. Gentry once lived behind the ornate facades of those tall, narrow mansions: rich cotton-merchants perhaps, or the owners of shipping companies, or the families of sea captains. Whoever they were, housemaids, cooks, scullions and boot-boys served them in rich, opulent surroundings under the eagle eyes of a butler.

As a school, the interiors of the converted houses retained little to remind the teaching staff and students of a history that embraced prosperous, quiet households, set in well-to-do surroundings. Only bricked-up fireplaces with ornate mantle-shelves, and finely worked, plaster ceilings remained. Large, wood-framed, sash windows, where lovelorn daughters may once have sat watching for their beaus, let in bright natural light to fill those ample rooms. Gone were the carpets and rugs from India, the lace curtains and velvet drapes, the plush furnishings, ancestral portraits and warming pans for silk-sheeted beds.

In 1951 the passages and stairs-ways echoed to the sound of rowdy, middle and working-class children, their foot-falls pounding upon the linoleum-covered wooden floors and stair treads; boys and girls schooled together again, in spite of the modern gender-separatism ideas found in Secondary Moderns like Gilmour and the Grammar schools of those times.

On his first day of term, feeling a little self-conscious dressed in his brand new uniform, and a tiny bit nervous too, Edward stepped down from the bus in Canning Street, the main road running parallel with the street where his school lay. He walked the short distance to report at the main Art College building, climbing the steps to pass, not without some veneration, through its imposing portals; just as he had on the day he sat for the entrance exams. He joined the other scholarship winners of the first-year junior students as they mustered in a beautiful, wood-panelled lecture room with its tiered, semicircular auditorium.

The eyes of Grecian and Roman, sculptured-marble faces, peered down upon the gathering from lofty niches as the room filled. Perhaps they wondered why the girls should tend to congregate on one side of the room and the boys on the other – a strange phenomenon born of twentieth century adolescence. The sound of teenage chatter grew as more new students arrived. Fleeting, veiled, half-hidden glances and whispered comment behind hands, fluttered amongst the two groups, reflecting a youthful interest and appeal, in a long denied co-educational arrangement.

On the stroke of nine the Head Master swept in, an umbrella hooked over one arm, an old-fashioned bowler-hat in hand and his unbuttoned Macintosh trench-coat flying wide at the hurried stride of his entrance. Silence fell upon the assembly. He welcomed the new students in a loud voice, before announcing instructions and calling-off names, under an alphabetical stream system similar to that of Gilmour's.

As it happened, Edward found himself in the 'A' stream again, as did Riley and Ferguson, the two boys seated next to him. They had already exchanged views about their 'new-school' misgivings and concerns, promising in a youthful pact that they would stick together come-what-may, even before the stream-allocation announcement.

Edward's new companions both carried the forename Peter, but after leaving Primary school, it was customary for boys to use their surnames with each other, just as the teachers did. Either way, the two Peters were his new confidants, with whom he would share his friendship and continued secondary education over the coming months, in form 1A.

Following his address, the Principal assembled the children on the pavement outside the college. He counted them meticulously before marching them in a double column up the road to their new school. The Principal led the way, conducting the procession rather in the manner of a military drum major, except that he used his rolled umbrella instead of a mace to direct his seventy-odd, marching recruits. When they reach the intersection of two roads at the top of the gentle rise, he detained traffic with convincing movements of his brolly, without anyone losing a single step. They crossed the intersection diagonally to reach the farther corner and the entrance to an access-drive that once would have carried horse-drawn carriages and cabs to and from the mansions of Gambia Terrace.

Still tramping in two straggling lines, the students followed obediently behind their triumphant, bowler-hatted Principle. They passed between open, ornate wrought-iron gates at the drive entrance, their feet crunching upon gravel as they passed ill-kept gardens of trees and shrubs that separated the drive from the main road skirting the Cathedral's grounds. Halfway along, they halted before the two mansions that were to be their place of study for the next two to three years. They submitted to another head-count, (a protocol that seemed quite unnecessary, since they had marched only a mere three hundred yards from the college) before receiving the privilege of formal entry through the school's front door for the second and last time, the first having been on interview day, accompanied by parents. Thereafter, they would use the rear entrance in the back yard.

Settling down into a new environment was not easy for Edward – with new timetables to copy down, new teachers to meet and new friends to be made. Very few of the students had the companionship of an acquaintance from their old schools. For the most part, each student was an individual, a singular scholarship-winner from his or her former school, thrown into a wholly new environment: a new school-life devoid of past friendships.

According to the class time-tables, studies in the arts were considerably more than the once-a-week routine Edward had experienced at Gilmour. Drawing, Design-in-Form, Painting, Craftwork and the History and Appreciation of Fine Arts were among the many specialised subjects on the curriculum. Metalwork came under the heading of craftwork and the curriculum dedicated a whole Friday morning to this subject, conducted in the metalwork shop at the main Art College building.

In time, Edward found great satisfaction in this pursuit and developed a comfortable working relationship with the metalwork master; a Kentish man who immediately recognised Edward's southern accent as not typical of Liverpool. Edward learned well; and, even after all the passing years, he retained several examples of the copper-work he made during those Friday-morning sessions.

He already knew the graphic-arts master, from Gilmour days. Nevertheless, it surprised him to see a master whom he already knew, on the curriculum timetable. He had not realised that his former master had left Gilmour to join the staff at the Art school. It was therefore a pleasing encounter on the first day of classes, especially since the master remembered Edward too, remarking upon his success in the scholarship exams.

Studies began in earnest once introductions were over and lesson-times set up. Homework, the teachers insisted, was important to the accumulation of knowledge. Sports did not carry the same significance however; probably because the school was city-bound and had no playing fields. Nevertheless, the history and geography master, an avid cricketer, managed to arrange a weekly excursion for a whole afternoon at Caulderstones Park, during the summer months. This was uniquely an activity for the boys, when the interested ones at least, were able to practice and polish their cricketing skills. The girls had a similar distraction at hockey or some such sport, but they went to Sefton Park.

Once a week in the winter months, the mixed class attended at the local public baths, for swimming lessons. Any other form of physical training, as with football, was a non-starter, and anyway there was little time on the curriculum timetable for any other form of sporting activity.

For those in the 'A' stream, studies prepared the students for pursuits in painting, sculpture, metal-work with copper and precious metals, design work, advertising, illustration, sign writing and lettering, etc. The girls studied design in millinery, dress and garments – training aimed at work in the fashion industry. Lessons in Lettering were of particular importance since the preparation of readable and aesthetically beautiful lettering touched all facets of artistic studies and prospective professions. In the 'B' stream classes, emphasis was in lithography, graphic arts and design studies. The boys trained for the printing trades: the girls, for design and shop window dressing.

All students studied the History and Appreciation of the Fine Arts; to Edward it was a loathsome subject if only because it was full of unpronounceable names and a million dates to remember. In reality, he did enjoy the in-between stuff related to method and techniques, developed by the various 'schools' of artistic thought and application.

The Head Master himself conducted these studies, beginning with early cave drawings, their significance and probable meaning; then onward through history, as 'man' became more civilised ...the Persians, Greeks, Romans, various European schools and so on. Architecture especially appealed to Edward, perhaps because it related to his technical interests.

While Edward settled into his new school environment, Freddy found good fortune in his selection for the post of Under Officer in his school's cadet force. He was in charge, so to speak, as a student officer. By then he had developed a more fitting physical build, suitable to the issue sizes of standard uniform. Indeed, the style of uniform had also improved. He had firmly set his mind upon the Army as a career and decided to enter as a volunteer when the time came for his call up. National Service Conscription was still operating but volunteers stood a better chance of further education, courses of specialised training and, perhaps of greater importance, more opportunity for promotion. The pay was better too. Freddy was then seventeen and fast approaching the 'call up' age. Besides his education, a keen interest in the cadets and his military ambitions, he had been Confirmed into the Church of England and taught regularly at St. Hilda's Sunday school in the village. He was a good teacher.

At Sunday school, David (the youngest of the three brothers) always insisted on being with Freddy's class, although he was really too young for the subject matter. No one could persuade him to sit with the children of his own age, in the appropriate class. Edward supposed that it was understandable, for the three brothers were very close, and David was only six, after all.

One Sunday, however, he finally agreed to sit in the proper class for his age group. Attendance at the Sunday school was quite large and the older children received their lessons in the congregational-pews in the nave of church itself, while the younger ones had their classes in the little hall at the back of the church. David trotted off down the pathway to his proper class and the older children settled down to classes in the church. Shortly, a breathless infant teacher, the same Miss Sharp who taught at the Primary school, came bustling in to whisper loudly to Freddy that David had pushed off after only a few minutes of the lesson in her class. Freddy delegated Edward to go after him.

David must have run all the way, because by the time Edward caught up with him he was quite close to home. There was little point in returning to the church, since classes were by then, bound to be over. David's reasons for taking off like that were a little unclear but it boiled down to the fact that he just preferred to be in Freddy's class. Who wouldn't – Freddy was a good teacher and took his religious work very seriously.

When David left Miss Sharp's class, he had sought to rejoin his usual place inside the church with Freddy's class but when he found the big, heavy church-doors closed, he thought that everyone had gone home without him, so he went home too. It was a fair assumption for a six-year-old.

Edward had reached the age to attend the vicar's once-a-week Confirmation classes at the church. The minister was an excellent teacher who understood children well. He had several children of his own. He made the whole course both interesting and instructive to his mixed class of boys and girls, preparing them for their first communion and confirmation into the Church of England.

The Minister was a Northampton man. He had served with the Royal Naval Volunteer Reserve during the war and seen action at sea before the war's end, before his ordainment as a church minister. As a Master of Arts, he often wore his graduate's gown over his ministerial cassock, and a mortarboard upon his head, when visiting the old and infirm people of his parish. This he did frequently, either on foot or by bicycle, his gown flapping and billowing behind him wherever he went but in spite of his eccentricities the Vicar commanded much respect in the district. He was a man's man, not averse to, though only rarely taking, a brown-ale with parishioners who used the *Greenall* pub in the village.

The King died in February 1952 and Elizabeth the II ascended the throne. There had been a period of mourning, then excitement, celebrations and a holiday in 1953, Coronation year. In that year, as with each year in Spring-time, the minister called upon his flock to help with the annual, church fete; a Rose-Queen celebration for the month of May. He knew that Edward had won a scholarship to art-school and that indeed both of the Weinel brothers had a flare for painting. With this in mind, he asked the boys if they would like to prepare banners with lettering for the various 'floats' to be used in the grand, carnival procession around the village, following the Rose-Queen crowning ceremony.

The Reverend, entirely determined to involve everyone, asked the boys' Dad to organise the collection of a nominal entrance charge at the gate to the garden-fete, the proceeds of which would go to a deserving charity. Fred Weinel agreed readily to the vicar's request for he shared the view of most in the parish, that Mr. Morgan was a likable fellow who had brought the people of his denomination much closer together during his tenure at St. Hilda's.

Village people and traders of the locality organised bring-and-buy sales, stalls for local horticultural produce, entertainment and games, all of which they gave freely. Folk dancing and sporting competitions were also on the list of things to entertain the visiting public, when they came to the primary-school playing field that would accommodate the whole affair.

Behind the scenes, members of the church committee accepted nominations of aspiring young girls of the district from which to choose a new Rose-Queen for that year. Selection must have been a difficult task for many attractive girls lived in Hunt's Cross. Competition was very strong. Meanwhile, Freddy and Edward spent several evenings a week after school and at weekends too, behind the scenes, painting and preparing the banners for the carnival-floats. When all was ready, they helped to decorate the lorries donated for the day by local traders, in readiness for the next day. Everyone prayed that the weather would hold, and it did.

Carnival day dawned with fine clear skies. It was the end of spring and the beginning of summer and Mother Nature had smiled upon St. Hilda's fete. For the boys, their part in the preparations was over and the bright Saturday was theirs to enjoy. For their Dad however, his participation was just beginning. The boys decided to help and went with their parents and David to the school, to prepare the entrance gate.

Two o'clock was the hour set for the grand opening. The stalls and games were ready, the committee members were in place on the platform, traders and organisers put finishing touches to their stalls, and people from all over the locality arrived in droves. The brass bands were already in full swing with a little pomp and circumstance music to preface the opening ceremony. All that remained was for the Vicar and his invited guest to arrive for the official opening of the fete. The Minister had managed to persuade none less than the local Lord of the Manor, whose Country Seat was the stately home of Knowsley Hall, not far from Woolton, to perform the opening ceremony.

Presently, a chauffeur driven limousine, carrying the Vicar and his titled guest, drew up to the gate. The gentleman was very large and emerged from the interior of the car with some difficulty. As he straightened himself and rearranged his attire, Fred Weinel stepped forward from his table with an engaging smile, his best Trilby hat brim turned down jauntily and his hand extended to shake-hands with both the Vicar and the dignified personage. After courtesies, Fred held out his hand again, palm upwards.

"That's two Shillings, please."

The Vicar became agitated and began to make animated hand movements behind the officials back trying to discourage Fred's insistence, with silently mouthed phrases. "Not necessary – official guest – opening fete."

However, Fred was adamant and pressed his point. "This is a

function dedicated to charity," he said. "It's a shilling for adults and sixpence for children," he went on, still smiling. "We've all volunteered our time and effort to help the fete but we've all paid for our entrance too," the boys' Dad concluded, or at least with words to that effect.

The vicar's guest nodded his concurrence, muttering words of agreement. "Quite right too, ahem, quite right," and he began to fish about in his pockets. The Reverend minister, somewhat taken aback by this unprecedented turn of events, flushed, as he too, dived into his cassock pockets to find some small change. Unfortunately, neither of them seemed to have a penny between them and so it was up to the Lord to resolve the problem in the most politic way. This he did by asking his Chauffeur if he had two shillings on his person, with which to pay the necessary charge to get through Fred's gate. Reluctantly, the chauffeur was able to oblige, and thus, with a Florin in his palm, Fred courteously stepped aside, allowing the two slightly embarrassed gentlemen to pass through. Edward wondered if the chauffeur ever got his money back and indeed, if the two-shillings that he donated were all he had, for he looked miserable and stayed outside the gate with the car, rubbing at its already perfectly shined paintwork and chromium.

As for the boys, they spent a splendid day sampling the entertainment and enjoying the sight of their banner work, amply displayed on the floats, as the newly crowned Rose Queen of Hunt's Cross toured the district at the head of her procession. However, it did not go without notice that some people had found another way in, at the far end of the field during the day's celebrations. By rolling on their backs through a gap under the paling fence, they avoided paying the entrance charge. Several were grown-ups too! It never ceased to amaze Edward how some people would try to evade paying, in spite of the charitable reasons for the very small gate charge asked, in return for an enjoyable day out.

15

Some weeks later, the Reverend organised a dance evening for the teenage-alumni of his Confirmation classes to be held in the church hall behind St. Hilda's church. He promised that there would be refreshments of sandwiches and bottles of pop, in various flavours. He would also take his aging gramophone along to provide the music by which to dance. For Edward's part, he had never learned the art, and, being somewhat shy in those days, he felt more than a little uncomfortable at the prospect.

The Minister insisted that his confirmation students should all attend, although Edward was apprehensive about his own participation; dancing seemed a bit girlish to him. Under some additional pressure from his parents, he put on his best suit, shirt and tie and, at the appropriate hour, made his appearance at the church-hall door. Most of the others were already there and Edward noted that all the boys had nametags pinned to their lapels. A few of the girls too, had tags pinned to their dresses but not all.

One of the Vicar's helpers greeted Edward at the door, fished about in a hat to find a girl's nametag for him. By luck of the draw his had ANNE printed on it. Edward's heart leapt as the helper pinned his own nametag to the lapel of his suit. There was only one Anne, and she was one of the prettiest girls in the village.

She was to be Edward's partner, certainly for the first and last dances anyway. He took the tag and went to seek out his good fortune, the lovely Anne. She was about the same age as Edward though not quite as tall. Her curly, golden-blond hair, and her big blue eyes that reflected her always-charming smile, had often attracted his attention recently. Edward had no difficulty in finding her, especially since an animated group of boys surrounded her, arguing noisily over nametags.

Amongst the evening's gathering, there were several girls by the name of ANN, without the E on the end. Some unscrupulous knaves had added 'E' to the end of their "hat drawn" ANN tags; and then tried to pass themselves off as genuine and legitimate partners to the fairest of them all: Anne, with an 'e'. Talk about bees around a honey pot! At least she, the real Anne, was observant enough to note that the other Ann tags also had the first letter of their respective surnames printed after the forename.

It was Edward who had the real tag and therefore he presented himself as the rightful partner to pin it upon her dress. Edward felt ten feet tall when she gave him a lovely smile and helped his trembling fingers with the tiny safety pin. It was rebuff for the others and joy for him, that is, until Anne asked him if he would like to dance. It was he who should have asked her, but he was in a dream and hardly heard her words.

"Oh dear! I never learnt how," he said.

"Well I'll show you, Ted." She grasped Edward's arm and propelled him to the dance floor. "It's really very easy. You just have to learn the steps and keep time to the music."

"Well, I'll try." What else could he say? He was on cloud seven – spellbound and looking at a wonderful future life. The gramophone struck up with a well-known waltz played by Victor Silvester's dance band. The record crackled and scratched, straining out the waltz at seventy-eight revolutions per minute. Anne was right. Dancing wasn't so difficult once you got into the rhythm and understood where to put your feet, not to mention your hands. Edward dared not think what state Anne's white shoes were in, and he dared not look, after his feet had trodden upon hers several times in picking up the steps.

Eventually the music stopped with a needle-scratching hiss and they broke to wait for the next tune whilst the Reverend resuscitated his aging machine, rewinding its clockwork heart and placing another disk on its turntable platen. As the dancers stood in readiness for the next tune, Edward noted a few angry faces across the room. It was clear to him that his luck of the draw might have upset a few of the other boys. One of them, a member of the boxing club, made threatening gestures and pugilistic movements with his fists. All this in the church hall too! Then, the ancient gramophone groaned into life again with more music. Around they went again. No one, yet, had challenged Edward's partnership with Anne, in spite of rude and threatening gestures.

Edward didn't do a very good job with the next dance. He couldn't get the hang of the Fox-trot; it was far too complicated. Why ever did they call it that, he wondered struggling with the foot work. Resignedly, Anne said, "Oh well, let's break for a sandwich." She was still smiling though, and she took his arm as they headed for the refreshment table.

Soon enough, other lads came to ask Anne for a dance. Edward submitted to the 'excuse-me' protocol and sought out those girls who had temporarily lost their predetermined partners. Several politely refused when he presented himself but one, not so politely, turned him down with, "What? Dance with you? You have to be kidding." The triviality of refusal did not worry him unduly, having at least done his duty. Instead, he was content to sit and watch the lovely Anne from a distance. Edward thought to himself what a fortunate fellow he was, to have been partnered with the fairest girl of his age in Hunt's Cross.

The evening turned out to be a contradiction of his earlier nervousness. He told Anne how much he had enjoyed it while she danced the last dance with him, before her parents came to take her home. She agreed that it had been very enjoyable and more, that she really had enjoyed Edward's company. Then, her parents were at the door, and she took her leave but not without a furtive kiss on Edward's cheek that sent his head whirling and his heart pounding.

Edward dearly wanted to see her again but he learned some little time later that her father had accepted a posting out of the district and she moved with her family to another part of England. At the dance, she hadn't mentioned that she was leaving Hunt's Cross, and Edward never saw her again after all. He didn't feel cheated; just a little disappointed, but in spite of that he felt a lot less timid about being with girls in more grown-up circumstances. He had Anne to thank for that, for she had made him feel very comfortable with the notion. It was all part of growing up.

Meanwhile, at Art school he was struggling with maths and French, both subjects being taught by a complete buffoon, in his view. He found little relevance in the blackboards full of calculation related to algebraic proportion and Lord knows, he was having enough problems with English grammar, let alone trying to learn the conjugation of verbs in a language at which only the French could hazard an understanding. The other subjects didn't come too hard and he was managing, somehow, with history and geography.

A lady teacher took English language and literature, but she had small regard for Edward's short-comings at spelling and grammar, and she did little to encourage him. As a result, his studies in these subjects did not prosper. He had little interest in Shakespeare or poetry and the understanding of grammatical intricacies left him cold, when it came down to interrogatives, past participles and participial phrases. Academic application and theoretical compromise didn't come easily to one accustomed to doing things of a practical nature, through his own hands, and observant eyes.

His only claim to fame in English classes came about when

she asked for an essay as a weekend homework project: one of many. The subject matter interested Edward greatly, if only because it related to history, a subject for which he had regained some regard. When the Spanish Armada set itself against the British mainland and lost the sea battle to Sir Francis Drake, some Spanish ships escaped. One managed to sail North to Scottish waters but foundered upon rocks in stormy seas to sink in a little known Northern bay. Story had it that the ship carried the gold coin and bullion with which to pay King Philip's invading Spanish army once they had over-run British land defences. Although the tale was perhaps more myth than reality, the idea of a sunken treasure ship was an intriguing notion.

Edward elected to write about how men of the times, with none of today's technology to assist them, might have attempted to recover some of the wealth that lay beneath the waters. He researched how men may have approach the problem in those days and discovered that the principles of diving-bells were not altogether unknown at the time of the Armada in the late 1580's.

The English teacher made it clearly known to her class that not only was Edward's essay well researched, but he'd managed to submit the project in good order – free, for once, of serious grammatical error and almost totally devoid of spelling mistakes. This won him a second-hand propelling pencil in recognition of his effort. However, thereafter Edward's interest waned again, and he fell back into his usual lax ways. The propelling pencil, by the way, had no retractable leads with which to write nor could Edward find any of the type it needed. It graced his blazer-pocket for a time, but ended up as a useless trophy by which to remember only a sadly singular moment of euphoria, once experienced in his art-school training.

At art, Edward excelled. Drawing, painting, studies in perspective, composition and figure drawing came as second nature to him. The Graphic Arts Master took his class outside the school on many occasions, to sketch and draw as exercises in observation. Groups of art students were often found seated at vantage-points in the Cathedral grounds to observe and draw pictures of the many statues or details from the structure itself, windows, arches, flying-buttresses and such-like architecturally interesting features.

Sometimes he took them to the Pier head to draw ideas from shipping, buildings or the riverside activities. Those were days to remember and Edward developed his skills in precision, accuracy and detail – skills that were at the very crux of his desire to be a good artist.

He fell afoul of his teacher's discerning and sometimes humourless character only on one occasion. The art master had set a homework project with the subject as an exercise in figure drawing from memory. He proposed 'Police on Duty' as the central subject. It was quite normal for discussion amongst the students on such occasions and it became clear to Edward that most of the others elected to present a policeman on point duty, directing traffic, a common enough sight in Liverpool's busy thoroughfares. Several students suggested a policeman giving directions to lost citizens, and one, Ferguson, the motorbike fanatic, thought he might do something like a policeman on motor-cycle patrol. All were good ideas, but Edward wanted to be different.

His submission, whilst gaining superior marks for draftsmanship and painting, etc., earned a reprimand for stating the illustrative-obvious. In his picture, a burglar, hotly pursued by a uniformed, whistle-blowing bobby, carried a sack over his shoulder as he ran before the long arm of the law. Caught almost in the act, the robber clearly had a head start. What annoyed the master was Edward's depiction of the policeman's quarry. The thief wore a black mask over his eyes: point one. He also wore a striped jumper, well displayed beneath the open front of his jacket, flying wide as he ran: point two. The strap of a cosh or was it the loose end of a stocking full of sand for head bashing, hung from a jacket pocket: point three. Finally, and most important of all, the bag, slung over the robber's shoulder, clearly displayed the word SWAG, just as if: point four. The art master described the mask, the jumper, the cosh and the word SWAG, as unnecessary appendages to identify the fugitive as 'wanted by the police'. He considered such visual identifiers were unnecessary labelling. The chase would have sufficed without such 'comic orientated, strip-cartoon' devices. They spoiled an otherwise persuasive work of art.

Edward accepted the criticism as fair, but at least his work was entirely different from the twenty-nine other presentations, showing rigid, unmoving, static, statue-like scuffers, all remarkably alike. Edward had achieved his aim – to be different in his approach to the subject, and to provide a picture full of movement. He noticed, however, that his teacher overlooked comment on the number Edward had applied to the policeman's collar: PC 49. Perhaps the art master did have a sense of humour after all. He probably needed it, as his students, at only fourteen years of age, were still young enough to listen to the popular radio police saga or read comics, were they not?

Time crept along until the school received notification of a visit by Her Majesty the Queen to the city of Liverpool. The Art School itself didn't figure on her itinerary but her visit was part of a nation-wide tour of the land, following her father's death and her accession to the throne earlier that year. The tour of Liverpool included a halt at the still unfinished, Anglican Cathedral, for which the city's council-workers erected a rostrum and platform in the roadway, immediately in front of the art-school, overlooking the cliff-like escarpment that descended steeply into the Cathedral's grounds. It was a wonderful vantage point from which to view the towering structure. The workers decked the structures with awnings, flags and bunting, in readiness for the grand arrival, due on the following day, a Saturday.

The Head Master opened the school on the day, to allow students to observe the proceedings from the upper floor windows. Fortunately, there was adequate window space for all to have a good view. It was suggested that students might use the colourful scene as a subject for an interesting exercise in drawing and painting; there was likely to be much activity.

Brass bands blared out their marching music, and flags fluttered in the light breeze as thousands of sightseers lined the streets. Her Majesty the Queen duly arrived with pageantry, in a procession of limousines, to view the cathedral and specifically to hear the very first ringing of the cathedral's fine peal of bells. Customary speeches followed, with tributes to the cathedral's architect, Sir Giles Scott; and to the noble city of Liverpool. Finally the bands struck up again as the cars moved off. Disappointingly, it was all over very quickly.

16

Appendicitis took Edward down in the early autumn of 1952. After a terrible day and half a night of sickness and pain, his parents decided to call in the doctor. Edward had a history of tummy-troubles, but this time Doctor McCullack expressed concern at the boy's clearly severe condition. Edward needed hospital treatment. It was just as well, because things had gone badly wrong inside. His appendix had perforated.

They whisked him away in the middle of the night by ambulance. His Dad went with him and the Alderhay Children's Hospital admitted Edward for immediate surgery. They placed him in a preparation room where nurses fussed with his pyjamas, changing them for a white gown and socks for his feet. They gave him an injection and made him count to twenty. He started but couldn't finish. Vague images of corridor lights flew above him as an Orderly wheeled him away into the depths of welcome sleep, and away from the gnawing pain in his belly. The surgeon told Edward's Dad later that a few hours more and it would have been too late.

When Edward came to, he was in the ward, his bed surrounded by screens. He had a drip-needle stuck in one arm, a bed-shirt garment about his body and a feeling of complete tranquillity embraced his whole being. Gone was the pain of yesterday. A nurse, preparing some other apparatus next to him, spoke. "Oh! There you are. I've been waiting for you." Edward fell immediately in love with her. "I want to put this tube up your nose and I want you to swallow a few times so that we can get it down into your tummy," she continued with a lovely smile. She was very beautiful and quite young – couldn't have been more than twenty. Edward was just fourteen.

"What happens if I sneeze?" He asked through his drowsiness.

She smiled and told him not to worry. Another nurse came into the roofless tent, to assist the object of Edward's new-found love. Together the two nurses manipulated apparatus into position and popped the tube into Edward's left nostril. He swallowed several times and, without too much discomfort, he was 'fitted'.

"What are you going to do with the other end?" He asked, more to hear his own voice rather with than a genuine interest in the extremities of the tube.

"You'll be fed through this for a few days; a liquid diet; you won't be eating solids just yet." The answer came from somewhere beyond a veil of mist. Edward felt tired. There was no pain and that comforted him. Sleepily, he asked if his Dad was still at the hospital.

"No, he popped in to see you after the operation but you were still asleep. We told him to go home and that everything was fine. You'll be making a good recovery, believe me."

The love of Edward's life chatted on, as he drifted away, down and down, into a deep, deep, painless sleep.

He woke much later, with bright sun-light all around. The screens were no longer around his bed. There was noise, rattling trolleys and crockery, children's voices and a baby crying somewhere. He felt stiff and uncomfortable. His mouth was dry. He was thirsty and his arm hurt. A nurse, passing by and seeing Edward awake, smiled and came to his bedside.

"Hi," she said, "how d'you feel?"

"Fine," replied Edward, "but I'm thirsty."

"I'll get you something to drink and then I have to give you an injection," she said, and hurried away.

Looking around, Edward could see that he was in quite a large ward with many beds. All the occupants were children. Someone alongside said, "Hello! What are you in for then?"

Edward turned his head on the pillow to see the owner of the voice. In the next bed, on the right, a boy stared and craned to get a better view of his bedside neighbour. Perhaps a little younger than Edward, he sat with a large book balanced on his drawn-up knees.

"My father's a bacteriologist you know," he went on, "works in this hospital too."

"Oh," said Edward in response, "that's interesting. My Dad works with aeroplanes. I've had my appendix out. What's wrong with you?"

"Same," the boy answered rather proudly, returning his eyes to the pages of his book.

Shortly, the nurse came back with something to drink and a tray of injection kit. She pulled back the sheets to expose Edward's legs. He wondered how long he was going to be in hospital, how much school he was going to miss and which days would be visiting days. "Roll over a bit if you can," prompted the nurse, "I want to pop this needle in your thigh." Edward moved over a little so that she could inject. It didn't hurt that much. He'd never had a ‘thing’ about needles anyway. The nurse massaged the area a little as she explained, "You'll have four of these each day for a while, but we'll have to alternate the legs, day by day, because this stuff's a bit thick," she said, rubbing gently at the point of injection.

The ward sister appeared at the foot of Edward's bed. At least he assumed that was who she was, her uniform was a different colour and the headgear she wore was large, triangular and stiffly starched.

She came in with a tall man dressed in a white coat and carrying a stethoscope. They discussed Edward's chart before the Ward-Sister introduced Mr. Foy, the surgeon who had operated. Edward liked him immediately. The surgeon spoke softly and with very gentle fingers, looked under the bandages and lint that covered the wounds in Edward's abdomen. It was then, with some shock that Edward noticed there was a piece of rubber tube sticking out of his belly. The surgeon observed the concerned look on the boy's face.

"That's just a drain tube we put in to get rid of the nasty stuff that you had inside," he explained. "We'll shorten it each day, little by little, until it’s out and before you know it you'll be right as rain."

He patted Edward's shoulder and asked the nurse to replace the dressings. "I'll see you again tomorrow, old son." He smiled and mussed Edward's hair before moving on to the next bed.

Edward's parents came in to see him the day after the operation, just for an hour. Poor Edith, clearly quite upset and near to tears when she saw all the paraphernalia and apparatus coupled up to her son. She smiled bravely whilst dabbling at her eyes. She soon recovered her composure, however, when she saw that in spite of the tubes and tape, hanging bottles and temperature charts, her son was smiling and animated. They had brought goodies, to put in the bedside cupboard – chocolates and sweets and stuff, but one of the nurses said that they would be a temptation there.

"We'll keep them in the office until Eddy's back on solids," she said. Edward never liked the shortening of his name to Eddy. He didn't know why, but Eddy smacked of the wise-guy spiv or one of Al Capone's hit men. Edward, Ted, or Teddy was more to his liking and more appropriate to his pacific nature. He made a mental note to remind the ward staff of this.

Along with the denied sweets, his parents brought some books and magazines too. These were very welcome since the ward library, brought round on a trolley from time to time, was full of nursery books more suitable for the younger children. At fourteen, he guessed that he must have been the oldest patient in the ward. Most of the others were quite young, some even babies.

So, for the next few days the routine was very much the same. Meal times came and went and the meal-trolley always passed by Edward's bed. His hunger grew with the passing days. He mentioned it to the Surgeon and the Ward Sister on their daily rounds but they just smiled and told him not to worry, he would be on solids again soon. In due course, they took away the tubes and promised Edward a proper meal. It turned out to be soft boiled eggs, just two of them, and fingers of soft white bread with the crusts cut off. Edward asked the meal-lady if he might have some tea.

"We don't serve tea to children," she said, though not unkindly, "you'll only get warm milk here, love." Edward grimaced, but accepted nevertheless... What I'd give for a cuppa right now, he thought, wading into the eggs, bread and milk.

The days continued to drift by. Edward's legs ached a bit from the injections but at least he was healing up well, so they said; he was getting regular meals by then, too. However, he was fed up with having to ask for a bottle at the call of nature, or worse, a bedpan that required him to do his stuff balanced precariously on top of the bed; no mean feat, when you are taller than the bed is wide. Very undignified! Yes, they put screens around, but at that altitude his head was always above the top and everyone knew what he was doing.

One morning before dawn, Edward needed to go so he crept out of bed for his first solo excursion. The nursing staff had disbarred him from getting out of bed thus far in his 'post op' recuperation, let alone walking about. "Your wound may reopen," they said. Well, he thought, I want to use the toilet so I'll go under my own steam. He did feel a bit wobbly on his legs and a bit tight around the belly but all went well. He got to the sluice room without incident and found the toilet. It was such a treat to sit in the accustomed manner, and Edward felt more than pleased with himself. His self-congratulation was short-lived, however.

On his way back to bed, he got caught. The night nurse came hurrying down the ward with a grim look on her face. "What are you doing out of bed!" She fired at Edward in a fearful stage whisper. Taking his arm in a vice-like grip, she accompanied him back to his place. "Don't you dare do that again Ted!" She exclaimed. "Not on my shift, not while I'm on duty!" She was angry.

Edward heaved himself into bed again while the nurse fussed about with the sheets and pillows, all the time muttering under her breath about naughty boys who wouldn't follow the rules but Edward smiled to himself because he did feel ...so good.

He was in trouble again a few days later, although they had allowed him out of bed by then, to amble around the ward and talk to the other kids. It was November the fifth: Guy Fawkes night with bonfires and fireworks. Outside, in the grounds of the hospital, the staff had built a big bonfire, complete with an effigy of Guy dramatically placed at the top of the wood and branches. A few of the smaller children in the ward couldn't reach up high enough to see over the window-sills to watch, when the revellers outside lit the fire so, with the help of a couple of the other patients, Edward rearranged some beds for the youngsters to climb on.

There wasn't enough room on the beds for one little girl, so he picked her up, so that she too could watch the fire and the sparklers. The Ward Sister was on duty that evening and she rushed over to give Edward a hard time for ‘lifting’. "I know your stitches are out, Teddy, but you'll pop that wound if you carry on like this," she said, taking the child from his arms. Reluctantly the little girl clambered up on to one of the repositioned beds and squeezed in, among the other children, to continue watching the festivities.

"And who moved these beds?" Sister went on. "You'll be in trouble with the Surgeon, you silly boy!” Silly or not, everyone in the ward saw the celebrations and Edward felt pleased about that.

Shortly after bonfire night, Edward was well enough to be discharged and to go home for an extended period of convalescence. Perhaps, by then, the ward staff at the hospital had grown well and truly fed up with Edward's rule breaking but he felt relatively fit anyway and was glad to be going home. His Dad came to pick him up, bringing day-clothes, a topcoat and a scarf. It was mid-November and the bright autumn weather had changed to the cold, grey, short days of winter.

Edward made his farewells to the Dragon-Lady Ward Sister, apologising for any trouble he'd caused, and he thanked all the nurses who were on duty that afternoon.

"What's the first thing you want when you get home?" His Dad asked, as he and Edward settled into their seats on the homeward-bound bus. He knew very well what the response would be, but he wanted to hear Edward say it, with relish.

"A pot of tea, Dad," answered his son, "a whole pot of tea."

Fred laughed and put his arm round Edward's shoulder in half a hug. "Done," he said, "your mother will have the kettle on, right now!"

After several weeks at home with his mother's ministrations and large helpings of her lentil soup, which remained a favourite thereafter, the doctor allowed Edward to continue with his schooling but it had been a very long absence and he'd missed too much at a critical time in his education. He found himself far, far behind the others in his class and, in spite of extra effort, he somehow knew that he would never catch up. The teachers seemed to give up too.

Edward took each day as it came, for over the passing years he'd discovered that there seemed little point in sinking into the depths of despondency. He could only do his best, and if that wasn't good enough, then so be it. He had developed a philosophical approach to life, though he often indulged in deep thought and self-analysis.

During their mid-day breaks for lunch, Edward and some his class companions frequently walked the streets of Liverpool in the locality of their school to peer through the motor and motorcycle show-room windows. There was precious little else to do after their sandwich-lunch. It was because of these excursions that Edward became interested in motor cycles, for there were many showrooms and shops in the district. Ferguson, one of Edward's friends, and a bit of a show-off, often described his jaunts on the back of his uncle's bike, a big, *Vincent Black Shadow*. The rest of the group would hang on his every word.

Bikes of the *Vincent* marque were expensive but they were (and still are) the essence of motorcycling. They had character, rather like the American *Harley-Davidson*: ageless. Apart from *Vincents* the boys ogled the *Nortons, BSAs* and *Ariels* on show but clearly their pocket money would never have covered the cost of such machinery in a month of Sundays. They were under-age too, of course, but dreams were dreams.

The end of term, and of Edward's second year at the art school approached, exam time was coming and he was still a very long way behind. Again, he'd lost heart and didn't really try hard enough to recoup his losses due to absence. He failed miserably at all the academic subjects. Final reports were emphatic about that. In the arts however, he did extremely well, gaining top exam marks in all subjects. Term class marks were of course uncountable because of his earlier long absence. There would be "No chance at all in the forthcoming GCE exams," as the Head Master wrote at the foot of Edward's final report. He would therefore leave school at the age of fifteen, obtaining solely a minor diploma, with a pass only in the second class.

Disappointing? Yes, but Edward felt that he had attended Art school to develop his artistic skills and having done so well at that, in spite of failure in the other subjects, he had at least achieved something. His hopes and dreams remained indomitable.

Part 6

To be an Engineer

17

In the summer of 1953, his creative aspirations dashed and possibilities of a career in design and artistic pursuit put firmly aside, Edward contemplated his future with some apprehension. Leaving school at fifteen didn't irk him particularly, since he couldn't say that he liked school all that much anyway. He took council with his father and they talked at length about what Edward should do for a career. There were several options open to him but clearly he would have to make up his mind and start work soon, to train for some sort of gainful occupation, something that could lead to a lasting career: the sensible path.

Finally, they reached a consensus that an apprenticeship in the aviation industry would be most suitable, if only because of Edward's latent interest in aeroplanes. Anticipating a positive response to this suggestion, Edward's father had already spoken with a friend who owned and operated the Flying Club at Liverpool's Speke Airport. Edward came to realise that his father understood his middle son's character precisely – better than he could ever have imagined.

Together, they went for an interview, to meet the Club's owner, Squadron Leader Gordon Wright (often known to his intimates as Wilbur for obvious reasons) and his Chief Engineer, Cliff Watson. They posed Edward a barrage of questions about his schooling and his keen following of aviation matters. Everyone seemed pleased with Edward's answers and agreed that he should start work on the following Monday. Wages were to be thirty-five shillings for five-and-a-half day week, and night school was an obligatory requirement of the apprenticeship articles. The Chief Engineer went to great lengths to explain that as a first-year apprentice Edward should expect only to do very mundane work to begin with: errand running, hangar-floor sweeping, cleaning chores and tea making. Technical stuff would come later in brief moments between the general duties. That seemed fair enough to Edward and he signed the documents of indenture, next to his father's signature.

On the weekend before starting at his new job, they bussed into town to buy overalls: two pairs of dark blue coveralls with pockets and shiny, tin buttons. They were a trifle big, but Edward's father explained that they would probably shrink in the wash anyway. On top of this, his son still had some more growing to do. The prospect of having a job, of earning a wage (little enough though it was) and of his expectations for a future in aviation filled Edward with excitement and pleasure.

His father gave him three thick books, volumes on aircraft engineering, maintenance and overhaul. This gift added to his pride at having gained even such a lowly position at the Flying Club. The books were supplements to a popular, pre-war aviation magazine published by Newnes and printed on fine, glossy paper; the pages collected and bound with care by his father, over the passing years. Edward cherished them and retained them throughout his career, referring to them frequently over the years, old though the technology between the covers may have been.

School-holidays were a thing of the past. Now that his full time education had come to an abrupt end, it was clear to Edward that adulthood and responsible activity must now supplant his childhood. A new way of life lay ahead of him. During the short time between finishing school and starting at his first employment, he studied hard at home.

He read everything to do with aircraft that he could lay his hands on. He ordered the weekly magazines, Flight and Aeroplane, from the newsagent's shop, in the hope that he could begin to understand the latest developments and the function of post-war civil aviation. Many older aeroplanes still flew in British skies during those post-war times in spite of the great changes then becoming evident in aviation. Light, civil aeroplanes built by *Miles, Percival* and *De Havilland*, for instance; the *Hawk*s, *Geminis*, under-powered *Proctors*, *the Falcons, Tiger Moths, Puss Moths, Dragons and Rapides* – all were frequent users of a less congested airspace than that of today.

Edward was keen to get amongst it, although as an apprentice he did not see his new situation as quite like that of being a full-blown working-man, and being paid for it, but he was, and he assented to his father's suggestion that he should give his mother half his earnings each week. The notion of making a living and at least paying some of his way, added to his feeling of stepping into the world of adults.

Edward's new adventure began when he cycled off to the airport at seven o'clock on a mild Monday morning towards the end of summer that year. The Flying Club and its hangar nestled on the western side of the airport, some twenty minutes cycling away. It lay quite close to the river's shore and opposite the threshold to the south-western end of the airport's main runway. He'd excitedly planned his journey the day before, the route taking him through Hunt's Cross, turning left into Woodend Avenue at the roundabout, over the red-brick railway bridge, and on towards Speke.

Turning right at the next roundabout, to pass the factories and housing estates, would set him on the road to the industrial estates and on to yet a third roundabout where the dual carriageways to the airport and Garston began. On reaching that point, he would take to the cycle-track and press on with the airfield to his left.

The route would carry him along the northern extremity of the aerodrome, crossing the reciprocal approach for the airport's second runway on the way. Long and curving, this stretch of road was very prone to the wind because of the flat nature of the landscape. It followed the airfield perimeter with its fence of six-foot metal stakes painted in a pale turquoise blue, the principle colour used by the airport authority of the time. A wide, flag-paved footpath ran next to the fence, with the cycle-track set between the path and the westbound carriageway of the road.

So there he was, on his way, full of enthusiasm and eager to make a good impression on his first day. At that hour of the morning, traffic wasn't heavy. Few people owned cars then anyway; most travelled to and from their work by bus, tram-car or bike, just like Edward. Occasionally a van or lorry passed him, and the odd double-decker bus to Garston and the City but the cycle-track was his!

An enormous timber yard and railway sidings dominated the view on the farther side of the dual carriageway, with huge stacks of wood standing in the open to season. Some piles were twenty or thirty feet high and more. The yard stretched back as far as the railway and its sidings, the same line he'd crossed earlier at the brick-built bridge, and reached from the roundabout at the south-eastern corner of the airfield, almost as far as the outskirts of Garston. In spite of the wide, dual roadway running between the airfield and the timber yard, Edward speculated at the possible hazard presented to aircraft where the woodpiles stood in the approach and take-off path at the second runway's end.

From halfway along the curving road, he could see, distantly, the control tower, the two large hangars and the main entrance gate to the airport's terminal building. Built before the war, at about the same time as his first school, the terminal bore strong similarities in structural design and style. The airport building, an almost crescent shaped structure, along with the two fine hangars nearby, displayed the same light-brown coloured bricks as those of the primary school. Indeed, much of Speke, in both housing developments and the industrial areas, had similar characteristics, quite modern in those times, with soft curves complementing straight lines, reflecting the contemporary geometry of pre-war architecture.

He peddled on, brimming with energy and growing excitement. It was a calm morning, with no wind to impede him and Edward made good time with little effort. It was no different from riding to school. The hour was much earlier of course, but the distance was no more than his former daily journeys to Gilmour. The airport trip was no farther: a twenty-minute excursion. Passing the airport entrance gates, Edward continued on towards the outskirts of Garston, where the dual road and cycle tracks merged into a wide two-way road leading into the town. To get to the south-western corner of the airfield where the Flying Club lay, he needed to turn left before Bryant and May's match factory, with the Gas works beyond, where the road flanked the Garston side of the aerodrome, crossing the approach to the north-western end of the airport's third runway.

At that farther side, there were more hangars, two black-painted, corrugated iron buildings with huge sliding doors at each end. The road curved to the right, to pass behind the gas-works and enter old Garston town but Edward's goal, however, was to turn left, off the main road and into a side road leading towards the river, past urban housing. Beyond, was the gate giving access to a service lane behind the airfield's earliest hangar and the stone farmhouse (Chapel House) that had once been Speke's 'station' building in the thirties. The Flying Club lay farther along, with its small hangar and brick-built workshops nestled at its landward end.

Edward made for the front of the building, to reach the hangar-doors on the airfield side. He propped his bike against the clubhouse wall and went inside, squeezing through the gap between the partially open hangar doors. Packed full of aeroplanes, the place was not at all like the day of his interview, when it was empty. Two twin-engine biplanes, *De Havilland 89 Rapides*, stood with their tails tucked into the back corners of the hangar, with a *Miles Gemini* parked neatly between them. The *Gemini*, a low-winged monoplane, also had two engines. It was smaller than the *Rapides*, and distinctive with its quarter-spherical, bubble-like windscreen, swing-up doors and twin tail fins.

Several types of *Auster* and a couple of open-cockpit, *Miles Hawk*s, filled the rest of the floor-space. Edward already knew that the *Hawk* carried the name of *Magister*, in military service; and likewise, the *Rapide* went under the service name of *Dominie*. This much Edward had already learned from the many aircraft 'spotters' books he'd accumulated over the passing years, to satisfy his latent interest in aviation.

To reach the Chief Engineer's office he had to crouch and weave among the tightly packed aircraft. Edward ducked his head to pass beneath the strutted, high wings of the Austers, and squirmed around the wing tips of the low-winged Hawks to make his way through the crowded hall of silent flying machines.

Birds squawked, twittered and whistled from their high perches up in the steel work of the roof, their cries echoing, bouncing and rebounding from the concrete floor to the roof and from wall to wall.

A loose roof-panel rattled timidly, flapping in the light breeze. Sunlight was everywhere, focusing in through two rows of translucent panels running the whole length of the roof. Having reached the far end of the hangar, and its row of office-doors, Edward found the one marked Chiefy, with its hand-drawn picture of a Red Indian, bedecked in ceremonial, feathered headgear. He tapped on the door but since no one answered his knock, Edward concluded that he was early: a punctuality habit that he would maintain throughout his life.

While he waited for his new boss to turn up, Edward wandered among the aeroplanes, looking through the windows into the cockpits and running his fingers over the smooth, fabric covered surfaces. The smell of cellulose 'dope' and the faint whiff of gasoline filled the still air of the hangar; particular and unique odours that he thereafter always associated with memories of his earliest days with light aircraft.

Dodging a strut under an Auster's wing as he ducked to move on, he heard footsteps in the hangar. Someone was coming. Edward could see only a pair of legs, from where he was, still ducking down under an aircraft's wing. "Morning." He volunteered, amid a torrent of echoes.

A surprised face appeared, joining the legs beneath the engine cowlings of a nearby *Hawk*. Edward almost laughed out-loud; he could see the humour of the encounter: two people stooping to greet each other with their heads at hip level.

"Hello, my word, you're on time aren't you?" The face said. It was the face of the Chief Engineer, who had conducted the interview a week ago. Beneath a shock of fair hair, the Chief appeared to be in his late thirties. He had a friendly face with a wide mouth that smiled easily. His piercing blue eyes showed a commanding character, to match his stocky build. "Come on then," he said, with a meaningful glance at his watch, "let me get the office unlocked and then I'll set you up with the others, when they get here!"

When the others did finally arrive, he introduced Edward to them as Ted. "This is Nick, our licensed engine-man. This is Ron, airframes. Here is Bernard, he's a senior apprentice and eats too much. That's Malcolm, he's a senior apprentice too, and currently looks after our stores." Taking a tall fellow by the arm, Cliff introduced Peter. "He's been with us for a couple of years and you'll be with him to start with. OK?"

Peter was about seventeen, very tall, more than six feet of gangling youth, with long sideburns to his thick black hair, plastered with *Brilliantine*, and combed into an enormous quiff.

There were others too, Cedric, Gerry (not Jerry, he insisted), George and finally, 'Harry the Banner,' a balding fellow sporting a handlebar moustache; all greeted the new boy with smiles and handshakes.

The sound of a car outside announced the arrival of the Club's Boss. The group dispersed as he squeezed through the space between the hangar doors, with a cheery greeting. Edward had already met him, at interview. A short, stocky man with a beaming, ruddy face and the owner of a beautiful, shiny, black, *Brough Superior* sports car. The Boss had flown *Mosquitoes* with the Royal Air Force during the war, as a pathfinder. It wasn't clear to Edward how his own father had come to know him but his sympathetic agreement to take Edward on as an apprentice, demonstrated that he was more than pleased to do a favour and provide the opportunity for a friend's son, to begin a career in aviation.

Introductions over, the Chief handed Edward over to Peter, the tall apprentice with the quiff and long side-burns. Peter lived not far away in the tenement-flats opposite the match factory. He came from Italian stock, reflected in both his looks and his surname. He had a terrible stutter; in sympathy for which, Edward soon found himself finishing sentences for him. Peter had a trick elbow too, and couldn't straighten his right arm completely. It soon became clear that the two would get on very well together.

As a first priority, Peter showed his new companion the Crew Room, so called because the hangar-staff used it for their tea and lunch-breaks. It was a small room, with a double casement window glazed with frosted glass and furnishings of old bus seats, littered with yesterday's newspapers. There was a small square table to make the tea on, a kettle, a teapot and a whole range of dirty cups and mugs. A dart-board in its cupboard hung on the wall, next to the door, though the room seemed hardly big enough to contain all the people Edward had met that morning, let alone mount a game of darts, too.

Peter showed him the tea making equipment and took a tin tray of dirty teacups through a second door, leading to washroom and toilet facilities beyond. "You'll have to clean this place up three or four times a day," Peter explained, "and make the tea at ten o'clock, lunch-time and again, in the afternoon, at three o'clock and five-thirty too, if there's overtime. At lunch time, before you make the tea, you'll have to do the Chip Run for the chaps," he continued, "I'll go with you for the first couple of days to show you where to buy the things, then you'll be on your own."

Peter grinned. Probably because, on Edward's arrival, he was elevated from junior apprentice to not-so-junior. As a result, he would be relieved of 'duty-run and chippy-run' errands. The two of them began to tidy up the crew-room and Peter, in his halting conversation, referred to Harry the Banner as they worked. He explained that Harry 'the Banner' had seen RAF service as a tail-end-turret gunner with *Lancasters* during the war, which was why, perhaps, he got into banner-towing with light aircraft, earning him the suffixed nickname. Peter went on to explain that banners, used for airborne advertising, trailed behind the aircraft, attached by cable to a strong point and release mechanism at the tail, although there the affinity with Harry's past ended.

Harry had learned to fly (as a pilot) after the war and flew the flying club's, banner-towing *Austers* frequently during the summer months. It was his specialty. Some said he was a bit crazy, with mannerisms rather like shell-shock, perhaps because of the many hours spent in the turrets of operational bombers: a cold, lonely and dangerous existence. Befriended after de-mob and taught to fly by the Boss, Harry had become a useful member of the Club staff in spite of his eccentricities. Nobody quite understood what the rest of his job was when he wasn't flying banners around the district and it wasn't unusual to see him sweeping the hangar floor sometimes or scrambling about in overalls under the boss's car, or wrestling with his Alsatian dog that he always brought to work with him each day, in his old *Morris* car.

The two lads had almost finished cleaning up in the crew room, when the call of a raised voice came to them distantly from the hangar. "On the doors then, you lot," it bellowed, "come on, fingers out. Two-Six!" The shout accompanied the rumble of rolling, iron wheels on steel rails, announcing the opening of the big hangar doors. Edward followed Peter out into the hangar, at the trot, down three steps from the crew room door, ducking under the wings of planes to throw their shoulders to the task of pushing the doors.

There were four doors to open, two overlapping each other on either side. As they rolled, more sunlight streamed into the hangar, for the doors faced almost eastwards. The other fellows scrambled about, removing wheel-chocks from the first aircraft and prepared to push it outside. Peter said, "You just watch for today, to see how it's done and where to push. You can help tomorrow."

With that, Edward stood aside and watched as three people carefully manoeuvred the first aircraft out of the confined space of the full hangar. They moved out onto the tiny hard-standing in front of the hangar, from where a narrow paved roadway led to the airfield's perimeter track or taxi-way. One by one, the chaps pushed each aircraft out of the hangar on to the grassed areas flanking the roadway, parking them into wind with the brakes on and chocks placed in front of the wheels.

The last to roll out, were the *Rapides*. Edward noted that these aircraft needed six people to manoeuvre the larger size. They pushed the first twin-engine, nine-seater biplane, nose-on to the edge of the open door-way. Then, tail first, they swung out onto the grassy field, with two men on the tail to guide the castering tail-wheel and steer, whilst the remainder pushed at strategic positions, one on each wing tip and one at each of the two engine nacelles. They parked the *Rapide* just outside the hangar, to one side of the open doorway, nose-on to the red-painted, refuelling pump that Edward had noticed when he arrived.

Whilst the men went back inside to manoeuvre the second *Rapide*, Peter caught Edward's diverted attention. "Ey'are! You can watch this too. We're going to do a DI on this kite," he stuttered, adding, "but first we'll check the oil and refuel it." The word 'kite' jarred on Edward's ears. He didn't like the reference, 'kite', to an aeroplane. It smacked of ‘Much Binding in the Marsh’ and pseudo RAF jargon to him. His Dad always referred to aeroplanes as aircraft, aeroplanes or machines – likewise did Edward. Anyway, he kept his mouth shut. It wasn't his place to make criticism or pass judgment over a silly word.

Instead, he asked a question, "What's a DI then?"

"Daily Inspection: we do one on each serviceable plane every day, before flight," said Peter. "We'll do the oils first though, then we'll top-up the fuel tanks, like I said." He drew Edward to a corner near the front of the hangar where a forty gallon drum rested on its side, supported on a tubular-steel trestle. Peter took a spouted-can and filled it with thick, yellowish-green oil drawn from the brass tap plugged into the end of the drum. Then, taking a key with a ‘T’ bar at each end, from a hook, and a hank of rag from a nearby box, they marched back to the *Rapide*.

At the trailing-edge end of the first engine-nacelle, a small opening gave access to a brass tank-cap with two protruding key-dogs. Peter inserted the smaller end of the T key and unscrewed the cap, revealing an opening from which he drew out a long dipstick. The blackened oil in the tank showed to be halfway down the stick markings, so Peter emptied most of the fresh oil from the can, into the orifice and re-dipped. "Spot on," he said, replacing the cap and introducing Edward to a new word as he added, "You don't have to 'graunch' the cap down too tight, its only made of brass and the key-dogs will shear off easily if you put too much pressure on them. Just nip the cap down gently."

To Edward, the word 'graunch' had a lovely ring to it. As a word commonly used amongst pilots and engineers of the time, it spoke volumes and described, beautifully, damage caused by ineptitude or over zealousness.

"Ey'are," Peter offered, again breaking into Edward's thoughts, and handing him the can. "You can do the other one."

"Wow!" thought Edward, my first technical job! I'll be an engineer tomorrow at this rate! Later, he watched as Bernard came over to help with the refuelling, but it was a disappointment when Edward didn't get to see the rest of the Daily Inspection. Instead, Peter detailed the two of them to finish cleaning up the crew room, before showing Edward how to prepare tea of sufficient strength and quantity to satisfy the tastes and large mugs of the men... quite some feat, with only one small teapot!

As they waited for the kettle to boil, Edward could hear engines running outside, propellers whirring, the crackle of exhausts and the resounding roar, as throttles opened. *Gypsy Major* and *Cirrus Minor* engines sounded very similar but the six cylinders of the *Rapide*'s bigger, *Gypsy-Sixes* with metal propellers fitted, sounded tremendous. He would loved to have watched that too but he was making tea and washing cups. Oh well, he thought, perhaps tomorrow.

When tea-break time arrived and all was in readiness, it only remained to call the chaps. Peter pointed to an old engine-cylinder hanging on a thick cord, outside, next to the crew room door. An equally old connecting rod dangled inside, its smaller end protruding from the cylinder's interior, rather like a bell clapper, for that indeed, was its intention. "Give that a good, long, ring, Ted," he said, "and shout TEEEEEA UP!"

Edward grasped the con-rod and waggled it vigorously. The clattering, bell-like chimes rang 'loud and clear', throughout the hangar. His adolescent voice broke as he shouted, but the bell's rebounding echoes from everywhere masked his call. It was Ten o'clock already, time for morning tea break.

"Roger Dee!" "OK!" "Coming!" Voices from the hangar responded to the bell. People appeared like magic from the four corners of the hangar, heading for the crew room. With steaming tea poured and sandwich packets opened, the crew settled down to tea-break chat, and reading newspapers.

The Daily Mirror was popular, with its girlie pin-up and Garth, Jane and Flook cartoons, whilst some read Flight and Aeroplane magazines. Conversation centred on sporting events, football pools and family matters among the married men. Bernard, he who ate too much, had two double chins and pudgy fingers. At eighteen, he really was overweight. His sandwich-pack was the biggest and deepest in the room. He opened it with relish and a smack of the lips.

He was a good-natured chap who smiled a lot – strange for a red-haired fellow, really. Edward never saw him angry or upset in spite of the 'stick' he often took about his weight and the size of his appetite. Perhaps it was just as well that he had a quiet nature.

A game of darts was usual at tea break. The room had just enough length for the throw but its width was very narrow. It wasn't unknown for the odd rebounding dart to end up in someone's knee, or worse. However, the game was always on. Three-O-One was the standard for tea breaks, the best of three. Five-O-One was the usual for lunch times. "Do you play darts Ted?" Nick asked, while waiting his turn to throw.

"Yes, but not regularly," responded Edward.

"Well get yourself some arrows, lad, and get stuck in."

Nick's throwing technique was to hurl the darts like spears, lancing them through the air, straight and fast, burying the points to the hilt. Ronnie's technique differed completely. He'd launch the darts upward in a curving trajectory, coaxing them to drop into the target, to sink only shallowly and precariously in the board's material.

"PFFOR! ArrOW!" Nick cried, distracted as his partner mopped up the current game with two darts and a double sixteen.

Peter took Edward back into the hangar after tea break and presented him with one of two yard-brushes that must have been over a yard wide. "Right," he said in a commanding tone, "Now we're going to sweep the hangar floor." The statement came out like, "Now we're ger-ger-ger- going to swer-swer-swer sweep ther-ther-ther-ther-ther..........."

Before he could finish, Edward sympathetically prompted him with, "Hanger floor?" In response to this and with a look of relief upon his face, Peter nodded, gathered his thoughts and blurted,

"Right!" It was the beginning of a good relationship between them. Peter took no offense by the timely, prompting intervention. Edward was beginning to get the hang of communicating with Peter. His stutter always seemed to get the better of him, but he endured. His impediment was serious but nobody took a rise out of him for it.

Pushing their brushes, they marched up and down the hangar, sweeping clean the concrete floor. Starting at the back of the hangar they worked forwards, towards the doors, closed by then, to keep the breeze out. Edward learned that keeping the brushes at an angle to their passage, ensured that the dust, grit and sawdust formed a straight-line pile down the length of the floor. It was rather like reaping a cornfield. Then, turning at the far end, they could follow the same line back, a yard farther on and a yard nearer to the goal. Eventually, the accumulating pile of sweepings ended up in one corner ready for shovelling into a bin. The system worked well but, even so, hangar floor sweeping was tiresome, for the brushes were not light.

While Peter and Edward marched up and down the length of the hangar, there was much activity outside. The lads took a moment's break and stepped out into the sunlight, in front of the hangar. From there, looking left, the airport's terminal building was visible in the distance, with its crescent frontage and central block supporting the control tower. The building nestled between two large hangars, on the farther side of a wide, paved hard standing beyond the threshold of the airport's third runway. Sweeping his eyes Eastward, Edward could plainly see the tops of the wood stacks in the timber-yard that he had passed earlier, on his way to the airport and the distant factory roofs of the industrial area of Speke: *Dunlop*'s in the old *Rootes* factory, *Bibby's* and *Lockheed*'s.

Almost directly opposite the Club, but far away on the south-eastern side of the aerodrome, lay the stately Tudor home of Speke Hall, almost hidden in its copse of trees. To the right, the river Mersey flowed, its broad surface rippled with tiny waves, drifting sedately seawards between the shores. The opposite shore was clearly visible at that widest part of the river for it was a clear, fine day. There lay the Wirral peninsular, separating the Mersey from the river Dee, on the other side. Liverpool was still a major seaport, handling millions of tons of cargo each year, with access to Manchester through the ship canal, still active and thriving.

On the river side of the Club's hangar there was another building of beige brick. It was low, being only single storey. Peter had told him that it had once been the original Clubhouse, but was taken over by the Radio Communications and Navigational Aid Department of the MTCA. Clearly, it was an important centre for airport control operations, frequently visited by the orange and dark blue painted MTCA vans and cars that circulated ceaselessly.

A semicircular (or rather half hexagon) extension jutting from the southeastern corner of the Club's hangar accommodated the firm's registered, general office. Its windows were large, providing an all-round, panoramic view of the airfield. It was from this office that the Boss and his staff conducted business affairs under the Company name of *Wright Aviation* Ltd. Next to the Boss's office, more windows on the flat frontage of the hangar provided for the Flight Office, inside.

Peter again explained that this was where student pilots, instructors and club member pilots prepared their flight plans and signed the necessary documents for their flights. He took Edward inside, showing him the flight detail chalkboard next to the door, the maps, weather charts, safety posters and all the other paraphernalia for a navigation and flight op's room. It was all so new and interesting; Edward metaphorically squirmed as he revelled in the technicalities of his new surroundings. What absolute joy!

Reluctantly, the lads went back to the job of sweeping until about half past eleven. At this point Peter advised that they should prepare for the lunch break, clean up the crew room, wash the cups and set up the lunchtime tea for those of the hangar crew who didn't normally go home for a meal. "We'll make the tea after we get back from the Chip-Run, but first we'll get the orders," Peter said. He produced a pencil and notepad, before starting his tour of the hangar and offices. "We'll do the offices first."

He wrote down orders for chips and fish, pies and cakes, cigarettes and chocolates against each person's name, postage stamps for the office, biscuits and cakes. "You'll need some more tea, sugar and milk too," explained Peter, "get Golden Stream tea, it's only ninepence a quarter; the money for that comes out of the tea-swindle box in the crew room. I'll show you." Each person on the order-list gave enough money to cover his needs: half crowns, florins, shillings, threepenny bits and sixpences. Edward's overall trouser pocket became heavier with every order. "Don't forget to put down the amount they give you, Ted, so you can work out the right change to give them," Peter advised.

With the list and the money in Edward's pocket, they took off on their bikes to secure the goods. Peter showed where they were to shop. The corner shop just outside the gate kept the tea, sugar, chocolate, cigarettes and matches, etc. They rode on farther, into old Garston, to the Fish & Chip shop that Peter always used. They had just opened when the boys got there. The first batch of chips had just gone into the boiling fat but there was already a queue of five or six people.

Peter took the list and calculated the order quantities under prices. For example: Four-penneth of chips, Six-penneth of chips, Eight penny-fish, penneths of peas, so many slices of bread, Seven-penny steak and kidney pies, Four-penny pork pies, Four-penneths of scallops and so on. In this way, they could place the order quickly and concisely, when their turn at the counter came.

When their turn did come, they moved up to the counter and Peter, unflinchingly, made the order which he had calculated from that day's list: "Four four's, Six sixes, Nine o'fish, Four pork pies, One steak-an-kidney, Six o'peas and Six o'bread; no scallops today." The order came out pat, no stuttering; an amazing feat. Peter grinned at his success.

The shop owner quickly prepared their order from the sizzling vats; each item in its greaseproof envelope and wrapped in a double page of last week's Daily Mirror. Four or five full pages from the broadsheet Liverpool Echo completed the assembly into one big package of perfectly insulated, hot food. The two errand-runners pedalled off again to the next shop, where they could buy cakes and biscuits. The smaller packages fitted neatly into the saddlebag on Peter's bike. "You'll have to bring a bag tomorrow Ted, or a satchel or something. Sometimes the order is bigger than this," Peter flung over his shoulder, as they cycled over the cobbles of Garston's streets. He had noticed that Edward's old saddlebag was indeed falling to pieces.

Peter's bike had dropped-handlebars and Derailleur gears, but he carried the newspaper parcel under his arm, cycling confidently with one hand. They arrived back at the hangar just in time to make the tea and dish out the chips and change to the hungry crew, before leaving the gannets to it.

For Edward it was lunch at home. He'd worked it out; twenty minutes to ride home, twenty minutes to eat lunch and twenty minutes to ride back again was a very orderly one hour lunch break. He'd bought two sixpenny bars of Pain's chocolate for himself on the chip run, one to eat going home and one to eat going back to work. This was the life – not a bit like the rigours of his school days.

Edith had lunch ready on the table for her boys as Edward arrived, in good time. Naturally enough she and David had a battery of questions about Edward's first morning at the Flying Club. They discussed his experiences at length while they ate. Edward couldn't wait to get back to the Club, although clearly he knew, he wouldn't get to work very much with the aircraft, for the time being anyway, but just being part of it was important to him. His new adventure had begun. What would tomorrow bring? Would he get to fly as well, one day?

With lunch finished within the twenty minutes he'd allowed himself, Edward peddled off again with enthusiasm. The afternoon brought more tasks for him to learn about. How to wash an aeroplane was not the least of these. His part was to clean the undersides of an *Auster*'s wings and belly. The wings were not too difficult, though of course he got quite wet as soapy water ran down his arms. The under-belly of the aircraft was something else, however.

With oil streaks from the engine and exhaust pipe stubs to contend with, the supine position, lying on his back on a wheeled, wooden duck-board and soapy water dripping all over him, it wasn't a job to relish; but when he'd finished, Edward felt quite pleased with himself. The engineer in charge complimented him on a job well done, before releasing his wet apprentice for other work.

Then it was tea break time again and back to the crew room to wash cups and prepare for the mid-afternoon recess. Later, Edward made the evening errand-run, and made more tea before ending the first complete day of his apprenticeship, tired and still not a little damp.

Each day, thereafter, followed a similar routine but every day too, for short spells between the mundane jobs, they put Edward with someone to learn something definitely related to the care and maintenance of aircraft, even if it was only holding tools and passing spanners.

18

One morning, after helping to push the aircraft outside, Bernard took Edward to the flying club's unique *Taylorcraft Plus D* aircraft. This, a very utility, fully aerobatic, high-wing aeroplane with a closed cockpit, two seats fitted with Sutton harnesses and very simple controls, had none of the upholstered comforts enjoyed by the later *Auster* variants. It was an aircraft of simplicity, having no real complexities, no wing-flaps on the high, strutted mainplane and a windscreen of flat, perspex panels. Powered by a *Blackburn Cirrus Minor* engine, driving a wooden, two-bladed propeller, the *Taylorcraft* airframe originated in America and its design preceded the newer *Auster* range, then built in England.

Bernard had Edward place the chocks in front of the main wheels, with the pull ropes placed from the inboard side, in front of the chocks and stretched out towards the wing tips. "Why like that?" Edward asked, as he followed the instructions.

"When the engine is running, the wheels press quite tightly against the wedge-face of the chock," explained Bernard. "When the rope is pulled to take the chock out, it turns the chock out, from under the tyre, rather than sliding it sideways. The rope pulls the inboard side of the chock towards the puller, making it easier to dislodge, especially if the brakes aren't holding too well." With that information absorbed, Bernard drew Edward to the cockpit door on the left. "You're going to learn how to swing a prop this morning, Ted," he said with a grin. Edward's heart leaped. "First, here's the cockpit drill. You MUST do this before you touch a propeller."

Bernard clambered into the pilot's seat, no mean feat for a chap of his build. Edward moved closer to watch the procedure through the open door. Bernard placed his heels on two little pedals on the floor of the cockpit, immediately below the rudder pedals, and by straightening his legs, pushed, to depress them. At the same time, he pulled on the ratcheted brake-handle that protruded from under the instrument panel. "There, that's BRAKES ON," he said. Then, directing Edward's eyes to a pair of switches mounted on the instrument panel, he continued. "These are the Magneto Switches, they MUST be OFF: OK?"

"Right." Edward responded.

"Just because they are off, doesn't mean the engine is not still alive," Bernard continued, "always assume that an engine IS alive when you're on the end of the prop: OK?"

Edward nodded, "OK."

"The petrol-cock is over here, and it should be OFF too," said Bernard, tapping the fuel handle with his finger. "This is the throttle." He pointed to a push-pull handle with a flat yellow knob, projecting from the centre of the instrument panel. "It should be CLOSED, that's right OUT, not IN: OK?"

"OK."

"Right then, that's brakes ON, petrol OFF, switches OFF and throttle CLOSED," he affirmed, as he cast his eyes over everything again. Satisfied that all was in order, Bernard climbed out of the cockpit and took Edward to the front of the aeroplane. He continued with the lecture of instruction, drawing attention to the grassy turf. "Before you swing a propeller, Ted, it's generally a good idea to have a quick look around on the ground, to make sure there are no obstructions or loose stuff that might trip you up, or anything that might be picked up by the prop-wash once the engine's running." He squatted briefly to look under the plane.

He stood up again and leaned around the nose of the aircraft, looking at each side. "Check that the cowlings are all closed and fasteners secure, also check immediately behind the aeroplane to be sure there's nothing there that might get blown away too, by the prop wash." Bernard stepped back a pace or two before continuing with the instructions. "Place yourself where the pilot can see you and catch his attention by calling out, BRAKES ON. Don't touch the prop until he responds to ALL your calls – none of that 'ALL THOSE THINGS' nonsense; the prop-swinger is in charge until the prop is swung and the engine has started. Remember that, OK?" Bernard went on, glancing at his pupil to be sure that he was listening.

"Right," said Edward, mentally struggling to retain the order of the instructions. Bernard continued with his lecture.

"The next call is PETROL ON, then SWITCHES OFF; make sure you get your responses. Now, this aircraft has a *Kigas* primer. It is a little hand pump on the instrument panel that shoots petrol into the manifold for better starting. The pilot does this from the cockpit." Bernard had shown Edward the primer-pump earlier, mounted, as he said, on the instrument panel. "So you call;

'PRIME, TWO SHOTS'; Get your response."

"Right!"

"Next thing is to get some of that priming fuel into the cylinders." Bernard moved towards the propeller. "You call, SWITCHES OFF, SUCKING-IN; the pilot should respond with SWITCHES OFF, SUCK-IN." Edward nodded again; thinking it was a lot to digest, all at one time. "Now you can touch the propeller. You'll see that it's at an angle as you stand in front of it, about twenty-to-two and it turns clockwise, viewed from the front."

Bernard twirled a finger to emphasise the rotational direction. "Grasp the top blade close to its tip with both hands." Bernard reached up to demonstrate. "Put your weight on your right foot and place your left foot forward a bit, off the ground." He followed his own instructions to show Edward how. "As you swing the propeller downwards, sweep your left leg back so that when the prop blade in your hands reaches about the twenty-five-past position, your left foot is behind you, on the ground. Swing your body a little to the left, move your right foot back as you release the blade. You'll probably find that your left hand will automatically leave the blade before the right hand, as you swing your body. That way, if the engine happens to fire, you are well out of the way. OK?"

"OK."

"You do this four times for the sucking in sequence; that's usually enough."

With that, Bernard swung the propeller a couple of times to demonstrate the footwork. "Next, you call THROTTLE SET? The pilot sets his throttle a little open and responds with Throttle SET. Then you call CONTACT; the pilot switches the magnetos to ON and responds with CONTACT. Then you swing for a start. OK, have you got all that Ted?"

Yes, Edward had it! It was a lot to remember but he thought he'd grasped it quite well. Bernard went through it all again anyway, to be sure. He made Edward repeat it from the beginning and nodded his satisfaction. "D'you want to have a go then, Ted?" Bernard asked, with a grin. "The engine needs to be run, so I'll get in and you can swing the prop."

Just then Nick, the engine man, joined them. "All set then? I'll just stand by here," he said, "to keep an eye on you. All right?"

Edward nodded as Bernard climbed into the Plus D's cockpit again. Not without a little anxiety, Edward stood by the propeller, in sight of Bernard, through the windscreen. Nick stood off to the side, near the wing tip, from where he could see and hear both Bernard and his apprehensive pupil. Edward called out his first request, "BRAKES ON?" His voice broke a little and he grinned at his own nervousness. They went through the whole routine until they reached the SUCKING-IN part.

Edward stepped forward to the propeller and grasped its cold surface – fingers clenched over the trailing edge, near to the yellow painted tip. At such close quarters, he could see where the black paint of the leading edge had eroded slightly. His heart was pounding and he was sweating a little too. Then, with his left foot forward and his weight on the right, he called, "SWITCHES OFF, SUCKING-IN!"

"SWITCHES OFF, SUCK-IN!" Bernard responded.

Edward swung downwards with his arms, swept his left foot back, turning his body slightly, and stepped back with his right foot as he released the blade. He heard the click of the magneto impulse starter as the blade turned under his thrust, and jerked through half a revolution. Three times more, then they were ready for a start. Edward wiped his sweat-damp palms down the sides of his overall, as he stepped to the blade again. He took a quick glance at Nick, who smiled widely and nodded vigorously.

"THROTTLE SET?"

"THROTTLE SET!"

"CONTACT!"

"CONTACT!"

This was the crucial moment! Edward swung the propeller, heard the impulse, and stepped back as the propeller-blade left his hand, snatched away rather than released, as the engine fired erratically, then burst into full song. Edward's newly learned foot-work, which reminded him of the church-dance he'd been to, when he danced with the delightful Anne, took him well out of harm's way and he moved obliquely away towards Nick, who, still grinning, put his hands together in an unheard charade of applause.

Bernard set the throttle for warm up revs, the engine-exhaust battering through the stub pipes under the cowlings. Nick put his arm around Edward's neck in a comic wrestler's grip and shouted into his ear, over the noise of the running engine and whirring propeller, "Well done, Teddy! Well done!"

At age fifteen, Edward had swung his first aircraft propeller to start an aero-engine, elevating his enthusiasm and growing love for aviation to new and even greater heights.

Thereafter, along with the other apprentices at the flying club, Edward performed the task frequently, whenever pilots, engineers or senior apprentices with approval to run engines, called, "Can someone give us a swing please?"

Edward felt proud of his achievements and the learning he'd acquired during the first few months of his new life at work and although he remained the most junior apprentice at the club, still expected to do the lowliest of tasks, there were many moments of tremendous happiness. Having gravitated to unsupervised prop-swinging, refuelling and performing small, technical tasks of maintenance, his life had become so much more fulfilled and enjoyable. He came to accept that each task given to him, however lowly, had a reason behind it. Sweeping the hangar floor, cleaning up the crew-room, making tea and running errands, all taught him the importance of cleanliness and tidiness around aeroplanes, orderliness and discipline amongst his companions, and tolerance of the many unpleasant but necessary chores that frequently came his way. He was on the road to becoming an engineer.

Part 7

Of Aeroplanes and Learning

19

Edward was very happy to be working among the friendly people at the flying club. His new environment wasn't at all like school. He was still under training of course, as an apprentice but he no longer felt like a schoolboy. Between mundane chores and errand running, his technical involvement on the aeroplanes began to take on interesting and instructive dimensions, although most of it was the dirty stuff to begin with. Cleaning engine-cowlings and drip-trays, scratching at corrosion in hard-to-get-at places, were typical routines for a first-year apprentice, along with watching and listening closely.

There were occasions when he had to clean cockpits or the *Rapide* cabins, after flights in rough weather. Apprentices received a special bonus of ten shillings for mopping-up the mess made by sickly passengers; a common example being a planeload of revellers returning to Liverpool, after a day-trip to Dublin and the Irish racetracks. Victims of over-indulgence and stomach churning air pockets over the Irish Sea, the hapless passengers invariably clubbed together to raise a few shillings with which to pay for the unsavoury task.

Edward's great delight was in working outside the hangar, if only for the sunshine, fresh air and the sights to be seen on the airfield. There were frequent aircraft movements at Liverpool's busy airport, and Edward loved to watch them. His affection for aeroplanes grew and deepened with every passing week. The daily visit by a *Spitfire* on meteorological duties (THUM Temperature and Humidity Flight) was the highlight of each morning. Its arrival always brought the fellows running to the hangar doors, to watch the landing. They never seemed to tire of seeing the *Spitfire*'s beautifully streamlined shape, with its elliptical wings and its slim fuselage, nor of hearing the sound of its throttled back Merlin engine, exhausts loudly muttering, as the pilot made his turning-approach over the river. Based at Woodvale, some distance north of Liverpool, the *Spitfire* performed daily weather-reconnaissance flights.

It stayed at Speke for an hour or so usually: enough time for the pilot to deliver his report and perhaps to have a spot of breakfast at the cafeteria in the terminal building. Later, the loud detonation of its engine-starter cartridge would herald the *Spitfire*'s departure. At this sound, everyone would rush outside again, to watch; hoping the wind was right for a takeoff from their end of the field. If this were the case, then the pilot would have to taxi his plane along the perimeter-track, passing close to the Club. That's just what the fellows wanted. It was like an addiction, to hear and see one of the hitherto most beautiful, and renowned aircraft in the history of aviation.

Edward had made several models of the *Spitfire* so for him to see the real thing at such close quarters was incredibly exciting. The erratic, bubbling mutter of its engine thrilled him as the plane made its zigzag passage along the perimeter track, to pass in front of the Club. Rolling on its narrow-tracked undercarriage, the fighter's huge yellow propeller-spinner, its long nose, its tiny windscreen and bubble-glazed canopy and its slim shape, plainly mirrored the power and speed at the pilot's fingertips. The pilot sometimes waved to the eager mechanics, from his cockpit, as he passed the small group of enthusiastic watchers at the club's hangar doors. Some of the engineers had worked with *Spitfire*s during and after the war, and so they knew the aircraft intimately.

Later, when the pilot opened the throttle for takeoff, the ex-servicemen groaned with sheer joy at the throaty roar of twelve cylinders as the air reverberated with the battering bellow of the exhausts and propeller. "OH! WHAT A BRAHMA! OH! YOU LITTLE BRAHMA," was one engineer's way of putting it, shouting to make himself heard over the din and close to tears of sheer rapturous joy. Edward could not help being drawn into the enthusiasm, along with the veterans of the club's engineering team.

As time passed, the Chief Engineer appointed Edward to work with George Parks, an airframe fitter who was somewhat older than the rest of the staff. He had also served with the RAF during the war, working with flying boats mostly, at Pembroke Dock and later, in Aden at the foot of the Arabian peninsula. George owned a very old, single cylinder motorcycle with a hand-change gear lever at the side of the petrol tank. He rode the bike to and from work each day, an hour's run from his home in the seaside resort of Southport. Whatever the weather, rain or shine, he always wore fur lined, flying gear, well-used but real leather: boots, trousers, jacket and gauntlet gloves. He didn't have a crash helmet; instead, he used a leather flying helmet and pilot's goggles. It took him a good fifteen minutes to get his gear on in the evenings, to go home.

George was a loner. He didn't become involved in the sometimes heated arguments at tea-breaks and preferred to remain on the fringes of general conversation. He worked mostly in the workshops, a small redbrick complex at one end of the hangar. Its windows over-looked the back of the clubhouse on one side and the narrow access road on the other. With so many windows, there was plenty of natural light by which to work. It was a good working environment.

A door at that end of the hangar, opening into a corridor with store rooms to the left and a few steps up gave access into the large workshop-room with benches and wide, sliding, exterior doors. Beyond, lay a smaller workshop with overhead lifting tackle and special stands for engine work. Edward joined his new supervisor in the larger of the two rooms, where George was restoring an *Auster Autocrat* airframe, removing corrosion, inspecting, changing structural parts and completely replacing the fabric. Stripped almost to bare metal, the wingless fuselage stood on strong wooden trestles. George gave his new apprentice the task of removing corrosion at the tail end, where, he told Edward, the structure is most vulnerable to moisture, particularly in tail-wheeled aircraft.

Work had already begun on recovering the first of the two wings. New linen fabric was already in place, stretched across the structure. George was in the process of sewing and stitching at the trailing edge, closing the material that enveloped the aerofoil like a bag. Nearby, also resting on trestles, the other stripped main plane waited for inspection and reworking before the fitting of its new fabric skin.

Under the guidance of his new supervisor, during the ensuing weeks, Edward learned how to use the curved needles and how to sew with the recognised stitches used in aircraft fabric work. He learned the arts of stringing, using long straight needles threaded with waxed cord, which they passed through the wing's new fabric skins from top to bottom and bottom to top, before knotting to hold the fabric close to the camber of each rib. He learned taping, doping and the names and applications of the various types of fabrics used in covering the surfaces of aircraft. There was so much to absorb, so much to remember, but Edward revelled in the work's technicality.

At the back of George's workshop lay two *Miles Magister* fuselages, resting together on the workshop floor, with their wings and tail units nearby. Painted yellow, they still bore R.A.F service markings. They were next in line for George's attention, waiting for repairs, inspection and reassembly, before being placed on the Civil Register. George explained that the Boss had acquired them very cheaply from an ex-WD auction sale, with the idea of increasing the Club's fleet of ATC cadet, pilot-training fleet. The two *Magister* airframes were George's upcoming jobs, after the *Auster*.

The *Austers* had very little wood in them. The wings contained wooden front and rear spars, separated by tubular steel compression struts, and mounted with pressed and fabricated aluminium ribs to form the camber. The fuselages had only wooden stringers attached to the welded, tubular steel framework and plywood floorboards in the cabins. Conversely, the *Miles Magisters* (known as *Hawks* in civil aviation) had semi-monocoque structures of wood, using metal parts only for the strong points and fittings, etc. Typically, the designers at *Miles* used this structural principle in most aircraft types built by that company.

It occurred to Edward that a capable person engaged in renovating and repairing aircraft constructed in these two distinctly different techniques, must have a very broad range of technical knowledge and ability. George was one such person; equally content in reworking with wood or metal structures. Yet he was a man of considerable experience, knowledge and capability who had never bothered with obtaining a license. When Edward questioned this, George simply shrugged and with a wry grin, he explained that after many years in the Air Force, the thought of renewed studying to gain a civil license, at his age, did not appeal to him. Nevertheless, George taught Edward much of his fundamental airframe knowledge.

During that time in the overhaul-shop, working along with George, there were two important, annual events in the aviation calendar. The first was The King's Cup Air Race, for which the Boss entered one of the Club's *Hawk* aircraft, with himself as pilot. The choice of aeroplane for this task had hung between the single engine *Hawk* or the twin engine *Gemini*. Although the Boss's real preference was to fly a twin in the races, because of his RAF background as a *Mosquito* pilot, Edward supposed but since a question mark hung over the reliability of the *Gemini*'s *Cirrus Minor* engines, the Boss decided that the *Hawk* was a better bet with the well-proven *Gypsy Major*.

The other event was the S.B.A.C. Air Show at Farnborough. Everyone hoped to attend on one or other of the two public days. The prospect of such a trip filled Edward with excitement for he had never been to the annual Farnborough show before but first, preparations for the King's Cup got under way. Before the races, it was necessary to prepare the club's entry, with meticulous attention to detail.

The club owned several *Hawk*s and the Chief engineer selected the one with least hours achieved, before delegating engineers and apprentices to the tasks of preparation. The Boss wanted to fly the aeroplane from the front seat, with the rear cockpit closed and faired in. He asked for the removal of the windscreen from the rear cockpit, since on this type of aircraft the rear screen was very large and rounded to accommodate a blind-flying hood. Removing this encumbrance would reduce drag and, hopefully, improve speed performance.

Normally flown only from the back seat for solo flights, the *Hawk* needed ballast to maintain the Centre of Gravity, when flown solo from the front seat. Therefore, for the race preparation they fitted bags of lead-shot in the rear cockpit, lashing them to the seat. The preparation-team fitted main-wheel fairings (streamlined, teardrop shaped cowls, often called spats, set around the wheels) to reduce drag, improving both performance and looks. Run-of-the-mill training flights didn't warrant the fitting of these because they were awkward and difficult to remove or refit when a wheel needed changing, which was often. They painted large, black roundels on the wings and fuselage to accommodate the allocated race identity number and the apprentices used up pounds of Simoniz wax polish, giving a lustre to the aeroplane's skin that would have done justice to any French-polished surface. Finally, the cream-coloured paintwork with its red cheat-lines and black roundels gleamed with a rare brilliance.

What preparation! Nick worked enthusiastically on the up-rated *Gypsy Major* engine. His objective being to squeeze every ounce of power out of it, to drive the brand new metal propeller that had arrived recently from *Fairey*'s. Air-test day proved that his work, and the rest of the team's, on the airframe, had paid off. The Boss reported a tremendous improvement in speed and the handling was perfect. The *Hawk* looked immaculate in its full race regalia. Everyone was proud of the achievement.

On the day before the races, the beaming Boss flew south in the early morning, accompanied by the *Rapide*, full of spares, tools and a couple of engineers. Tomorrow was to be the big day. For the rest, there wasn't much work done next day; there was too much excitement. The staff waited impatiently for news of their plane's progress through the race heats. The office staff was in constant contact and eventually they informed the impatient engineers that the Boss had won his heats, qualifying for the finals. To this news, a loud cheer went up. Joy and noisy discussion covered the possibilities of a win. However, an unhappy race-crew returned to Speke two days later. Having done so well in the initial stages, ultimately the *Hawk* did not have the speed to match that of other finalists' entries. Nevertheless, the Boss, though very disappointed, was philosophical about it and thanked everyone for their efforts.

The King's Cup races were over for another year. Next on the agenda was the Farnborough show, the singular and most prestigious event in British aviation's calendar, capturing the hearts of industry, operators and enthusiasts alike. Although put on to display the industry's wares, sell aeroplanes and aviation related equipment, this show drew hundreds of thousands of people to the Royal Aircraft Establishment, as it still does. The last two days of the seven-day event allowed the pubic to savour the excitement at first hand.

The year of 1953 was no exception. It was the time of the new jet era in aviation, the time of breaking the sound barrier, the time of experimental ideas, and of the 'V' bombers, Victor, Valiant and Vulcan. It was the time of Neville Duke and John Derry and of old established companies re-emerging with the coming of new, innovative thinking. The race was on, for higher speeds and higher altitudes. British companies wanted to be world leaders again. Like a magnet, the S.B.A.C Air Show drew anyone with an interest in aviation, as with all at the Flying Club.

The Boss very generously made one of the *Rapide* pilots (Peter Tare) available to take most the Club's technical people down to Blackbushe. A hired bus would take them onward to Farnborough. Each paid his own entrance charge of Five Pounds, and of course they bought their own lunches and refreshments, etc. Edward had saved like mad over the preceding weeks to put enough cash together for the trip.

The flight south, to Blackbushe, was very rough. The overcast weather in the northwest improved the farther south they flew but it was seat belts all the way and the crew felt somewhat relieved to get their feet back on firm ground again. Once inside the Farnborough airfield gates the party divided into smaller groups. Since Edward had never been to the show before, Nick and a couple of the others took him along with them to enjoy the sights, and to watch the flying display later. There was so much to look at. Edward had never before seen so many aeroplanes and exhibition stands all in one place. The static display was simply marvellous, for there were aircraft of all types on show.

Strolling among the show-tents, display-stands and aeroplanes, for what seemed like hours, Edward and his companions absorbed the latest aviation wizardry and new technology like sponges. When the air display was due to start, they managed to find a good place from which to watch, quite close to the front line of spectators and on a small rise.

The Ranald Porteus demonstration in his *Auster Aiglet* was close to the hearts of the fellows from the club. He amused the crowds humorously as he made his aeroplane hop from one wheel to the other on landing but by far the most magnificent sight, for Edward anyway, was the beautiful, all-white delta wing. The Vulcan made its low fly past, swiftly followed by the roar and crackle of the four jet engines buried in its wings. It was a remarkable sight as the huge aircraft flew at no more than a hundred feet above the runway. Now there was a 'Brahma' if ever there was one.

That year's Farnborough Show was one to remember and it was a tired but happy crew, which flew back to Liverpool in the dark at the end of a splendid day. They all felt well compensated for their failure to do better at the King's Cup air races. Perhaps next year they could win.

Any kind of air-show was popular in those days. Wing's Day was typical. It was celebrated on just about every R.A.F station throughout the UK, in memory of 'The Few' who defended the British Isles so bravely, during the Battle of Britain. There were two squadrons of Meteors based across the Mersey, at R.A.F Hooton Park, often seen operating in the local airspace. On Wings Day, they put on a show for the public. The club people didn't attend, but they saw the activity from their side of the river; the comings and goings of the aircraft based there and those of visiting display aircraft.

Sadly, an accident marred the Hooton celebration that year, when one of the Meteors crashed on the airfield, early in the show. The resident squadron's display included the demonstration of a low level, line astern, fly past, in which each aircraft by turn would complete a three-sixty degree, hesitation roll, in front of the crowds. The accident happened when one of the Meteors suddenly pulled-down during the inverted sequence, plunging nose first to the ground. At the club, they heard the explosion and saw the pall of smoke. The pilot died in the accident. It was tragic.

20

For Edward, still in his first year of apprenticeship, the days passed in an eventful kaleidoscope of interest and learning. However, the learning was sometimes by the hard route. Not far from the club, across a wide expanse of concrete hard-standing in front of Chapel House, the next hangar accommodated RAFVR unit, No3 Port Squadron. From time to time, the senior NCO, Corporal Morgan, popped over to borrow the odd spare part or a tool, or just because he had nothing better to do and wanted to chat.

One morning, Gerry asked Edward to pop over to the R.A.F hangar to borrow some Frog-gob nuts. For his part, Edward had already received instruction on the subject of nuts, bolts and washers, etc. As a result, he was well acquainted with the many types of hardware generally used on aircraft and how to identify them: the Plain nut, the Simmonds self-locking nut, the Castellated nut for use with a spit pin or locking wire, and so on. However, his instruction had not included Frog-gob nuts, and he questioned it. "What do they look like?" After a moment's hesitation, Gerry smiled and changed his mind.

"Tell you what, Ted, never mind the nuts, just pop over and ask Corporal Morgan for a long stand. You'll need one of those today."

As always, keen to please, Edward cheerfully responded. "OK, I'll go now."

Having quite unwittingly avoided a wild goose chase for frog-gob nuts, Edward willingly agreed to nip over to the MT Section for the long stand, though what it was and what he should need it for, that day, remained a mystery to him. As he walked across to the next hangar, Edward wondered about the frog-gob nuts, too. Although he'd mentally questioned it, he was still green enough to believe that perhaps there was such a thing; perhaps it was something overlooked in his aircraft-hardware tutoring.

Corporal Morgan greeted his visitor with a grin as Edward stepped into his little office. His smile broadened when the club's youngest apprentice explained his errand. "Come with me lad, this may take some time," he said, swinging his booted feet from the top of his desk. Edward followed him into the middle of the hangar, where the corporal told him to wait. Time went by slowly – perhaps half an hour. Eventually the Corporal came back. "You've had it today, mate," he said shaking his head, "you'd better cut-along now."

With a shrug, and in total innocence, Edward pushed off back to the club to find and inform the fellow who had sent him on his strange and fruitless errand. He assumed that the MT Section had run out of 'long-stands' or that perhaps they were all in use, whatever they were.

"Did you get one, Ted?" Gerry asked with a grin, as the willing apprentice stuck his head into the cockpit where he worked.

"No, the Corporal said, 'you've had it today,' and told me to come back here," Edward answered.

"Did you enjoy it then?" Gerry questioned, by then unable to contain a chuckling countenance.

"What?" Asked Edward, mystified.

"The long stand. You've been gone for hours, did they keep you 'standing' for long then?" How green can one get? The penny dropped, finally. Edward felt the colour rising to his cheeks.

In retrospect, he had wondered why the RAF chaps had grinned and whispered among themselves, as he stood like an idiot in the middle of their hangar. What a fool he felt. But he realised that it was true what they said about living and learning. Dim-witted or not, Edward was certainly warier after the episode.

On another occasion he embarrassed himself in front of every-one in the hangar, although he saw the humour of it later. He was on a lone, floor sweeping sortie; marching up and down the hangar with his yard-brush. The place was clear of aircraft and resounded with the noise of hammering and banging at the benches as people worked to repair components of wood and metal. The radio blared from its high perch on the hangar wall, adding to the bedlam of noise.

As Edward turned, close to the benches, aligning himself to make another straight-lined sweep towards to opposite end of the concrete floor, he noticed the Chief engineer standing outside his office at the far side of the hangar, hands on his hips and looking directly (Edward thought) at him. The chief shouted something but the din in the hangar drowned the words and Edward cupped his ear to catch the repeated phrase. He was always eager to comply with instructions and requests. Harry-the-Banner's dog, Peter, lay sprawled in the middle of the hangar floor, halfway between the Chief and Edward. The dog, too, had pricked up his ears.

Dropping his broom Edward turned to pull the plug on the radio. The music stopped. This caught the attention of the other fellows, banging and hammering energetically at the benches close by. "Can you all stop banging for a second, the chief's shouting something to me, and I can't hear, for the noise," said Edward, as hammers froze in mid-stroke and the banging ceased.

All eyes turned to him and then to the Chief, as the echoes died away. Silence reigned in the vast, empty space of the hangar. Cupping his ear again to hear Cliff's words, Edward marched off to the centre of the hangar, towards the Chief. "What was that Chief. I didn't hear you for the noise?"

In turn, Cliff cupped his hands to his mouth to form a speaking trumpet and enunciated very clearly and concisely, for all to hear. "Not you Ted, I was addressing the dog." With that, he turned and went into his office with a shake of his head, leaving Edward to stand bewildered, as someone switched the wireless on again and the hammering resumed, drowning the laughter and chuckles. Edward went back to his sweeping with a red face and the dog moved away to a distant corner, eyeing him with disdain as he went.

As the days grew shorter with the approach of winter, business at the Club declined. There weren't many Club-member pilots or students who wanted to fly in the autumn months. However, the ATC cadet training programs continued, when weather permitted, and there were a few charters from time to time.

One grey, windy day in late autumn, a request came from a newspaper or magazine photographer who wanted to fly over, and around, the big power station just down the river from Garston. The chief pilot decided to use the *Taylorcraft* for this mission since it had side windows, hinged at the top so that they opened outwards and upwards. The photographer's large camera needed space and the capability to take pictures, unimpaired by Perspex windowpanes. He wanted unrestricted aerial views of his subjects.

The wind was blowing more than half a gale out of the north at the beginning of the day but, by mid-morning, it was up to gale force. When the photographer arrived, the Chief Pilot briefed him on the difficulties of the flight, because of the wind, in a futile attempt to postpone the trip. Indifferent and unconvinced, the newspaperman insisted that the flight should go on. He had a deadline to meet. Delay was impossible.

Because of the raging wind outside, the pilot and the pressman (with all his equipment) climbed into the aircraft while it was still in the hanger. The hangar crew opened the hangar doors just wide enough to push the aeroplane outside, closing them again quickly afterwards, for the wind threatened to cause havoc inside, among the other planes parked. Eight people manhandled the *Taylorcraft* to a position on the grass, in front of the hangar as the wind continued to rage and shriek. Even then, faced into wind, with all hands hanging on grimly, the aircraft bucked and pitched about frantically. Someone detached himself from the group to swing the prop.

Once the engine was running, warmed up and mag checks done, the pilot waved the chocks away. The wheels were double-chocked as a precaution. The pilot opened the throttle and the 'hangers on' at the struts and wing tips moved forward with the aircraft trying to keep it stable in the teeth of the gale. Before they knew it the plane was airborne, going up like a lift without much forward speed until, at about fifty feet, it was apparently flying backwards. The wind remained steady and very, very strong. Even so, the *Taylorcraft*'s wings wagged about quite a bit as it gained height. Then it turned cross wind, flying briskly sideways, before turning downwind towards the river. The plane went very fast downwind and someone said that it would be a nonsense trying to get back. There were questions about the wisdom, or lack of it, in making the trip at all, in such weather. The ground crew themselves could hardly keep their feet; the wind was so strong. However, the pilot was ex Fleet Air Arm, with carrier deck-landing and take-off experience. He was one of the club's most skilled pilots and it had been his decision to make the trip.

It wasn't long before they heard the *Taylorcraft*'s engine at full bore, it seemed. The pilot was attempting to get back. Then they saw the aircraft, well over on the up-wind side of the airfield, aimed at the control tower, across wind and flying sideways towards them. By the time he'd reached the field, his position was well to the West of the flying club, at some three hundred feet. The *Taylorcraft* turned into the teeth of the gale and slowly drifted backwards with a fair amount of throttle on, until it was nearly opposite the clubhouse, but on the farther side of the peritrack.

By then the pilot had managed to get the altitude down to twenty feet or so, still pointing directly into wind and using the throttle, managing somehow, to keep his position as the plane came down. The *Plus D Taylorcraft* had no flaps and the wings were waggling about quite dangerously. The watching ground crew heard an increase to the sound of the engine as the pilot opened the throttle and his plane started to make headway into wind. Edward prayed that the wind did not drop. Slowly, almost imperceptibly, the aircraft lost height with next to no forward ground-speed, in spite of a high throttle setting. The plane wagged about and bucked dangerously until the wheels were almost on the ground.

The ground crew had already detached from the clubhouse

and stationed themselves to grasp the wings, struts and tail end quickly, when the moment came. This they did as the main wheels touched the ground in a near-perfect 'wheeler'. The tail wheel came down with a thump as the pilot closed the throttle and pulled the stick back. It took all eight men and boys to hold the plane steady and manhandle it across the grass to the hangar. Edward had never before seen such a flying feat, nor did he ever see such a thing again. Needless-to-say, there were no pictures taken that day! The pilot decided that wind conditions over the city were far too dangerous after all. The photographer became very poorly, filling several sick-bags and ruining his raincoat. Later, Edward made an extra ten shillings for cleaning up the mess in the cockpit.

It was evident that general aviation business at Speke had begun to prosper, although the Flying Club was going through a lean spell. Several companies had taken over facilities on the airport soon after the war finished. Two large hangars, with brick-built walls supporting arched roofs, stood on either side of the terminal building. They were very beautiful in shape, not at all indicative of the purpose for which they were built. A company dealing in heavy construction equipment, bulldozers, cranes and such like, occupied the farthest hangar at the Eastern end of the airport buildings. This hangar was inaccessible from the airfield, since the company had no real involvement in aviation activities.

On the western side, the second hangar housed an aircraft maintenance contractor, *Airwork Services Ltd.* contracted with N.A.T.O. to refurbish military aircraft before redistribution to the new European Air Forces. Parked nearby, dozens of aircraft awaited space in the *Airwork* hangar – *De Havilland Vampires* and Venoms with their strange, twin tail-booms, along with a few American *Shooting Stars*. From time to time, more aeroplanes arrived to fill the empty parking slots, as they became vacant. Together with these movements, the arrivals and departures of regular scheduled flights of civil aircraft and the club's flying operations, Speke was a very busy airport.

Just the other side of the second runway and close to the *Airwork* hangar was a group of smaller hangars, of the same type as used by the RAFVR unit next to the club. These housed a Liverpool-based operator, *Starways*, operating more than a dozen *DC3 Dakota* aircraft. The company carried out all their heavy maintenance in the four hangars. They were always very busy because scheduled movements of their aeroplanes were frequent. Charter contracts were their main source of work, not least of which were regular pilgrimages to Lourdes in France. The *DC3 Dakota*s were (and still are) very reliable. They served in most theatres of the war, carrying parachutists, towing gliders and hauling war equipment and materials to wherever needed.

The famous Berlin Airlift saw extensive use of the Dak, after the war, when the Russian occupation of East Germany cut off the city from the rest of free Europe. Many small, emerging airlines bought and used *DC3*'s because they were cheap and readily available. The size and load capacity suited the type of routes that smaller operators sought to serve. It was with one such aircraft that an incident occurred at the beginning of a *Starways* flight to Lourdes. The incident demonstrated single-engine reliability of the type. It was not an unusual sight when the aircraft took off with a full load of pilgrims in maximum density seating. However, that day proved to be out of the ordinary.

As the plane revved-up and slowly accelerated for take-off. Someone suggested that the plane carried an overload, squeezing in a few extra seats to cater for the larger number of pilgrims, who had booked up for the trip that year. In those days, the main runway at Liverpool had a hump in the middle and aircraft needing a longish run to get airborne would go over the brow and down the slight incline on the other side until lift-off. This flight was no exception.

The club staff, sitting in the sunshine outside the hangar at tea break, casually watched as the plane topped the brow and then sank beyond the hump. Only the aircraft's white-topped fuselage, fin and rudder were visible, as the take off run continued on the downhill stretch of the runway. With its tail well up, the plane lurched into the air, clawing for height, to climb slowly skywards, main wheels tucking up, one retracting undercarriage lagging behind the other, so typical in that type of aircraft. Soon the Dak was almost out of sight as it made a wide turn, gaining height to set course to the south. The chaps paid no more attention since it was quite a routine sight. However, within fifteen minutes the aircraft was back overhead again. One engine was not running; its propeller feathered. The Dak was returning with an engine failure, to make an emergency landing.

The aircraft droned around the circuit to use up some fuel and eventually lined up on the main runway, while everyone at the club lined up outside the hangar to watch. Normal approaches to this runway were from over the river. The plane began its approach from a long way out, still over the Wirral. The pilot was making sure of alignment and height. It looked as if the aircraft was too high in its approach and indeed this turned out to be the case. When it arrived above the threshold of the runway, with too much altitude, the pilot opened the throttle of the healthy engine, to go around again.

The plane didn't seem to make much height and skimmed over the brow of the hill halfway down the runway, then sank a bit. The remaining engine was all out, at full bore, as the aircraft approached the end of the airfield. They watched the white top and tail of the Dak as it gained height, very slowly, over Speke. Phew! Not a few sighs of relief emanated from our gathering of concerned watchers.

The *Dakota*'s second attempt was a success and the pilot managed to get the main wheels on the ground just a few yards before the threshold. The club people went over to see the divots in the grass afterwards. Fortunately, the ground was hard. Such were the activities at Speke during the early part of Edward's apprenticeship. Although his position at the Club was of low stature at the time, he felt very much a part of the activities. He looked forward with enthusiasm to a bright future in aviation and he felt happy and contented.

By now, winter had closed in with shorter daylight hours. Edward bought a new leatherette jacket for use on his bike. Belted and hip length, it had a deep collar and large side pockets. It didn't have a lining but it was large enough to wear a thick, heavy sweater underneath. The coat certainly kept out the cold wind and rain, providing the perfect answer to his continued use of the bike during the winter months.

Edward enjoyed riding after dark. At the end of a day in the hangar, with night falling he pedalled home to the whine of the bike's dynamo, driven by the back wheel. He bought two extra lamps to mount on the front forks and wired them into the dynamo's power. These gave him additional light by which to make the journey homeward in the darkness. Street lighting was very sparse beyond Hunt's Cross. The bike also served him for going to night school. He cycled to Allerton several times a week to attend technical evening-classes at his old school, Gilmour.

One wet and windy night he set out for school and had just passed through Hunt's Cross, taking an ill-lit back road, when he ran into the back of a parked car. Visibility was low owing to the rain-spots on his glasses and he had his head down anyway, battling into the driving rain in a strong wind. The vehicle had no lights on and Edward ran headlong into it. The collision unseated him. He sprawled across the boot, hitting his face on the rear window of the car. He broke a front tooth and almost lost his glasses, too. To cap it all, after picking himself up, he discovered that the front wheel of his bike, bent by the impact against the car's fender, would not rotate.

Since he was closer to home than school, Edward decided to call it a night and carry the bike back home on his shoulder. He staggered into the house much later with a bloodied mouth and in a shocking temper. His astonished parents did their best to calm him down with sympathetic words and strong tea. He was wet through and half frozen after his long walk with the bike on his back.

Edward took the bus to work next morning and had to do the chip-run errands on foot at midday. He stayed to eat fish and chips with the others, making a change from the normal run home and back at lunchtime. It was fortunate that the bus services past the Weinel home had improved over the years since the war, and the Liverpool Corporation had opened new bus routes to the City through Garston. Next day, the Number 77 took Edward directly and very conveniently to the airport main gates. A long walk around the perimeter track got him to work in good time. Two days later, he was able to buy a new wheel for the bike, to replace the badly bent one, and by Monday morning, he was back on the road again with his independence.

Over the months, Edward had begun to assemble a tool kit of sorts. His father had given him a large wooden toolbox, with a hinged lid and hasp for a padlock. Slowly, over the months that followed, Edward was able to put aside enough money to make individual purchases. He bought his first set of open-ended spanners, some screw drivers, a pair of pliers and wire cutters with which to carry out the assembly and disassembly jobs that the engineers delegated to him. Piece by piece, he added woodworking tools to the kit, improving his overall capability.

The winter months passed. There wasn't much flying but this provided opportunity to catch up with other important, though less urgent work. The Stores needed reorganisation and ground support equipment needed attention in readiness for the coming season's flying operations. Christmas and New Year celebrations came and went. The weather remained mild, but wet and Edward continued his learning in the workshops with George.

By his sixteenth birthday in March 1954, the *Auster Autocrat* they had been working on was almost ready for final inspection. With the airframe-work finished, only final assembly remained. They moved the fuselage to the hangar so that they might fit the wings and rig the flying controls. Nick installed a fresh engine and propeller and the finished aircraft soon stood ready for air test. George and Edward looked forward to an early start to the year's Club activity and ATC flying season, with the proud inclusion of this additional aeroplane. Spring was not far off.

Completing close to eight months in the first year of his apprenticeship, Edward looked to the future with satisfaction and enthusiasm. School days seemed to be a long way off, buried in the past, and he felt very glad that he'd taken his father's advice to make a career in aviation.

Part 8

The Orange Raincoat

21

By 1954, some changes had occurred in districts of Hunts Cross, Halewood and Speke. During the years of post-war austerity, the trappings of a nation at war had disappeared. The tank-traps and air raid shelters had gone and the barbed wire and sandbags at important locations taken away. New housing estates had sprung up and people had begun to re-adapt to peacetime life.

During the war, the city, main-line railways, factories, the airport and important docklands of Liverpool were frequently targeted by enemy bombers. Many people were the victims of the blitz or mothers of children whose fathers had never returned from the war. Some were childless wives who had waited, in vain, for soldier husbands lost in a foreign land. These poor souls were building new lives from the heartache and rubble of the past. For Edward's part, he reckoned that he was one of the luckier people. His family had survived the conflict. Now, he had a job and his future looked bright.

His daily cycle-journeys, to and from the Flying Club, took him over the broad, redbrick bridge on the road to Speke. Arching over the four-track railway cutting, the bridge had a long, smooth, downhill gradient on the far side, running southwards into the industrialised zone of the district. The bridge's descent carried traffic down to a large roundabout, where two major roads crossed at the north-eastern corner of the airfield. In those times, the industrial area of Speke extended northwards, from the river shore to outskirts of Hunt's Cross, and from the airport, eastwards, to the beginnings of Halewood.

Although there were many factories and warehouses established before the war, much of the area remained as vacant lots. Fenced off and deserted, they waited for better times when industry would one day occupy them. They had become overgrown in the intervening years, abundant with tall grasses and thickets of bushes and shrubs growing wild, perfect places for roosting birds and small creatures of the field.

By contrast, the extensive and busy *Lockheed* Hydraulic-Systems factory, visible from the foot of the downhill run, stood next to *Bibby*'s chemical factory on rising ground to the left, dominating the upper reaches of the locality. Camouflage markings, the paint fading and flaking, remained vaguely visible on the walls and rooftops of the buildings. They were ghostly souvenirs of the war and reminded Edward of his childhood days, taking those Saturday-morning walks with his brother Freddy and their Dad.

Farther along on the Eastern side of the airport, well beyond the big roundabout and closer to the river, *Dunlop* had taken over the old *Rootes* factory where *Bristol Blenheim* and *Handley-Page Halifax* bombers were produced. The factory had become one of the largest employers in the district. Hundreds of council-houses, built in estates upon nearby urban developments, provided dwellings for employees of the many factories in the zone. The rooftops and chimneys extended southward, filling the sloping shore-lands of lower Speke, towards the river's edge. The broad concrete roads of the neighbourhood were still in excellent condition, in spite of the peeled tar coating that had survived the many years of wartime inattention, the inclemency of weather and the passage of heavy traffic.

Edward enjoyed cycling, whatever the weather. High winds were a nuisance of course but if they were against him going to work, he could at least look forward to a tail-wind on the return trip. By the age of sixteen, Edward had grown much taller. He was still lean, but he felt very strong and full of vitality. The twenty minute bike ride was good exercise before the day's work. He always felt alert and wide-awake when he arrived at the Club, ready for the chores and very happy with his lot in life.

It was on one of those fresh mornings, in the early spring of 1954, when Edward's young life took a turn for the even better. It happened on his usual way to the airport. As he cycled over the brow of the bridge, with its wide footpaths and redbrick parapet walls, he noticed a young girl, a teenager like himself, walking up the incline of the bridge on the farther side. She wore an orange-coloured raincoat, which struck Edward as rather a strange hue for such a garment. Oddly enough, the colour seemed to suit the girl very well. It was probably the coat, which first caught his notice.

From a distance, Edward observed with keen adolescent interest that the girl was very attractive. She wore no hat to cover her short-cropped, blonde hair. The belt of her raincoat, cinched tight at her slim waist, emphasised and exaggerated the broad flaring of the knee-length coat, a style so popular in those times. She walked upon low-heeled shoes and carried a large handbag slung over one shoulder, to hang at her side, with an arm crooked through the strap. They passed at the crest of the bridge, she on the pavement and Edward on his bike.

Sensing his intense stare, the girl looked in Edward's direction and smiled. So fleeting was the moment that he hadn't time to react, for they'd passed and he had begun to accelerate on the downhill stretch from the crest. He flung a glance over his shoulder, but by then the girl was waist-down on the farther side of the bridge and she wasn't looking back at Edward anyhow. Her smile so excited him that he almost hit the kerbstone with the front wheel of his bike in his anxiety to look back. He grinned at his own ineptitude as he straightened his path. That would have been a sight to see – base-over-apex for a smile. Edward applied himself more diligently to keeping on track.

It was spring, a girl he didn't know had smiled at him, birds were singing all around and the sun shone with the promise of a fine day. The old adage about a young man's fancy turning to thoughts of love in spring crossed Edward's mind, and he whistled a merry tune as he pedalled on. He arrived at work in a very happy frame of mind indeed. He felt wonderful, contemplating his daily chores with sublime contentment and unconcealed joy, indifferent to the mundane aspects of the tasks. The girl's smile had made his day perfect.

In the hangar, everyone worked at the many tasks concerning the overhaul of "Peter Tare," the *De Havilland Rapide*. It's Certificate of Air-worthiness renewal had fallen due. The stripped aeroplane rested on trestles in flight-attitude, occupying a corner of the hangar near the crew-room. The apprentices, all tasked with anti-corrosion work on the airframe, scrubbed away at rust on metal components of the structure with wire brushes, re-protecting the cleaned areas with green-primer paint afterwards.

Edward had already learned that corrosion was a major source of concern in aircraft maintenance. The engineers explained that both ferrous and light-alloy metals used in structures are subject to corrosive action, especially in atmospheres where moisture and salt are prevalent. Sudden variations in temperature sometimes cause condensation to form upon metallic parts, having similar effect and producing corrosive action. For these reasons, strict and frequent inspections follow a programme of checks called up by the Approved Maintenance Schedule. Rectification work is always necessary.

The boys made small wire brushes from short pieces of narrow-bore aluminium tube threaded with a length of steel control-cable, crimping the tube's end to hold the cable in place.

Fraying the cable's protruding end opened the wires to form stiff bristles. The little hand-made brushes were ideal for getting into the tight corners, and they were less aggressive than emery-cloth for removing rust.

The bracing wires, strung diagonally between the bi-plane's wings were amongst the most vulnerable to corrosion, as were the fittings around the tail end, tail wheel area and at the main wheel areas behind the engine nacelles. In particular, and partly due to acid-vapours, corrosion was considerable around the bolts securing the battery and its tray, at the aft end of the cabin floor. The floor panels themselves, at that point in the cabin, tended to get damp, soggy and inclined to disintegration, being made from a compound material of compressed cork, wood-chippings and resin, fixed in metal fittings. Work on the *Rapide* was intense, for an aeroplane sitting-idle on ground cannot earn its keep, as the Chief had so often said.

The aircraft stood with areas of the fabric stripped from the airframe providing access for the extensive repair work needed to restore portions of oil-soaked woodwork in the wings and rear fuselage, again particularly around the engine nacelles and the tail wheel. The under-belly fabric-skin of the aircraft hung open like the bomb-bay doors of a bomber. Unlacing a cord from two rows of hook-like buttons along the under-side, centre-line of the fuselage, allowed the fabric to fall open in two longitudinal flaps. This gave access to the flying control cables, pipe work and electrical wiring under the floor of the cabin.

Engineers and boys went about their tasks to the strains of Music-While-You-Work, blaring from a radio perched high on its shelf at one end of the hangar. Edward thought about the girl on the bridge, and her smile, as he scrubbed vigorously at the corrosion around the *Rapide*'s tail wheel fittings. He felt very happy and strangely alert. His life seemed full of promise. Next to the *Rapide*, the flying club's *Miles Gemini* aircraft also rested on trestles. Several engineers were addressing undercarriage problems that needed repairs and retraction tests.

On this type of aircraft, normal retraction and lowering of the landing gear is by means of electric motors, driving screw jacks. In an emergency, such as electrical failure, lowering the undercarriage relies upon an alternative mechanical system, operated manually from the cockpit. The emergency system depends upon very strong coil-springs, compressed and latched only whilst the aircraft is on the ground. By pulling a handle between the two front seats, the pilot can operate the system to release the springs. This action separates the screw jacks from the landing gear and allows the springs to force the wheels into their down and locked position. Once released, subsequent retraction is impossible, until resetting of the mechanisms and springs. Quite fool proof on the face of it!

Nevertheless, one of the club's newer pilots, returning from a charter job, had reported ‘no green lights’ over the airfield. He suspected a minor electrical fault to do with the indicator lights, but the control officer advised that his landing gear was still up when he made a low pass with his aeroplane over the tower. The pilot spent a little time over the airfield and the river, pointlessly diving and swooping in a futile attempt to shake the gear down but unlike some hydraulically operated systems, rotating screw jack systems will not shake down. Finally, the pilot pulled the emergency handle. The wheels came down and locked with green lights but the pilot ended up with a couple of yards of loose, steel cable, plus the handle, in his hand. He had yanked out the whole thing in his anxiety to get the undercarriage down.

The root of the problem lay in a failed solenoid switch, controlling current to the electric motors that normally drove the screw jacks. The job then, was to reeve a new cable from the cockpit to the spring release mechanisms, change the solenoid switch and then test everything for proper function. As if there wasn't enough work to be going on with, one of the *Miles Hawks* lay in a sorry state too, at the back of the hanger, waiting for major repair-work after a crash. Some weeks earlier, an ATC cadet had flown his aeroplane into the ground during his first 'touch-and-go' solo, reluctantly ‘ploughing the fields and scattering’ in a very convincing incident. The accident virtually wiped-off the main undercarriage legs, and it was questionable whether the wings were within the constraints of economic repair.

The cadet concerned had taken much longer than the usual twenty to thirty hours of instruction before going solo. Eric the instructor had shaken his head more than once over this student's aptitude. "A nice-enough lad, and keen, too, but I don't know if he has the makings really. Even yet I don't know if he's ready to go solo." However, after fifty hours, sort of second time around, Eric said that he thought the lad just might be ready.

That day, Eric and his student had taken-off in the *Hawk*, intending to practice circuit-and-landing procedures. The wind was from the west. Landings were therefore possible on the grass, with an approach over the trees of Speke Hall, touching down towards the club hangar but on the far side of the airfield's main runway, pulling up well before the pavement. It was a fine day, about mid morning, and the hangar crew took their tea break outside in front of the hangar, to sit in the sunshine and watch the student's progress.

After several impeccable landings, with equally impressive, rolling-take-offs, the instructor dismounted. The young student was about to go solo. He turned the aircraft and backtracked to the farthest end of the grassed area. Turning again into wind, he started the take-off run with textbook precision. With one notch of flap and at full throttle, the aircraft accelerated obediently to the thrust of its invisible propeller. Its tail rising and its wheels bounding over the bumpy grass, the *Hawk* lifted smoothly, before leaping heartily into the air to carrying the cadet on his very first solo-flight. The hangar crew gave a little cheer.

Eric began to walk back towards the club, halting briefly at the main runway to watch his student's progress around the circuit. The cadet was doing fine. Having climbed out over the gas works, he made his first ninety degree turn across wind. Eric took a green light from the runway control van and continued his march across the runway and the dew-wet grass, back to the club.

The aircraft by then, was well into its down-wind leg, letting down gently before turning crosswind again. Eric had reached the watching group, and was taking his first sip of hot tea, as the student made his final approach. He had done splendidly, thus far. With flaps down, his height good, the cadet brought the *Hawk* swooping in over the copse of trees close to Speke Hall, flaring a little above the grass, throttling back, floating and flaring, wheels skimming, then touching the ground with just a trace of bounce; it was a good landing. The student wouldn't have heard the murmurs of approval from the gathering, outside the distant hangar, but he might have felt pretty pleased with himself as he opened the throttle wide, to take off again.

The *Hawk*'s wheels bouncing as the speed gathered over the bumpy grass, the bounding plane momentarily lifted off – but what then? The student retracted the wing flaps to fully up, losing precious lift at the critical moment. The *Hawk* sank fast and the tail was too high. The main wheels ploughed into the ground, doubling back under the wings as the aircraft sank onto its belly. On impact, the cadet must have snapped the magneto switches off an instant before the propeller tips touched ground, for the *Gypsy Major*'s roar ceased abruptly as the *Hawk* slithered to a halt, on its belly, just short of the paved, main runway.

Eric and two or three of the engineers piled into the 'Tilly' wagon that they always kept parked outside the hangar, ready for such emergencies. They rushed off across the airfield. An MTCA car and one of the airport fire engines swiftly followed, careering down the peri track from the control tower building, also heading towards the scene of the incident. There was no fire and the cadet had scrambled out of his cockpit unhurt (apart from his pride) to stand dejectedly some little distance away, surveying his crippled bird that now lay silently upon the grassy field.

When the 'Tilly' brought the humbled cadet back to the club, Eric took him away into the instructors' room to debrief, privately, over the incident. Poor chap: he really looked appalled by what he'd done. Meanwhile, the airport authority provided a tractor, a low-loading-trailer and a winch with which to retrieve the wreck. Even so, it took the rest of that morning to get the damaged *Hawk* back to the hangar. There she lay, forlorn, silent and without wings, still waiting for repair.

As Edward scrubbed and cleaned at corrosion on the *Rapide*, club-flyers made good use of the fine spring weather and the *Austers* flew regularly. Welcome distraction from his corrosion-control work occurred frequently as the intrepid pilots called for a prop-swing or for someone to pull the chocks. Periodically the planes needed refuelling; another job for the young apprentices of course.

Between times, Edward's mind returned to the girl whose smile had so enchanted him earlier that day. Who could she be, he asked himself. Why hadn't he seen her before? Would he see her again? Another shout for a prop swing distracted him, as he happily mused and fantasised over the morning's romantic occurrence.

Back on the job, he mentally dallied with his prospects as an apprentice with the Company. The Flying Club had contracts with the MoD for Army Co-operation flights using the *Rapide* and the *Gemini* for those lucrative flying tasks. On occasion, this gave the apprentices an opportunity to fly and Edward took every opportunity he could get. Two or three of the other, senior fellows were already taking flying lessons for their private pilot licenses, but at three pounds an hour, such a desirable spare-time activity was beyond Edward's current financial capability. Several of the engineers had gained their PPLs, already and they flew for an hour each weekend during the summer months, to keep their hand in. Perhaps Edward would follow their lead in due course. Until then, any kind of flying experience was highly desirable.

When the boys knew that there was an up-coming Army co-op program, it was only a matter of seeking permission from the chief pilot to go along as an observer. Most times, such flights were at night. Take-off time would be around eleven, with a flight duration of two hours or so. The pilot would climb the aircraft to near maximum ceiling, around eleven thousand feet, and then fly on a set triangular course over the given area. The Army would then train their radar and gunnery pointing equipment, using the aircraft as the target. There was not much to see on those night-time flights, depending on the cloud cover and it was always very cold at that altitude. Warm clothing was essential. This Edward learned, to his cost, from his first such flight. He could hardly stand up, after getting back from that first ill-clad trip but it WAS flying, and that was important.

On subsequent flights, Edward was sure to wear two pairs of

long trousers and thick sweaters under his overall, gloves and a borrowed pilot's-helmet. Some pilots were kind enough to share their thermos of hot coffee or tea and perhaps even a sandwich or two.

22

At weekends, during the summer months, they used the same aircraft to make daytime pleasure flights over the thriving dockland of Liverpool, and around the power station. Speke airport had public terraces overlooking the apron hard standing, and this is where Edward spent many of his weekends, selling pleasure-flight tickets. On Saturdays and Sundays, people came mainly to watch the arrival and departure of aircraft at this busy little airport.

There was always plenty to see. *BEA* had regular flights in and out of Speke: *Aer Lingus*, too. There were also regular visits of *Bristol* Freighters from both of these companies, bringing milk etc. from Ireland. Most air-traffic movements in those days were *DC3 Dakota* aircraft, although military aeroplanes came and went, frequently – sometimes singly or, more often, in flights of four or five. *Vampires, Meteors, Sabres* and *Shooting Stars* were commonplace during those times.

Edward's secondment to the pleasure-flying duties, albeit on overtime, was a pleasant job. He enjoyed it. He wore splendidly white overalls to impress the public. On the airport terraces, he had a desk with a lockable drawer, a sign saying 'Pleasure Flights Here' and a book of tickets with counterfoils.

At a price of one pound for adults and ten shillings for accompanied children, the fifteen-minute trip was good value and, for Edward selling the tickets, quite easy to count. Business was normally very good if the weather was fine but if people didn't come to the desk, Edward would take a little walk-about, to entice the timid to take a trip. This worked, more often than not and he frequently managed to sell enough tickets in this way to fill the waiting *Rapide* – eight or nine people.

After gathering a full load of passengers, he would take them down the stairs to the aircraft on tarmac hard standing, board them and make sure they were all strapped in properly with their seat belts. He also had to ensure that they knew where the sick bags were too, for in spite of the ten-shilling bonus, he didn't much care for cleaning up after sickly passengers. Edward looked forward to another summer season on the terraces and returned his attention to the job in hand, scrubbing at corrosion.

He realised that it was Friday and as he painted over the areas of bare metal that he'd been cleaning, his thoughts again drifted to the vision of that morning, on the bridge. He surmised that if she (the girl in the orange raincoat) was on her way to work, when Edward saw her, then he would see her again probably on Monday or maybe even tomorrow, Saturday. Had he been early that morning, or late? What time was it when they passed, on the bridge? Edward turned the questions over and over in his mind, against a background of wild fantasy, as he scrubbed at obstinate corrosion on the aircraft's metal fittings. He glanced at his watch; it was time to make tea for the chaps who were working-on that evening. Oh well, he thought, back to the drudgery, make the tea, run the errands but then, with a happy feeling inside, he'd be off home on his trusty bike, to wait impatiently for the next morning to dawn.

The minimum working week in 1954 was forty-four hours, so Saturdays were half-day, working until lunchtime. Edward resolved to keep to his normal timetable and be sure to keep his eyes peeled at the bridge. He felt like doing something completely mad. He wanted to make an impression. If only he had the motor cycle he so badly wanted. Alas, stuck with his pushbike, the only thing he could think of, to catch the attention of the girl in the orange raincoat, was to be doing handstands on the saddle as they passed. Silly thought: but he grinned at the very prospect.

When he wasn't at work or night school, Edward's spare time obsession was making model aircraft, both solid, balsa wood models and free flight, flying models. He spent hours at this 'busman's' hobby and produced many models to hang on threads, suspended from the ceiling of his room. It was his room now, since Freddy had joined the Army. David still had the little room, over the front door hallway. However, on that Friday evening, Edward just couldn't concentrate on any of his hobbies or studies; his hopes were high for another glimpse of the girl on the bridge, in the morning.

Saturday dawned. Edward set out from home, pedalling eagerly, intent upon the possibility of seeing the girl in the orange raincoat again. Alas, what a disappointment! He reached the bridge but no girl! His mind raced... Shall I wait – has she passed already? No, surely I would have seen her! He took the downhill stretch, from the bridge-crest to the roundabout, very slowly, peering ahead to catch the first movement of her ascent on the farther side of the rise, but he saw no one, not a single soul!

He reached the wide, circular roundabout and cycled its circumference several times not knowing which of the approach-roads the girl used. He glanced at his watch. Time was getting on, and he would be late for work. With a final circuit, he pushed on, joining the cycle track on the westbound road towards the airport. He felt completely disheartened. What a miserable morning he spent. At least, he wouldn't have to do the chip run that day, since the half day ended at lunch-time.

Edward wondered if there would be Pleasure Flying in the afternoon. He didn't really care one way or the other. He felt totally deflated. Would he really have to wait until Monday morning? Would she smile again? Perhaps it wasn't him she had smiled at, that first morning on the bridge. Perhaps his shirttails were hanging out, or his hair was sticking up, uncombed; perhaps she saw something amusing about the tall, lanky lad on his old, lady's-bike; perhaps she was laughing at that. Perhaps – perhaps what? Perhaps he was just being daft!

Just as he anticipated that afternoon, the Chief asked him to look after the pleasure-flight ticket-sales again, starting immediately after lunch. Edward cycled home to eat, wash and change, returning to the airport, resplendent in his newly laundered, special, white overall. The weather was fine. It would likely be a busy afternoon. He collected the tickets and moneybox from the office and hopped into the back of the *Rapide* as the pilot prepared to taxi up the peri-track to the terminal building.

It was a beautiful sunny day. Edward and the pilot sat next to the desk, basking in the sun. They didn't have long to wait before people began to drift onto the terraces. The airport authority supplied, without charge, folding, metal chairs on the public terraces and thanks to this, visitors tended to stay longer – some, all afternoon. Tea, sandwiches, cakes, soft drinks and ice cream were all available from the little cafeteria, downstairs in the main hall. It made for a good afternoon out.

Before long, the first tentative customers came strolling along to the table. It was strange, Edward noted, because most people seemed to approach the idea of flying quite timidly to start with. However, many came back for more once they had taken the initial step. Edward often wondered how many children, who in later life might have made good pilots, didn't, because they didn't take that first flight.

Soon, there were eight anxiously waiting pleasure-flyers. Dicky, the pilot of the day, said, "Let's go; Eight's enough. Let's get airborne." There was room for one more but since there wasn't another taker at that moment it was better not keep the others waiting, so off they went. In fifteen minutes the *Rapide* was back again, disembarking eight, happy, smiling people; a good sign because Edward had also noticed, on previous weekends, that potential customers would often wait to see the reaction of the first load. Indeed, it wasn't long before Edward had another group of nine lined up. In reality, he had a long queue of people at the table, to whom he had to apologise for his momentary absence, getting the second group boarded and seated.

The afternoon passed extremely well, so much so, that Edward hardly had a moment to reflect upon the girl with orange raincoat and his disappointment of that morning. Towards the end of the day, business was a little less hectic and Edward was able to daydream just a little. He had one more load of nine waiting, plus an additional man, his wife and their son. Edward explained to the man, that unless at least four more passengers came forward, he would have to return the money and ask the family to come again, another weekend. Edward could not send just three people. The minimum was seven. The family were content with this and said that they would wait, to see if more passengers came whilst the penultimate flight was away.

The flight turned out to be the last of the day, because no more people did come forward. Edward returned the money to the man, who promised his disappointed son that they would surely come back another day. Unfortunately, Edward forgot to retrieve the tickets, and later realised that the moneybox and the ticket-book counterfoils didn't tally. This mistake cost him more than a week's wages. The Club's cashier wasn't sympathetic to Edward's mistake and insisted that he make up the balance. Edward reproached himself with the thought that he must have been thinking about bridges, or fleeting smiles, or orange raincoats, or something, to make a silly mistake like that.

He woke on Monday-morning, feeling that it was special. All weekend Edward had ground his teeth over the loss of two pounds and ten shillings through his own stupidity but at least Monday promised to fulfil his hope to see the girl again. With quick farewells to his people at home, he shoved off to work on his bike, trying not to pedal with too much haste. He was just about on time as he got to the top of the bridge. His heart leaped. There she was, coming up the other side, orange raincoat and all. Edward slowed down a little and found himself whistling one of Dicky Valentine's more popular songs. He tried not to stare or look too interested. He wanted this to be as casual as possible. He looked straight ahead, trying to use peripheral sight to catch reaction.

Then they were level. Edward looked towards the girl. There! She had been looking in his direction, he could tell. At the movement of his eyes, she quickly averted hers, to look straight ahead. She smiled though, and then looked down at her feet for a moment, as if the sole had fallen off one of her shoes. It hadn't, and Edward smiled too. Then they had passed, the moment was gone, but far from forgotten, ever. Edward took a quick glance over his shoulder, being careful about kerbstones, manhole covers, drain-gratings and the like.

What he expected, he wasn't quite sure. That she would be running after him, tears streaming down her face, arms outstretched, appealing for him to stop? Huh! Chance would be a fine thing! She didn't of course. Edward's adolescent imagination had run wild again and by then, anyhow, she was over the brow of the bridge and gone, whilst he was gathering speed down the incline on the other side.

His day made (and hopefully all his tomorrows), Dicky Valentine's song sprang to Edward's lips again. He sang heartily, all the way down to the roundabout. 'Life is delicious', he affirmed, out loud, to no one in particular and to everyone at large!

Later, at work, the radio never sounded so good, as Dennis Lotis sang, 'As you walk in the garden, the garden of Eden'. The voices of Dicky Valentine, Diana Decker, Jimmy Young, Lita Rosa brought the sweet melodies of romantic ballads to accompany everyone at their work. The hangar echoed to Edmund Hockridge's rendition of, 'Your eyes are the eyes of a woman in love' and 'Oh, how they gaze into mine', as Edward worked at his chores. Drowning in the music, he was oblivious to all but her, the girl with orange raincoat.

Edward's first job that morning, after cleaning up the crew room in preparation for the morning tea-break, was to mix up some Casine cement; a glue used frequently for structural repairs on the wooden aircraft. This came in powdered form, for mixing with water until its consistency reached a workable, liquid paste. Normally he would have done this by hand, in a cup, using a small, wooden spatula, not unlike those things the doctor uses to press down your tongue, to examine your throat, making you retch, and want to vomit all over his carpet.

It took time for the water and the powder to amalgamate when mixed by hand, so Edward decided to mechanise the operation. He found a very large split pin, (some people call them cotter pins) bent it about half an inch below the head, at ninety degrees, and slipped the still-closed legs into the chuck of a vertical drill. It rotated rather like an eggbeater when he switched on the machine. Placing the cup of powered glue and water under the rotating, bent pin, he lowered his invention into the cup of un-mixed ingredients. It worked very well to begin with and Edward soon had a mixture close to perfection.

To his astonishment however, the cup began to oscillate as the consistency of the mixture became thicker. In a minute, he'd lost control. Most of the contents, from the violently wobbling cup, spattered about for yards around. Walls, overalls, faces, the length of the bench-top and other people working at the same bench; all suffered the consequences of Edward's wild desire for mechanisation. There was stickiness everywhere, much to the amusement of onlookers who were out of range.

Edward got a rollicking, and under stern advice from a sticky, and not so happy, engineer, he used the prescribed, doctor's vomit-stick for the next batch!

Repair-work on the crashed *Miles Hawk* continued; which was why Edward was mixing glue. The Chief had placed him with Ronny, the licensed airframe engineer, to learn about major repairs in wooden structures. The wings having been removed earlier, the Hawk's fuselage rested on trestles, for inspection and repair by George in the workshops. Primary damage to the aircraft was in the main plane centre-section stubs, which provide attachment for the main undercarriage struts, as well as the wings. The struts had doubled back under the wings in the accident, seriously damaging the under-surfaces and main spar woodwork of the centre section stub-wings where Ronny and Edward worked together on the repairs.

When the undercarriage collapsed in the accident, the aircraft slid on its belly for some distance. Fortunately, sliding on grass was not as drastic as it would be on the concrete or tarmac of a runway. Therefore, damage under the fuselage was minimal. At one point, they thought the wings might not be recoverable. The inboard leading edges, on both wings, had suffered some damage as far back as the main spar. However, George, in the workshops, was very confident of getting the wings back into service within a few days. Ronny and Edward concentrated on the fuselage and centre section. They prepared new spruce pieces for scarf-jointing in, and new spruce-ply skinning, to replace the damaged stuff. Work progressed well. Ron was a specialist in woodwork and Edward found himself learning many more new techniques. He was in his element, extremely happy at work and joyful over his interest in the girl.

23

One day, as the repair work neared completion, the Chief sent Edward on two errands. The first, to collect a can of hydraulic oil, ordered from the *Lockheed* factory in Speke and the second to pick up a can of aircraft paint from the *Cellon* warehouse, not far from the railway bridge that Edward crossed each day. He shoved off on his bike to cover the first errand, planning to visit the *Lockheed* factory first, since it was the farthest away from the airport. He'd collect the paint on the return journey.

Situated on rising ground, a little farther on from the big roundabout at Speke, the *Lockheed* factory had an extensive frontage. Edward needed to locate the General office. He found the car park entrance, and with great pride parked his *New Hudson* bike, next to a large, shiny Humber Snipe car. The car stood in front of a little sign on a stick. The sign displayed the letters GM and Edward wondered about its meaning, while he searched the long frontage of the building for a door.

Red brickwork and windows were all he could see until a smiling young lady, behind one of the panes, prodded the air with her finger, pointing in the general direction to Edward's left. The main entrance looked much like all the other windows, except that it was a glass door and had a pathway leading to it, a couple of steps up and a brass plaque next to a bell-push.

His forefinger on the button summoned a doorman, who grumpily opened the door for him with an air suggesting that this visitor should be using the tradesmen's entrance. Grumpy directed Edward to a hatch with a counter to one side of the foyer, looking into an office where three girls, rattling away at typewriters, looked up from their work. Quite spontaneously, they stopped typing in unison. How strange, thought Edward, as the rattling ceased and the room filled with smiles and fluttering lashes.

He explained his errand to all of them, trying to determine to which he should address himself. Two of the girls reopened the recital, a duet for two Imperial upright's, as the third detached herself and approached the counter to take the company order-form from the visitor. She was the prettiest of the three, and Edward felt the colour rising to his cheeks. He'd noticed of late, that this would happen to him, on occasion, especially when faced with a young female of the species. "It's the Sales Office you want." She said, after glancing at the form, "I'll take you if you can hang on a jiff." She returned to her desk and spoke into a telephone for a moment, at the same time fixing her dark-eyed gaze upon Edward.

An attractive girl, shapely in her pencil skirt and white blouse, she flashed a smile at him as she put down the phone. The other two girls had reached a crescendo in their recital, and hammered away without looking again in the visitor's direction. "I'm so sorry," said the girl, "the salesman who deals with this kind of thing is at a meeting. Can you wait for half an hour or so?"

"Yes, but I have another errand to run afterwards, so I hope he won't be too long," Edward affirmed, though it crossed his mind that perhaps he could do the other errand first, and come back later, but the girl caught his attention again with a question.

Would you like some tea? I'm about to make some for us anyway," she asked with another charming smile.

"That sounds good to me, yes please," Edward replied, looking at his watch. It was tea-break time in the hangar. "I would be making tea myself, back at the Flying Club, but for sixteen people.

"Well this will make a pleasant change then," rejoined the girl with a laugh, "do you take sugar and milk?"

"Yes please."

The other two typists continued to hammer away at their machines. Edward sat in the foyer, from where he could see the girl, through the hatch, as she prepared the tea. Eventually, she brought a small tray to him; the tea in a china cup with a saucer and a spoon, unlike the thick, half-pint-mug he used at work. There was a plate of biscuits too.

As Edward munched and drank his eyes drifted frequently to the girl's face, just visible above the counter top. Their eyes met several times across the polished wood of the counter and she held his gaze each time, with a whisper of a smile. Edward felt a bit awkward. For here they were, playing a furtive eye-game, observing one-another through a hatchway at almost eye-level with the counter.

Grumpy, the doorman, sitting behind a tiny desk near the entrance, glared at Edward whenever their glances crossed. Edward noticed that the fellow didn't get a cup of tea. Perhaps that was why he glared. Edward finished his tea and stood up to take the tray back to the hatchway. The girl, raising her eyes at his movement towards the counter, came to take the tea-things from him.

"Thanks very much for the tea." He said.

"You're welcome; would you like another?"

"No thanks, but I enjoyed that."

The girl walked away with the tray, throwing a fleeting, smiling glance at Edward over her shoulder as she placed the tea things next to the kettle on a tiny side table. Edward stayed by the counter, it was more comfortable, leaning on its polished wood, rather than sitting at eye level with it and anyway, from that position he didn't have to tolerate the glares of the doorman.

The girl returned to her desk and rejoined the cacophony of typewritten script. Edward pretended to occupy himself, flicking through the pages of some *Lockheed* product-brochures that lay upon the counter. It was a pretence that allowed him to observe the girl discreetly, from time to time. She had a beautiful face, angelic and clear. Her natural smile, reflected in her brown eyes, was captivating. The ends of her short, dark hair curved down to touch her cheeks, moving gently as she returned the carriage of her machine.

Perhaps she sensed Edward's stare, for indeed, he was staring. She glanced at him from beneath her dark eyebrows; the smile reappeared as she looked down again at her work, with a modest quality that was strangely appealing. She seemed very quick with the typewriter, sending the carriage back deftly at end of each line, with a flourishing sweep of her manicured hand. The telephone on her desk rang. She took the message, nodding at Edward to indicate that the salesman had returned from his meeting, and was ready to see him. The girl told the others that she was taking their visitor to the Sales Office, to which they murmured responses without looking up from their keys.

"Come along then, I'll take you now," she announced as she lifted the flap and opened the half-door in the counter. Edward flushed again as he thought... Damn, I should have done that for her. Whatever are you thinking about, Ted? "You're from the Flying Club then," said the girl, "I've a brother who flies, he's in the Air Force."

"Yes," responded Edward, answering the question, "I'm an apprentice, training to be an engineer one day." The click of her high heels on the hard floor resounded through the polished corridor, lined with closed doors. "Sometimes I sell the pleasure-flight tickets on the airport terraces, weekends mostly," ventured Edward, adding, "Have you been up, in an aeroplane?"

"Oh, do you? No, I've never flown, is it fun? Maybe I'll come to the airport one day and take a flip," she said. "How much is it, to go up I mean?"

Their conversation seemed to be full of mixed statements and questions, compounding articulation. "Well, it's a pound for adults and if you do come, I might be able to arrange the best seat for you, where you'll be able to see more of what you're flying over." Edward's words seemed a bit impetuous, and he surprised himself with the spontaneity of his suggestion. She didn't answer.

They had reached the door marked Sales Office and with her hand at his back she ushered him inside, seating Edward in front of a desk, with a gallon-can of oil standing upon the blotter. Hydraulic Oil, it said on the label. Must be mine, he thought, as the man behind it, scribbled away at an invoice, humming a little tune to himself. He should be with the recital, in the front office, Edward mused, inwardly.

The girl slipped off with a tiny wave to accompany her charming smile. Edward smiled and waved back. His face flushed again. Damn and Damn! Why did it keep doing that? The salesman didn't speak for a moment or two, until he'd finished filling up the form. "Did you bring cash or shall I bill the Club for it?" The salesman asked.

"Well, they didn't give me any money, so I suppose it'll be a bill," said Edward, with a questioning look at the frowning man.

"They haven't paid the last bill yet you know, and that's from months ago. Better tell them, Eh?" The salesman scolded, his frown deepening. The man stood up, telling Edward he'd be back in a minute, and strode away in the direction of his supervisor who sat at the far end of the long office. The office contained a dozen or more desks, with people sitting at them, writing, or talking into telephones. No one took any notice of Edward, so he contented himself with reading the label on the tin.

They needed the oil for the damaged *Hawk*'s new landing gear struts. The oleo-struts contained seals made from pure rubber, requiring special oil of a type not used with synthetic rubbers. The can contained a gallon of the stuff, about ten pounds in weight. How was he going to carry it? Edward pondered the question, as the frowning man came back to his desk, with an invoice.

"Right, here's the invoice and there's the oil," he said, indicating the tin Edward had been contemplating. "Please tell your office people, to make good on the outstanding invoices, as soon as possible or my name's mud. Bills really must be paid on time you know, within the thirty days allowed."

Edward thanked the man, pocketed the folded invoice and, picking up the can, he made his way back to the front office. It wasn't difficult to find; the Imperial recital was still in full swing. The clattering typewriters and intermittent bells were clearly audible all the way down the corridor. Edward stopped at the front office hatchway to thank the young lady for her help. She sprang up from her desk and came to the counter, along with her lovely smile. Above the din of her two companions Edward said, "Thanks very much again, for the tea, and for your help."

"You're welcome. I might see you on Saturday then," she declared, leaning her elbows on the counter-top and clasping her hands together beneath her chin, her smiling eyes gazing at him intently, "at the airport, OK?" She raised her eyebrows to emphasise the point. The glaring doorman, anticipating the young visitor's departure, already had the door open and stood beside it, tapping his foot, hands on hips and impatiently staring at a far corner of the ceiling.

"Oh, right – on the terraces," answered Edward, lamely, "Saturday, I'll look out for you then."

She nodded, pursing her lips, her eyes smiling and watched as he turned to leave. "B'bye."

The girl's smile fascinated him, along with that little wave of her hand but her announcement took Edward completely by surprise. It was less than an hour since they'd met. He hardly knew the girl. She was older than he was too, by a year or so anyway. Nevertheless he felt strangely flattered by her proclamation.

As Grumpy closed the door behind his unwelcome young visitor, Edward wondered if the girl really would come to the airport on Saturday, and would she come to see the aeroplanes, or to see him? His heart skipped a beat.

The can of oil was quite heavy. Being square shaped, with only a finger-grip handle at the top, it was awkward to carry on a bike. Edward found a piece of string from his saddlebag and tied it to the can's handle. In this way, he made a loop by which to hang the tin from the handlebar, so that it rested against the frame over the front wheel. He un-parked from the space that his bike shared with the Humber and peddled off to his next assignment, reflecting over the unusual occurrences of that afternoon. Then, quite suddenly, he thought about the girl in the orange raincoat. He felt guilty, but he could not think why.

His next port of call was the paint factory, just next to the railway bridge where the orange raincoat girl walked each morning. Edward had been to the place once before, months ago. The building was long and low, standing with loading bays alongside the railway tracks. The roadside frontage was very narrow and painted white, with the company name blazoned in big letters. There were no windows or doors at that end of the building, so access to the offices and goods unloading area, was from the side. Edward made his way to the wide gate, on the railway side of the building. It was open and he cycled straight through, down the slope, and on to the office door, halfway along the building.

The place smelt heavily of paint; its pungent, penetrating odour pervaded everywhere, drying the throat and filling the nostrils. He thought about his old gas mask from the wartime days; it certainly would have come in handy for this errand. Perhaps he should have kept it. Edward already knew the counter-man in the office. He didn't like the fellow much, but they exchanged greetings as he entered. He explained what he'd come for, and with a nod the brown-smocked man poked his head through a hatchway behind the counter, calling for someone to bring the ordered paint.

He and Edward chatted for a moment until a girl came to the hatch with the tin. Edward's pulse raced. It was her, the orange raincoat girl, except she wasn't wearing it right then. Instead, she had on a light green coverall, but it really was her. He couldn't take his eyes off her as she spoke to the counter man in low tones and showed him the paperwork. Then, as if sensing Edward's eyes she looked in his direction and smiled that smile. His face flushed with the adolescent affliction he so hated, as their eyes met. He wanted to say something, but he didn't know what. See you tomorrow on the bridge, leapt to mind, but no, that's stupid, I wish the counter-man would go away for a minute, Edward thought. Perhaps a simple hello and a smile would have sufficed.

It was a moment he'd always remember with clarity. The girl held his gaze for a long moment, smiling, until she looked away again as the damned counter-man drew her attention to some problem with the documents. Then, looking at the red-faced apprentice, he addressed Edward with an exaggerated wink. "We'll soon get it right, son," he said, "spot of bother with the paper-work: no approved certificate." The girl hurried away without looking again in Edward's direction.

Meanwhile, having spotted his youthful customer's reddened cheeks, the counter-man grinned wickedly, adding to Edward's already embarrassed countenance. "Nice girl," he ventured, "only been with us for a short while, still learning the ropes, left school just a couple of weeks ago. She's only fifteen," he continued, with another theatrical wink, but built, Eh?" Propping his clbow on thc hatch-counter he leaned back to watch her departure over his shoulder, through the opening. Dirty old man, thought Edward. He's old enough to be her father.

So now, he knew. She had been on her way to work, there at the paint warehouse, when Edward had seen her on those other occasions, on the bridge, and she was about his age. Somehow he had to find a way of making contact – verbal contact. She must have a name. Edward could have asked the counter man but he didn't fancy hearing the sort of reply that the man would probably have made. Edward had to make a plan for tomorrow, on the bridge.

The girl came back to the hatch with the corrected paperwork but there was another woman with her, probably her supervisor. They discussed the problem with the counter-man and then disappeared again, into the interior without looking at Edward again. He took the paint and the documentation, leaving the office to make his way back to work. The paint tin had a wire handle that fitted well, over the other handlebar, and it too, rested next to the bike frame, balancing out the weight of the oilcan, hanging on the other side. He didn't remember anything of the ride back to the Club. He was far too happy, having discovered where the girl worked, and having seen her twice in one day. Now, he fretted only about tomorrow.

Next day saw Edward at the top of the bridge with a punctured tyre. It wasn't really a punctured tyre but he needed a plausible reason for being stationary at this spot. He didn't want anyone to think he was just hanging about, waiting for someone. He felt a wee bit wicked about it, even so. The bike was upside-down at the edge of the broad pavement with a few spanners and tyre levers littered about to make the situation look at least credible. A dog trotted past, had second thoughts, returned and cocked his leg on the bike's inverted saddle. This distracted Edward for a moment whilst he chased-off the hound with a few round remarks.

He didn't see the girl with orange raincoat until he turned his attentions back to the bike, with its imitation puncture and wet saddle. At the same instant, there was a hail from the roadside. It was Donald, a fellow with whom Edward had schooled at Kingsthorne and at Gilmour, although they hadn't kept company in several years. He'd pulled up at the kerb on his own bike to ask if Edward needed any help. Edward looked from Donald to the upturned bike, and from the bike to the girl. She had stopped and she was smiling, though not at Edward, but at the dog, who had by then run to her, probably to complain about Edward's harsh words.

She stooped to pet the creature as Edward, in disarray, quietly told Donald to sod off. Donald pushed off in high dudgeon and the girl, still talking to the dog, walked on down the other side of the bridge, without another look in Edward's direction. This left him with a puncture that wasn't, unused tools to clear away and a wet bicycle-saddle to mop-up. What a day it had turned out to be. Edward's plan screwed up by a dog with a weak bladder, intent upon making his mark, and an idiot ex-school-chum. When finally Edward got to work, he was still furious. He made in-roads into profanities that would have made the devil blush, and he spent the rest of the day slamming around and cursing his luck.

He continued to pass the bridge each day after that, sometimes seeing the girl, sometimes not. When they did pass, she didn't look at him. Rather, she looked at her feet to see if that sole had finally come off her shoe, or over the high parapet, or sometimes up at the sky to see if there was a hint of rain. Edward might just as well have been invisible. She never smiled at him again. He tried a couple of 'Good Mornings', but she didn't respond. He never saw the dog again either, and wondered if it might have been hers. In retrospect he felt a little embarrassed over the way his bridge-ploy had panned out. Thus ended a relationship that never began. His silly fantasy waned and he felt thwarted.

One morning, not long after the orange raincoat episode, the sound of many aeroplanes in the circuit filled the air, as Peter and Edward worked on some jobs outside the hangar. A number of *DH Tiger Moths* flew over in loose formation. They broke and formed up in-line-astern to make their approach for landing. The lads counted eight. They also noted that the planes were in R.A.F colours, complete with roundels and yellow bands painted on the silver-grey wings and rear fuselages.

The *Tigers* stream-landed one by one on the grass area next to the main runway, and taxied across towards the Club. Someone shouted from the open office windows, asking for them to be marshalled to parking spots in front of the Club House. Peter and Edward ran to take up positions, showing the pilots where to park as the first group of aircraft headed in their direction.

The pilots avoided the paved, main runway and taxied very carefully across the peri-track because the *Tiger Moth* has a tailskid, instead of a wheel, designed for use on soft ground or grass fields. Such a device wears quickly when in contact with hard pavement – not to mention the damage caused to the pavement itself. The skid also aids steering when taxiing, since the *Tiger Moth* has no wheel brakes. As each aircraft taxied into position, it was clear that they were in terrible condition. Fabric strips were hanging from some of them, and the bracing wires didn't look tight enough on any of them. The paintwork appeared to be flaking badly and two of the planes seemed to have massive oil leaks, judging by black slicks on the inboard leading edges of the wings, and on the rear fuselages.

The boss came over to greet the pilots as they climbed out of their cockpits. What a motley crew they were. Some had beards. All had flying overalls and parachutes strapped to their backs. In those times, most open cockpit training aircraft had contoured seats to receive the bulk of a packed parachute.

Peter and Edward looked the *Tigers* over, and noticed that the front cockpits contained dinghy-packs, strapped into the seats. This meant that they were likely to fly over the sea.

The lads conjectured over the destination of these wild men in their decrepit planes, as the pilots took off their parachutes and piled them next to the *Tigers*, along with their flying helmets, map cases and gloves. The boss took them into the Club House where the bar had just opened. Good timing! One of Edward's jobs that morning had been to clean the place up after the night before. The usual cleaner didn't normally arrive until close to mid day. Clearly, the Boss had expected the arrival of this band of brigands.

It turned out that the *Tiger Moth* pilots were Irish. They had flown up to Liverpool from an R.A.F station in the Midlands and their final destination was Dublin. The idea was to take a short rest at Speke, refuel (and oil), before making the longish trip across the Irish Sea, justifying the need for dinghies. The parachutes were an obvious necessity too, because the aeroplanes were in shocking condition. Protective red-lanolin still covered the engines. Sold as they stood through ex-WD auction, the *Tigers* could fly out of the country, (to Ireland in this instance), on a one-time-only special certificate that any buyer of such equipment could easily obtain. Edward thought, rather them than me.

In due course (and for sure, after several beers or Irish Whiskeys), the Boss and his guests emerged from the Club House in good humour. Laughing and jesting, the pilots put on their chutes and helmets in preparation for the onward flight. In the meantime, the hangar crew had refuelled each plane, checked and topped up the oils, and cleaned the tiny windscreens.

They helped the pilots to strap in, for the encumbrance of parachutes attached to ones back and backside makes it difficult to reach around to pick up the seat and shoulder harness straps. Then, with hearty farewells to all, they prepared for start-up.

Just for the fun of it, the prop-swingers tried to achieve a simultaneous start: all eight together. Six they managed, but the other two needed ‘blowing-out’ before they would start. Finally, all eight engines were running and the pilots taxied their aircraft away for takeoff.

The light breeze had veered. They zigzagged downwind towards the wooded grounds of Speke Hall, the pilot's whooping and shouting to each other as they lined up in two flights of four, on the distant grass.

The takeoff was impressive. Two flights of four planes each, belted over the grass towards the club. The first flight lifting off as the second flight, behind, got their tails up.

The pilots held their planes down, skimming the field and the paved main runway, gathering speed until, not far from the peri-track, they pulled up hard, to zoom skywards, passing overhead at a good height to clear the hangar, the propellers and exhausts battering the midday air with a din that sent birds scattering in all directions.

Waggling their wings and with hands lifted in fleeting farewells, the Irish flyers set course for home. As he watched the planes fly away westward, into the distance, gathering in loose formation like a flight of migrating geese, Edward experienced an emotion of another kind, validating his growing love for aeroplanes.

Part 9

The New Bosses

24

The *Tiger Moths*' visit to the club was certainly an unusual occurrence, but at least it was interesting and fun. It made a welcome interlude, diverting Edward's thoughts away from emotional turmoil. His life had become a little confused over recent times, since his self-inflicted foolishness over the girl with the orange raincoat. He was still smarting over that. The distraction created by the visiting Irishmen with their dilapidated flying machines had provided an agreeable change from the normal routines of club's activities.

The late summer days turned wet and cloudy with the early approach of autumn; and the pleasure-flying season came to an abrupt end. Edward's weekends were therefore free, unless there was a need for extra hours to cover the needs of ardent club flyers, intent upon braving the persistent overcast of cloud. He turned his attention to concentrating on his studies, and to developing new ideas in his obsession with model-making. However, it was on one of those weekends in October when Edward did work, to cover a last rush of club flyers who descended upon the club to take advantage of a break in the poor weather cycle, that he made a terrible error.

Weekend duty for prop-swinging, chock pulling, refuelling and putting the aeroplanes away in the evenings was completely voluntary for the apprentices. They normally agreed between themselves who would work and who would take the weekend off. There was no set roster for them but the Chief engineer always delegated one or other of the licensed engineers to be available for technical matters. A staff pilot would be on duty too, to ensure that the club pilots followed flight-office procedures correctly.

That Saturday morning flying programme had been busy, and the forecast for the afternoon and Sunday looked promising.

Edward agreed to stay on to cover the afternoon work with two or three of the *Austers*. From time to time Edward swung propellers, pulled chocks and refuelled, as the need arose. The latter involved notations in a refuelling book, kept by the petrol pump, against a check of the fuel pump's counter. During the late afternoon Edward noted, from his annotations and calculations in the fuel-book entries, that the underground storage tank was getting low. Both the engineer and the staff pilot had gone home by then, after helping to push the aircraft into the hangar, leaving Edward to lock up. He knew there would be a heavy flying program next day if the weather held and thinking that he should anticipate the need for more fuel, he called the Shell-BP service at the terminal building to asked for replenishment.

The duty Shell bowser-man asked if it was really essential, because they were short. Edward told him that his ground-tank was almost empty and that the Club had Sunday flying, and the following week's ATC cadet training operations to contend with. Edward also knew of two important *Rapide* charters scheduled for Monday with Army Co-op flying in the evening with both the *Rapide* and the *Gemini.*

The petrol-truck duly arrived, pulling a trailer-tanker and the driver gravity-fed its contents, one thousand gallons of aviation fuel, into the ground tank. This was the usual quantity delivered on such occasions in dire cases of necessity. Towards the end of the operation as they chatted good-naturedly about the change in the weather, the tanker operator asked Edward if he had ‘dipped’ the ground tank before calling for replenishment. Edward flushed and admitted that he hadn't; it had not occurred to him. The man looked concerned. By then, the one-thousand gallons had already passed into the tank, underground.

Together they took the dipstick and dropped it into the ‘T’ shaped hole under a cap, next to the replenishment connection. When they drew it out, it was clear to see that the tank now contained just over two-thousand gallons! The Shell-BP service-man shook his head and glared angrily at Edward. His hitherto friendly attitude changed to one of disdain. He shrugged off the fact that he hadn't dipped the tank himself, before off-loading the precious cargo from his trailer.

"Do you realise that I've just used up the last of our reserves. I won't get any more until Monday. All I've got is what's in the big tanker and that'll be used up on tomorrow's *BEA* and *Aer Lingus* schedules."

Edward shrank. What could he say? He'd made a mistake and they couldn't pump the petrol back into the bowser either. "That's not allowed," the bowser-man told him, "here, sign this." The man spoke impatiently, still overlooking his own failure to dip the ground tank. "Next time, dip the tank first, you silly twerp."

Edward signed the receipt and apologised for his oversight, but the Shell-man was unforgiving and stalked off to tow his empty trailer back to the terminal. Edward shuddered. Another mistake: perhaps he'd really get the sack this time. Things weren't going at all well for him in recent times. It was all his fault and in scanning the refuelling book he soon realised that he'd made a serious error in his subtraction calculations, losing one thousand at the stroke of a pencil. How stupid he felt; how stupid too, not to dip the tank before calling the Shell people. On Sunday he mentioned his mistake to the licensed man, Ronny, as they prepared the *Austers* for the day's flying but he just shrugged and he too called Edward a silly twerp.

Monday morning came, and Edward took the fuel-receipt to the office to explain his mistake to Mrs. Van, the office manager. She was normally very pleasant, sympathetic and kindly. However, on this occasion she wasn't. Bawling Edward out, she asked him who he thought was going to pay for the unnecessary, unwanted thousand gallons of petrol. Edward flushed, and felt close to tears under the tirade. He left the office feeling irresponsible and very small indeed. Ronny, meanwhile, had very unkindly told the lads in the hangar about the weekend episode and they teased Edward unremittingly until it was time for him to do the errands at lunchtime. He passed a very uncomfortable day altogether but at least he didn't get the sack.

Colder weather brought with it the need for heating. The wooden clubhouse, resting on a concrete foundation plinth, had thin walls and little insulation against the cold airs of winter. Its bar held licensed hours, rather like a public house, and it was therefore important to have the place warm by midday. Even though poor weather placed limitations on flying, many of the members came to the club to tell stories of their aerial exploits over a pint or two.

A small furnace to heat water for the radiators in the club house became Edward's responsibility. Cleaning-out, setting and lighting the furnace, took priority over his other duties each morning. This was one of Peter's jobs before Edward joined the company. In his boiler house handover, Peter had demonstrated the best way to get a good fire going quickly and where to find the kindling and coal-coke fuel. Then Edward was on his own with his new obligation.

Turning to, early on the first day of his fire-lighting duties, he prepared the club's small, water-heating furnace. He set paper, wood kindling and coke but the wood was damp. Peter had explained that in such cases there was no harm in throwing a little paraffin over the prepared fire, as an aid to ignition. This Edward did, and put a match to the paraffin soaked paper. It lit-up well and in readiness for stoking up, Edward popped over to the fuel store, in another boiler house, close by, to get some more coke in a bucket.

When he got back, the fire seemed to be out. There was just a wispy cloud of white smoke drifting up the tin chimney. Cursing under his breath, he struck another match and stooped to re-light the wretched thing. Next moment the shed erupted in a rush of flame that threw him headlong backwards, out of the open door, heels over head, to end up in a sitting position on the pathway outside. Edward's face felt warm and tight as he watched in dismay whilst the tin chimney slowly collapsed into the boiler house. The chimney fell in a series of disjointed jerks, amid a cloud of soot and rust-particles. The smoke he'd seen was vaporising paraffin: highly volatile. Edward's second match had done the rest. The force that threw him out of the shed also burst the aging and rusty chimney pipe.

His first thought was, Oh Lord, what have I done now? The fire was out, the chimney and some of the roof were down, his face and arms were almost hairless and it had begun to rain. He would surely get the sack for this. Later, Peter helped him to repair the tin chimney and roof, and to get to the fire going again. It wasn't the end of the world after all. Edward only lost the front of his hair and his eyebrows.

In the hangar it was cold and damp and without some warmth in there, doping the fabric covering over the plywood skin of the *Hawk*'s fuselage was impossible. Edward learned that in these conditions it was customary, although strictly illegal and extremely dangerous, to set coke-fired braziers around the hangar. As the most junior apprentice at the club, preparation of these fires was also Edward's responsibility. On damp mornings it was up to him to dispose three or four glowing braziers, in strategic places about the hanger.

The engineers showed him how to make the braziers from empty, five gallon cans, with the tops cut out and rough holes punched through the sides. A piece of thick welding rod formed a bucket-like handle at the top, by which to carry the braziers into position with a hooked, metal pole. Each fire stood on three house bricks, suitably spaced on the floor of the hangar, and fire extinguishers positioned not too far from each one, ready for use in case of accident.

In this way, they warmed up the atmosphere of the hangar, dispersing the damp air around the aeroplanes sufficiently to apply the 'tautening' dope, without it 'blushing' or 'blooming'. Under these improved conditions, work continued with good progress and since there wasn't much flying activity, there was plenty of uninterrupted time to bring the repairs close to completion. On those cold days, Edward served tea around the warmth of the fires, for the crew room didn't have heating, apart from an old paraffin stove that he'd turned up too high one day. The stove smoked so much that the cloud-ceiling was down to twelve inches. It took Edward a week of scrubbing to get the black soot off the walls and seats. They never used the stove again.

Soon enough the early snows arrived. On the first white morning, Edward managed to get to work on his bike, without incident, although the snow had drifted in open places to form deep undulations against hedges, fences and buildings. Peter arrived soon after Edward and between them they cleared the hangar frontage of snow and prepared the coke fires to warm the cold, damp air inside. The tea was steaming hot as the rest of the chaps arrived. There wasn't much to do on days like that, so they started a day of tidying up and cleaning tools: making-do and mending.

At teabreak, someone questioned why the Boss hadn't turned up. The snow was hardly enough to prevent his morning journey to the club. However, the staff was completely unaware of what was going on in the background, and why the Boss was absent. As it happened, they didn't see him ever again. The facts, revealed later in the week, explained why. The Flying Club was bankrupt and everyone was on a week's notice.

The Official Receiver had moved in to take inventory, and to advise that employment for practically everyone would cease at the end of the week. Two engineers would remain, to be kept on temporarily until other arrangements became evident. Meanwhile the club would conduct no further business during the foreseeable future. This news dismayed everyone, and some disappeared for the day to look for other work.

Edward pondered the situation. He wondered if his error over the petrol, last Autumn, had anything to do with the bankruptcy. Perhaps the cost of a thousand gallons of unnecessary fuel had been the financial last straw to break the back of the flying club's fortunes. Perhaps it was all his fault. Would anything go right, for once?

Several of the engineers and a couple of the apprentices secured fresh jobs immediately, to start the following week with a newly formed company that had recently taken over a small hangar close to *Starways* and the terminal building. The new company had acquired two *Avro Ansons* and a *Rapide*, with the intention of using them to fly mushrooms from Ireland to Liverpool. The new company was appropriately named *'Federated Fruit Company'*, being formed to supply a chain of green-grocery businesses in the Liverpool area.

Edward was distantly acquainted with the Managing Director, having met the man several times before at the *Wright Aviation* hangar. A prominent businessman, Mr. Gates had been an ardent club member, flying club aircraft regularly. He'd converted to twin-engine aircraft, with the *Gemini*, presumably so that he could fly his new company's *Anson*s, as well as directing their air-freighting interests. But when Edward approached him to ask if there was any chance of a job, the man explained that he had all the staff he needed for the time being. Those from the club, who had joined *Federated Fruit*, had clearly been negotiating for employment over several weeks prior to *Wright Aviation*'s demise. It seemed, to Edward, that perhaps they had known about Mr. Wright's business difficulties beforehand, and said nothing to the others.

As a very junior apprentice, Edward had no real understanding of the apparent linkage between the Flying Club and the *Federated Fruit* Co. However, he learned later that *Wright Aviation* had been asked to operate the new company's aircraft as a joint operation but it seemed that the Flying Club's backers withdrew their support at that point. All Edward knew was that he would be without a job at the end of the week.

As for other possibilities, Edward hadn't been lucky on that score either, in spite of several visits to the Youth Employment Bureau in Garston, and the only other established, general aviation companies at the airport, *Airwork* and *Starways*. There seemed to be no place for him. He became very despondent, as did the other fellows who had been no more successful than he had. They went about their chores with little enthusiasm during those final days at *Wright Aviation*.

Later that week, the Official Receiver's temporary manager called everyone together. He seemed unusually happy and guardedly pleased. He explained that important visitors were expected, and that they wished to meet with the technical staff and pilots. He asked all the staff to stay behind after work, until the visitors arrived that evening. This news was food for new discussion and the remaining staff tried to guess at the reasons for such a visit. Some said that this was likely to be a good sign. Rumour and conjecture reigned.

It was after dark when the visitors arrived. They found the staff in the hangar, waiting, huddled around the last remaining coke fire that Edward had been told to keep alight close to the crew room, in anticipation of the meeting. There were four people in the group, not counting the temporary manager. One wore a black overcoat and walked with the aid of a stick. He seemed to be the senior member: young, with sharp, inquisitive eyes. He was bare headed, but carried a Homburg hat in his hand. Of the other three, one was short, thickset and sported a large handlebar moustache. His heavy, dark-blue raincoat revealed a dark, pinstriped suit beneath as he held the coat open to capture some of the warmth from the fire. Another was taller, somewhat older and balding, whilst the third was large, hunched of shoulder – a heavy set man, well into his fifties. He also wore a black overcoat and a Homburg hat.

The scene reminded Edward of an American gangster movie his father had taken him to see as a birthday treat, years before, and in which Edward G. Robinson had featured. With near-darkness all about them, except for a couple of lamps at the benches, the glare of flaring, red-hot coals in the fire, the figures threw long shadows upon the hangar walls, grotesque in shape and accentuated by the fire's glow. Heavy coats and Homburg hats: it all fitted. It was like a meeting of the mobs to discuss territorial rights in Chicago's gangland disputes of the nineteen-twenties.

Edward's imaginary illusion evaporated as the visitors came to the fireside. Everyone got up from improvised seats and stood respectfully aside, so that the visitors might share the warmth of the glowing coals. The staff pilots and instructors joined them from the flight-office.

The temporary manager opened the meeting by introducing the owner of the stick as the leader of the troupe. He was indeed the senior of the quartet. Captain Guynan owned the handlebar moustache. The big man with the hunched shoulders was introduced as the first fellow's number two, whilst the fourth gentleman apparently had no name and remained a discreet pace or so behind the others. Perhaps with a Tommy-gun hidden beneath his coat, Edward thought, with misplaced humour, his imagination running riot again.

The leader took up the conversation with a friendly smile and a cheery greeting, as he cast his inquisitive eyes around the assembly. "Good evening gentlemen; it's certainly warmer in here than it is out there," he said, tilting his head towards the hangar doors through which the visitors had just passed. He continued by explaining the purpose of their visit to the Flying Club, and that he and his companions represented an interested buyer – it turned out that he was on the Board of the new *Hunting Clan* airline, a subsidiary of the *Clan Shipping* line. He explained that if the deal went through, the Flying Club would continue as before, but under new management and, with additional aircraft added to the fleet. These would be larger, possibly the new four-engine *DH Heron* type, with potential use for feeder line services up and down the country.

Guarded excitement began to fill the faces of the attendant pilots, engineers and boys alike. Could this be their salvation? Since there was still a day or two before the end of the week, everyone was asked to give some thought to what he'd said and if they were agreeable to reinstatement with a newly formed company, they should affirm this with the Official Receiver's temporary manager before the coming weekend.

During the following days most of the guys made their affirmations and by the end of the week, they found themselves with employment again. It was a weight off Edward's mind for he'd had no success in his job-search with the other companies on the airport. The Chief Engineer together with Gerry and Malcolm, had already committed themselves to leaving the Club to join the new fruit-freighting company up the road. Clearly, they would therefore not be with the rest of those who had decided to remain with the club under the new arrangements.

Another week passed by. Crew room conversation reflected upon the new company's plans, the new aircraft and the need for a replacement Chief Engineer. Meanwhile Capt. Guynan returned to take up his post as the resident Boss, and began settling-in.

The ex-*B.O.A.C* senior captain, along with his handlebar moustache, became a frequent sight in the hangar as he got to know the fellows of his technical staff.

The name of the new company was to be *Dragon Airways* (there was a Welsh connection) and the boss's announcement for new colours gave fresh inspiration when he asked for work to begin immediately on re-spraying the aircraft. The colour scheme was to be dark-blue, white and silver. The dark blue for the fuselage under-surfaces and sides, white for the fuselage top surface and registration letters, silver for the wings with dark blue registration markings, top and bottom.

They called in the sign writer whom the Club had employed for lettering work in the past, and he gladly accepted an assignment for painting registration markings and insignia. Times were lean for him too. Bill the sign-writer was very skilled at his work. It never ceased to amaze Edward how quickly and neatly he worked with his long-bristled brushes. He'd mentioned to Edward one day, as the apprentice watched his work, that he had trained at the Liverpool Art College. They shared a common background in that respect: small world. The Boss gave Bill a few ideas for insignia design, so that he could come up with a suitable company emblem to paint on the aircraft. In due course, Captain Guynan chose Bill's red dragon with spread wings, clasping a shield, and surmounting a scroll with the words *Dragon Airways Ltd.* printed upon it.

Within a month, the colour conversion was complete on all the aircraft. They looked splendid, lined up on the grass in front of the clubhouse in their new livery. Everyone felt very proud. They lived in hope and expectation for a bright new future, and for the arrival of the new aeroplanes.

25

Winter came suddenly that year, and one night it snowed heavily. Getting to work posed a problem because the snow was deep. There were no buses running but nevertheless, Edward decided to try the journey by bike. The soft, white snow parted to the front wheel easily enough and he found the going quite smooth in spite of having to press his feet into the snow on each down stroke of the pedals. No traffic marks showed the way in the snow and the white blanket hid all the kerbstones and other possible hazards. Edward's navigation was strictly by familiar objects above ground level, but in spite of hitting a kerbstone occasionally Edward managed to get to the Club without serious incident.

The snow had drifted in exposed places. In front of the hangar it was very deep. Edward was first, as usual, and when the others arrived they set about clearing the small hard standing in front of the doors. It was doubtful if there would be any flying but they made the effort anyway. The next priority was to get some heat into the place. Edward prepared a few coke braziers, taking a few hot coals from the overnight, banked-up furnace that heated the Club House. The new Boss didn't turn up, but staff knew that he'd gone away from the district for a few days to interview applicants for office-staff posts, and to select a new Chief Engineer.

With all the aircraft serviceable after the colour scheme conversion, and with no flying to speak of, there was little technical-work outstanding. Edward occupied himself by making a desk for the Chief Engineer's relocated office. They'd cleared the old hangar storeroom out, redecorated it and built a partition to provide an inner sanctum beyond a technical-office, where log-books and schedules would reside. An Operations Manager (filling a new, additional post in the organisation) had taken over the old

Chief's office on the other side of the hangar. Now they had begun to get really organised.

It was a quiet time and the days slipped by slowly. One morning, at tea break, a shout from outside the hangar doors sent Edward and Nick to investigate. They opened one of the centre doors a little, to see what all the fuss was about. This was a big mistake. For no sooner had they squeezed through the narrow opening, a barrage of snowballs took them completely by surprise. The Club was under attack by the R.A.F unit from the next hangar. They'd decided to make a morning raid.

Being well prepared, and armed with piles of pre-made snowballs, they soon had Nick and Edward diving for cover, backing-up swiftly through the open door. Once inside again, the two managed to get the door shut. Nick warned the others, grinning widely. Edward pelted off through the hangar to the workshop corridor, calling to Peter and Bernard as he went. "Come on, the RAF lads are out front spoiling for a battle. Let's get round the back and take 'em from the rear."

This was a shade of the games Freddy and Edward used to play in the fields and at school, when the snow fell in their younger years. Behind the workshops, snow had drifted deeply and Edward knew there was abundant material with which to arm against the attackers. This was war. Edward and his companions quickly prepared as many snowballs as they could carry and made their way quietly, to creep up on the flank of the men in blue greatcoats.

The rest of the chaps, with Nick at the head, had gone out through the other end of the hangar, through the boiler-room used for office heating, and around the far side, hopefully to catch the R.A.F fellows from the other flank. As Edward and his team crept through the cold air, between the club-house and the workshop walls, they could see the group of giggling airmen in front of the hangar busily scraping up more snow to replenish their depleted ammunition, and cat-calling at the closed hangar doors. Nick and his chaps appeared on the far side and the two defending groups advanced to pelt the enemy with a hail of snowballs from two sides.

It's hard to imagine grown men getting so excited about a little snow but there they were, a sight to behold indeed; whooping and hollering and advancing like the good soldiers they thought they were. They drove the airmen back under volley after volley, for they'd caught the attackers completely unawares just as the RAF chaps had done at the beginning of the exchange. The men in air-force-blue berets hadn't expected a counter-attack especially not a neat pincer-movement in which they now found themselves.

The snowball fight went on for quite some time. Eventually it broke down to hit and run tactics, in smaller groups. Edward got caught coming out of a workshop corridor window, trying to do a solo action. At mid-point, climbing through the high window, his feet on the sill, knees doubled up and his head ducked down beneath the lintel, two of the MT Section were waiting for him in the snow outside. A barrage of snowballs came at him from ground level.

Edward was helpless. Balls of snow smacked into his face, chest and knees. Should he drop back into the corridor again or should he drop forwards onto the snow beneath, on the outside? His glasses went, whisked away in the pelt of white missiles, so he followed them down, into the snowdrift below. Landing on all fours, he managed to retrieve the lost spectacles by which time the two airmen had run out of ammunition. They made off, giving Edward the two-finger sign and giggling like a couple of schoolgirls.

At last the two sides agreed to a cessation of hostilities. They sat around a glowing brazier to drink hot tea and analyse the finer points of their battle. There were no winners and they warmed themselves near the fire in a state of happy exhaustion.

The weather remained very cold, early snow coming and going periodically throughout the months of November and December. At home, preparations for another Christmas were well under way, not least of which were Edith's mincemeat and pudding; well matured after the making, months before. Christmases were always happy times in the Weinel household and that year, a turkey would grace their dinner table for the first time since the war.

Freddy wrote from Germany to say that he might not be able to get home for the festive season due to special manoeuvres scheduled for that time. His advancing army career had won him Staff Sergeant's stripes by then and he struggled with studies and exams to attain his long sought goal, to become an officer. The courses were difficult, and the standards high. In a peace-time army, places for promotion from the ranks to commissioned officer were hard to come by.

Nevertheless, the seasonal preparations at home went on. Edward helped with the holly and decorations, and the preparation of the front room as usual, ready for the Christmas dinner and evening revelry. The family's excitement grew as the celebrations approached. David wrote his letter to Santa Claus, and with his Dad's supervision, sent it off, up the chimney. He watched his behaviour carefully in anticipation of the coming season, being especially careful when in sight of the flag-pole that stood tall at the end of the garden. The reason for this was in deference to the watchful seagull that sometimes perched at the top.

It probably wasn't the same gull each time but since these birds are so alike, it appeared to be the same one. In jest, the boys'

Dad suggested that the gull might be Santa's messenger, convincing David that the bird was keeping an eye out for naughty children. Well, why not? If his belief meant peace and tranquillity in the home for a spell, then long may the gulls continue their vigil, for David was an energetic boy, full of pranks and mischief.

Soon after Edward started his apprenticeship with the flying club, David's interest and ambitions had switched dramatically from farming to aviation. He became almost fanatical about learning to fly and to become a pilot. At the time, Edward thought it was brotherly hero-worship, though no doubt much of his brother's interest must have surely come from the major topic of household conversation. The boys were very close and aeroplanes were most certainly an integral part of the family's fortunes: the best and, at times, the worst.

Christmas 1954 heralded a new future for the Flying Club. It had a new name, *Dragon Airways Ltd* and it would operate under the flag of Hunting-Clan, its parent company. They expected delivery of the *De Havilland Heron* aircraft in the early months of the new year. Plans were already afoot to open feeder-services between Liverpool and Newcastle with direct flights to continental Europe also under consideration. Three s were on order, one, a Mark 1B, with fixed undercarriage and two Mk 2's with retractable gear. Some of the engineers would be lucky enough to attend type-training courses at *De Havilland*'s factory at Hatfield and enthusiasm grew as an air of anticipation flourished among the few who'd signed up to continue with the company.

Before long, new office staff joined, and a new Chief Engineer arrived, together with a few new faces in the hangar. The office staff comprised three girls: Nina, and two others, both named Sheila. Their arrival at the hangar doors drew forth immediate and appreciative attention from the chaps inside. The boss ushered them in and courteously showed them to the main office, glowering at the wolf-whistles, emanating from various corners of the hangar. Later, Edward had the enviable privilege to be the first of the technical staff to meet them. Since he was still responsible for the errand running, it was his duty to call in, as usual, at the offices, to collect orders for the chip run.

Even Captain Guynan enjoyed his fish and chips out of a newspaper. "Gives me the chance to catch up with the news," he would often say with a generous smile, when Edward took his order. Edward liked the Boss. On this day, however, Captain Guynan teased Edward a little, over the lunchtime order-taking from the new staff in the general office. "Watch your step next door, Ted. The place is full of pretty girls now. Mind your language and keep your hands to yourself, laddie."

With a knock before going in, Edward faced the happy smiles

of three gorgeous girls. His cheeks burned as he introduced himself and announced that he would take orders for anything they needed for lunch. Nina, a specialist in office management and accountancy, was the senior, perhaps in her mid-twenties with dark hair worn short. In her grey two-piece, jacket and skirt, she looked very smart and very efficient, and exceedingly attractive. She occupied the big desk that Mrs. Van had vacated when she left the company soon after the old club went bankrupt.

When Nina had passed through the hangar, Edward noticed that she carried a briefcase, one of those slim, zippered ones, normally carried under the arm. This added to her efficient appearance.

The first Sheila looked after secretarial work for the Boss and his ops manager. She too, wore a two-piece outfit, but she was shorter than Nina, and plumper; perhaps well rounded is a better expression. Almost blonde, she wore her hair in an enchanting, unruly, curly style to match the bubbling personality of one who clearly enjoyed life. Her smiling, backward-glances at the earlier wolf whistles bore that out to a tee.

The second Sheila was beautiful – at least Edward thought so. She was quite tiny, petite, and gazed at the world with very large, amazingly deep-blue eyes. She was younger than the other two and smiled most of the time. Her job was typing, file-keeping and mail.

They all placed orders for fish and chips, cakes and cigarettes and being responsible for outgoing mail, Sheila II also ordered postage stamps for the letters. As he added to the order list and pocketed the money, it dawned on Edward that his good fortune at being the first from the hangar to meet the new trio of girls, served to create an extra three names on his list. That meant more to carry and more change to dish out, but he didn't care. Sheila II had become his latest youthful fixation and she knew it.

He tried, not very successfully, to control his adolescent, facial flushes in her presence. When she leaned close to look over his elbow at the list, correcting her order from six pennyworth of chips to four pennyworth, Edward's senses reeled in the fragrance of her perfume. He was sure she'd noticed how his hand shook as he wrote and his reddened face. When she spoke, she spoke with a trace of accent. The words lilted in that familiar Manx brogue that he'd heard before on the his school's camping trips to the Isle of Man. Edward realised that he was in danger of becoming besotted. He'd have to watch his P's and Q's with this girl, especially when she turned those beautifully blue eyes upon him, and he still didn't care.

The arrival of a new Chief Engineer created a sense of intrusion to begin with. Sid was somewhat older than his predecessor. His hair was almost black, swept back and neatly parted close to the middle. He had piercing blue eyes, set in a humour-lined, weather beaten face. His well-manicured hands were large, spider-like, having those jutting thumbs, with pointed joints that obtrude when applying pressure. He wore a heavy, grey-green, tweed jacket with leather elbow patches. The lovat-green, gabardine trousers that were so popular in those days, hung with neat creases over immaculately shined brown shoes. Sid's engaging smile won over the majority of the crew at their introduction, though Edward couldn't say that he liked the fellow much at the outset.

Sid lived near Manchester and drove a small, dark-blue, two-door, Standard Eight car. The staff soon learned that he was ex *De Havilland*'s and already had the *Heron* on his license. He was one of the original A, B, C, D and X licence holders and he'd spent some time overseas, in East Africa.

After introductions, the Boss took Nick and Ronny along to show Sid around the place and to discuss current and future work. Since the company intended to expand and ambitious plans were in the making, Edward supposed that the new technical boss wanted to get organised quickly. Later, in conversations around the crew room, Sid made frequent references to his East African experiences. Edward thought he was a bit of a line-shooter and that his mind for dirty jokes seemed like an ever-open catalogue. Edward blushed more than once, at age sixteen, listening to the never-ending and increasingly lewd recital of bar-room jests. Edward was to meet Sid again many years later, in very different circumstances.

The new chief engineer didn't like the desk Edward had made for his office. It ended up in an outer office, to be used for filling out the aircraft inspection worksheets and Technical Log Book entries. Fair enough, there weren't any chairs high enough to use it as a regular desk anyway, Edward had made it too high, but at least, it found good use.

Old George from the workshops had gone when the Club went bankrupt. He found a job closer to home. As a result, the new Chief put Edward to work with one of the new people, a youngish fellow specialising in sheet-metal work. He was also an accomplished ARB approved welder, and old George's workshops became his responsibility. Cliff was about thirty, recently married and expecting to be a father quite soon. He walked with a rolling gait rather like a sailor, and Edward never saw him without a neck-tie; just like Ronny, their airframes-man, in that respect.

After the formalities of introduction Cliff and Edward started on reorganising the workshop areas to meet the forthcoming, additional demands upon the maintenance facilities. Between them, they moved the two ex-R.A.F. *Hawk*s to another storage area, to make room for more benches and welding equipment and it was under Cliff's guidance that Edward began to learn more about metals and the techniques of welding. Meanwhile of course, he continued with his lowly jobs – sweeping the hangar floor, making tea and the twice daily errand running, which still took him away from the technical training with monotonous regularity. However, without a fresh intake of younger apprentices who might take over the mundane tasks from him, he endured, and developed his own day-plans for technical work and errand schedules.

26

Two incidents occurred during those hectic early days of *Dragon Airways*. They were poignant and remained as clear memories with Edward for all his years. The events concerned errand-running and Edward's perception of changing attitudes within the hitherto close working relationships between the hangar staff.

The first was the day when he fell off his bike with the lunchtime fish and chips. Depending upon the size of the package, most days would see Edward cycling through Garston with the wrapped food under his arm. The roads of Garston, paved with small square cobbles, became very slippery in rain. So it was, on that particular day, when his bike's front wheel skidded on the slick, uneven surface and Edward went down with a crash. The wrappings for the chips and fish burst open and the contents spread all over the road. Several passing cars put the finishing touches to the incident, mashing the food well and truly into the cobbled roadway.

After he'd picked up himself and his bike, Edward had no option but to go back to the Chippy to reorder. The fresh purchase cost him more than a whole week's wages. Fortunately he had enough money in his pocket to cover it, but the thing that amazed Edward more than anything, was the attitude of the people when he got back, and the disdain with which the crew met his explanation. He was late for a start, and he got a rollicking for that, as well as missing his own lunch at home. Furthermore, there was no sympathy and no offers of either full or even partial reimbursement for the extra cost incurred by the accident.

Edward thought it would have been all the same if a bus had knocked him down, or if a bolt of lightning had struck him as it did the proverbial postillion; no one would have cared much either way, it seemed, so long as they got their bloody fish and chips.

Nick and Bernard helped but the rest just laughed and told Edward to be more careful next time; it served him right for being thick enough not to look where he was going. How unjustly heartless they were!

The second occurrence came as a shock, and placed the new Chief Engineer several points lower in Edward's assessment. He was on a special errand one morning, cycling up the peri-track to the terminal building to deliver some documents and log books to the ARB offices. On the way back the front tyre of his bike went flat with a puncture. It was raining hard so he hopped off and wheeled the bike, at a trot, back to the Club. It was almost time for the lunch-hour chip-run so he hurriedly began to make the necessary repairs. Mid-way through, Sid passed by the workshop, saw Edward mending the inner tube and told him to stop.

"You don't do bike repairs in company time," he said, "get back to your proper work." Sid wouldn't listen to reason and so Edward went back to the hangar to start his order-rounds for the chip-run. If he'd repaired the puncture out in the rain, on the other side of the airport, Sid wouldn't have been any the wiser. As it stood, Edward would have to do the chip-run on foot.

Of course he was late returning that day, too. He missed his own lunch at home again and, to boot, he spent the rest of the day in very wet clothes. Edward was furious and refused to do the evening chip-run because his bike wasn't serviceable. He mended the puncture after hours, in his own time. Angrily, he decided to have it out with the Chief-bloody-Engineer, next day.

He advanced on Sid's office the moment he arrived. Sid was smiling and chatting with the chaps at his office door when Edward asked him for an interview. The Chief consented and ushered the lad into his office. "It's about the rollicking I got yesterday." Edward explained, as Sid closed the door behind them. "Getting to and from work is my own responsibility. I could use the bus; but I prefer to use my bike; it saves me hard earned money." Sid moved around his desk to sit, as Edward continued. "It's taken for granted that I use my own bike for errand running, the chip run or delivering documents to the AR-bloody-B." Sid raised his eyebrows at the vehemence with which Edward spoke, but nodded his agreement with the statement without comment.

"If a tyre gets punctured on a company-duty run, then I'd expect to mend it in company-time, if only to be ready for the next errand." Sid frowned, but allowed the blustering apprentice to go on. "Yesterday, I took your package up to the ARB and I got a puncture on the way back. It was raining and I was already late for the lunchtime errands, so it had to be mended quickly but you stopped me. I had to do the chip-run on foot and I got into more trouble for being late getting back." The Chief eyed Edward with his piercing stare, but he didn't interrupt. "If I stopped using my bike and came to work by bus, I'd have to do all the errands on foot, that would mean that I'd spend much more than half my working hours away from the hangar, and the remainder of the time making tea, sweeping the hangar floor and cleaning crew-rooms and offices."

Edward continued. "It's not fair; I'm an apprentice and I want to be an engineer, not a full-time bloody errand-boy."

The Chief still said nothing.

"It's not that I mind running errands, but if I'd wanted, I could have found a delivery-boy's job with the Co-Op or Irwins the grocers. They would have provided me with a bike too, and they pay a lot better than what I'm getting here."

Sid spoke at last. "So what are you saying Ted, that you won't do the errands?"

"No, I just want fair treatment. I don't like getting a rollicking when I'm doing the best I can with my own bike. I take my job seriously and if errands are part of it, fine, but I can't ride about with flat bloody tyres." The conversation ended abruptly, as one of the engineers came in with an urgent matter for Sid's attention, punctuating Edward's complaint.

With his little tirade over, Edward turned and walked out of the office. He didn't expect an apology, nor did he get one, but Sid never again questioned the sincerity with which Edward treated his duties. Edward found himself beginning to learn a few things about people, as well as about aeroplanes. The Club was not the happiest of places in the world in spite of the bright new future they all hoped for. New faces and regimes had changed what had been a very close-knit and friendly working environment.

Amusing interludes did, however, sometimes come their way.

One such situation arose during the winter, when the weather was cold enough for lighting the office-heating boilers. As the most junior apprentice, the daily furnace and fire rituals remained Edward's responsibility. One morning he decided to get the coke-braziers going first, before lighting up the boilers. He set about preparing the fires as usual, outside the boiler-house, on the edge of the road running behind the hangar. Within minutes he had several braziers blazing away. Just then, an airport security policeman rode up on his bike and stopped next to the flaming fires. Edward knew most of the airport policemen; many often stopped on their rounds for a chat and a warm at the fires.

However, this fellow was new, and he wasn't smiling but Edward greeted him anyway, in the customary manner, with a cheery, "Good morning."

"What's this then?" asked the uniformed officer, still astride his bike; the friendly greeting left unanswered.

"Fires," Edward answered.

"What are you going to do with them?" The police-man asked, eyeing the lad suspiciously.

"They're to warm the place up, inside."

"Don't you know it's illegal to light fires on the airport?" The officer's question was aimed at drawing an answer that would condemn the errant Edward, out of hand.

"Well.. I.." Edward began, weighing up his position carefully.

"Don't you know it's illegal and dangerous to put fires in a hangar, with aeroplanes," the policeman insisted, "what's your name and who's your boss then?"

At that moment one of the engineers poked his head out of the boiler-house window to ask if the fires were ready to take inside. The policeman dismounted, leaned his cycle against the hangar wall and took out a note book. Licking his pencil point he repeated, "What's your name, and his?" Indicating with a lift of his chin, the absent head of the engineer, who by then had ducked back inside. It was Ronny who'd been at the window, and Edward called out to him through the closed boiler house door, to bring the Boss. Ronny's muffled response affirmed Edward's request.

Within moments Captain Guynan's face appeared at the window and he asked Edward what was going on. Then he decided to come outside himself. The boiler house door wouldn't open. It was rotten and had stuck fast. The Boss reappeared at the open casement saying that he would come out by way of the window. At this, the policeman frantically scribbled away in his note book. He and Edward watched as two hands grasped the window-frame edges and two well-shod feet appeared on the sill. The boss heaved himself up, until his head, shoulders, hands, knees and feet neatly fitted into the frame of the window.

One might have supposed that his idea was to leap out onto the ground below, but as he prepared for this athletic feat, the window frame detached itself from the structure and fell out. The boss landed on all fours, on his hands and knees, with the window frame still surrounding him. The glass in the open casement window broke with a crash. What a situation! The policeman consumed another page of his note book. By then, his lips had taken on the purple hue of his pencil-point, wetted many times. He made no move to help Edward assist the boss to his feet.

Apart from the broken window there was no real damage done. The boss was furious of course and dusted himself off with his hands as he approached the security man. "What's the problem here, officer?" he asked in a commanding voice. Nice-one, Captain, thought Edward.

"It's the fires," the policeman said. "You can't have fires on the airport." There was no hint of respect in his voice, rather, his tone was intimidating. "You can't put fires in the hangar, it's illegal and dangerous."

"What did you tell this fellow, Ted?" the boss asked, turning to Edward.

"I just said they were to warm the place up." Edward answered. The boss nodded and addressed the policeman again.

"Right officer, the hot coals are to be tipped into the boiler furnace. It's quicker to light the things that way, to get some heat into the offices. Who said that the fires would go into the hangar?" Edward thought again, it's no wonder the boss is a Captain, his quickness of mind has turned the tables on our arrogant critic.

The man looked from Edward to the boss and then to the fires. He tapped his teeth with his pencil and glanced meaningfully at the broken window frame, lying in the grass where it fell. "What about the lighting of fires on the airfield and the damage to MTCA property?" he ventured, his eyes lighting upon the crumpled window frame and broken glass.

The boss quickly took up the challenge, his handlebar moustache waggling angrily. He was furious but kept up a very calm front, speaking quietly but with clear authority. "If you look up the road towards the building behind you, officer, you'll see that it contains an incinerator. Quite often we see unattended fires burning up there, but next to the building, not inside it." The policeman looked up the road to where, indeed, someone had lit a fire to burn rubbish, against the wall of the incinerator building.

The boss continued, pre-empting any response from the blue-lipped scuffer. "It's doubtful if the incinerator furnace has ever been used because it's always locked up. Here, at any rate, we have a responsible person looking after our small fires, so I don't see a problem." It pleased Edward to hear the reference to himself, as a responsible person; he was beginning to like the boss a whole lot more. "Furthermore, you can make a note for the attention of the MTCA building maintenance people, to come down here and do a bit of maintenance. The place is falling to pieces and I pay good money on the lease." At this, the policeman coloured up, snapped his note-book shut and put it away.

Capt. Guynan turned and walked stiffly away, towards the front of the hangar. Clearly, he must have hurt himself coming out through, or rather, with the window but he put up a remarkable effort to show that he hadn't. The newest addition to the airport security-force swung his leg over the saddle of his bike and viciously kicked the pedal backwards to bring it to the top of its stroke. He pushed off with the other foot, throwing a look at Edward that spelled daggers. Edward shrugged his shoulders, and the policeman rode off up the road to where the unattended rubbish-fire burned against the wall of the locked-up incinerator building.

During the same week on a dull overcast day, a shout from the flight office brought people running to the hangar doors. Two squadrons of *F86 Sabres* were on their way to Speke; coming from West Germany. Apparently they were low on fuel and had asked for preference at Liverpool. The airport was on alert with other incoming traffic held-off or diverted to Manchester.

The company's new radio engineer, had fitted up some HF and VHF equipment in the flight office, so that the pilots could listen to air-traffic control and communicate directly with the company's aircraft. They'd picked up the *Sabre*'s call to air traffic, and quickly arranged postponement of Club flying activity, on instructions from the tower, until the alert was over. Within moments of the caution they heard the jets overhead. The first group had arrived. They peeled-off and one by one made their approaches for a stream-landing. As each aircraft touched-down safely, they quickly tried to clear the runway by taxiing fast to the other end, or pulling up sharply to turn off at the intersection. Only two of the first group were able to taxi to the hard-standing. The remainder lay scattered about the airfield, out of fuel. The pilots did a good job though, and the main runway was totally clear for the next group.

The second flight arrived soon after, all landing safely. Of this group, only one managed to get to the hard standing. One tried to turn off at the intersection but ended up on the grass, axle deep in the soft turf, and the *Airwork* people had to use a crane to effect recovery. They had to tow in the rest from various places all around the airfield. It was another exciting day at Speke.

It was during those times at *Dragon Airways* that Edward added another achievement to his experience. He was working with Nick; pushing the *Hawk*s outside ready for the day's cadet-training programs. Nick asked Edward to give him a swing on the first one, for the customary morning warm-up, and to climb into the rear cockpit once the engine was running: an unusual request.

"Pull the chocks away before you do that though," he added. This also, was unusual.

Edward swung the propeller and the engine started first time. He pulled the chocks out, placing them well away near the wing tip, before clambering into the rear cockpit as Nick had asked him to do. Nick set the RPM for warm up. Five minutes later he checked the mag-drop and ran up to full power. Edward had sat-in on the morning warm-ups before, and even run the engines himself occasionally, under supervision, so he was quite familiar with the procedures.

After the power run Nick throttled back and leaning around from the front cockpit, he shouted to Edward that he was about to do a taxiing test. One of the engineers had changed the brake shoes on this aircraft, the day before, so the obligatory taxiing test was not an extraordinary formality. It was bedding-in the new *Ferodo* linings. With that, Nick opened the throttle a little, and moved off towards the peri-track. Once rolling, Nick zigzagged the aircraft, with occasional blips on the throttle to keep it moving. Up and down they went, in front of the Club. Finally Nick pulled the aircraft up. He leaned around again and said, "Want to have a go, Ted?"

Edward's heart leaped, just as it had at his first prop-swing. He nodded and put his feet on the rudder-bar pedals. He pulled the stick back into his tummy, holding it there with his right elbow and grasping the throttle handle with his right hand. The throttle was next to him on the left, so his right arm rested across his body, to hold the stick back and to manipulate the throttle.

He'd seen the procedure many times before. With his left hand he grasped the handbrake lever, situated below the throttle on the left. On this type of aircraft, there's a ratchet button on the top of the brake lever, used to lock the handle in the brakes-on position. To release the brakes, a tug backwards pops the button, and lowering the freed handle releases the brakes. One can then move the brake lever up or down freely, at will, to apply braking as required.

The rudder bar has differential connection to the wheel-brakes, so that, even with full left rudder, no braking occurs until one raises the brake-handle a little. This action applies braking only to the left wheel, allowing the right wheel to orbit, pivoting about the left and thus turning the aircraft. The same principle applies to turning right: full right rudder and a touch on the brake handle. The aircraft's rudder itself, at the tail, also helps to make the turn, provided the movement of air from the slip-stream or prop-wash, over its surface, is sufficient to force the tail around.

For uninitiated beginners, the problem with taxiing aeroplanes with a real, centrally-pivoted rudder-bar, instead of individual pedals, is that one tends to treat the bar like the handlebars of a slow-moving bike. Push right to turn left and push left to turn right. That doesn't work with aeroplanes. One must push left to turn left and push right to turn right. At first it's unnatural, but the knack soon comes. Apart from that, one can end up going too fast with too much throttle on, or turning too sharply with too much brake applied, or turning the wrong way for not thinking about the action of the rudder bar. Coordination is the key. Nick guided Edward through the exercise patiently until, finally, the boy was doing it without aid.

Nick, in the front cockpit, raised his hands above his head to show everyone watching at the hangar doors, that Ted, for the first time in his life, was taxiing the aircraft solo. This was yet another memorable milestone in Edward's embryo career; another high point to be well remembered. He discovered that, sometimes anyway, the hopes and dreams of life do mature. The point proved itself again some weeks later when Cedric asked Edward if he would like to go flying. Cedric had, by then, gained his private pilot's license and flew regularly whenever he could afford it. Having booked one of the *Hawk*s for that evening, Cedric intended to fly for an hour or so. Edward jumped at the chance. He'd flown in *Rapides* and *Austers*, on earlier occasions but he'd never flown in a *Hawk*.

Edward stayed on after work that evening, grabbing a flying helmet from the flight office when Cedric called him. Cedric climbed into the front cockpit and Edward swung the prop before clambering excitedly into the rear seat. They mover off as Edward closed the tiny drop-door on the left of the cockpit-combing and strapped himself into the Sutton harness as they taxied to up the peri-track to the threshold of the shorter of Speke's two runways. After power and mag-drop checks, a green light from the control van gave them permission to take off and they took to the air.

Communication between pilot and pupil (or passenger) used a simple speaking-tube system called Gosport tubes, as with most open-cockpit aeroplanes of that time. Each helmet carried a stethoscope-like appendage that plugged into the open end of a flexible pipe, the farthest extreme of which protruded from the instrument panel in front of the occupant of the other cockpit. They'd been airborne for only a few minutes when Edward heard Cedric's voice in the ear-pieces of his borrowed helmet.

"We'll just do a couple of circuits and bumps, then you can tell me where you live and we'll go and see what it looks like from the air, OK?" Edward leaned forward to place his mouth close to the Gosport-tube mouth-piece and found himself shouting in response.

"Right, from Hunts Cross, follow the main road to Halewood, I'll tell you when we're over the house."

"Roger!"

They circled the airfield and made a couple of standard circuits, approaches and landings before Cedric climbed the plane to gain height. Within minutes they were high over the village and following the highway towards Lunty's farm. Then, there it was: the row of houses where Edward lived.

"There, below us," called Edward, "the third house along from the end where the orchard is, on the right of the road." Cedric circled widely, lost a little height and pulled a split-arse elevator-turn almost directly over the house. Banked over and turning steeply on a wing tip, Edward could see neighbours in their gardens gazing upwards. Some waved. He waved back before Cedric levelled the *Hawk* and climbed the plane away towards the river. Cedric's voice murmured over the Gosport again. "Want to have a go, Ted?"

"Wow, do I?"

"Just put your feet on the rudder bar and take the stick in your right hand." Edward complied. The *Hawk* wallowed. Cedric spoke again as he stabilised Edward's crude attempts at controlling the plane's flight. "Not too hard, let the plane fly itself, you just guide it, keep the wings level and try to steer a straight course. Line up on something very visible ahead, or watch your compass heading."

Edward slackened his grip on the stick, tucking his right elbow in to rest his forearm on his thigh. "Right. Sorry, I didn't realise how sensitive the controls are in flight."

"That's normal at first, try to relax." Clearly, with ambitions to becoming an instructor, Cedric emulated his own instructor's advice.

Edward relaxed and picked out a suitable land-mark on the evening horizon, close to the heading they were on. He prodded the rudder bar gingerly to bring the *Hawk*s nose to bare. After a moment or two he briefly switched his eyes to the turn & bank indicator, mounted on the panel in front of him, then to the instrument indicating rate of climb-and-decent and finally to the altimeter. All remained stable. Cedric had the elevators trimmed perfectly for straight and level flight. It was just a matter of keeping the wings level and the nose headed in the right direction.

The slipstream battered at Edward's helmeted head behind the large curved windshield. The plane's engine roared and Edward could feel the very being of the *Hawk*'s capacity to fly. It was exhilarating. Cedric's voice broke in again: "OK, Ted, I have it, we'll have to join the circuit again now, my hour is almost up."

Edward withdrew his feet and hands from the controls feeling more than pleased with such an unexpected privilege. All too brief though it had been, it marked another notable moment in his, thus far, short career.

That experience was the first of many more unofficial flying lessons.

Part 10

Valerie, a Precious Encounter

27

Sweeping low over the hangar, the sweet sound of four *Gypsy-Queen* engines brought everyone running to the hangar doors to catch a glimpse of the aircraft whose slipstream had rattled the tin panels of the roof. And there she was, climbing steeply into the circuit again after her low swooping pass over the Flying Club. The company's first *De Havilland Heron* had arrived. They all knew she was coming, but they hadn't expected quite such a dramatic, aerial proclamation. The Boss was normally discretion itself, when it came to flying procedures and discipline. Mature ex-BOAC captains didn't normally beat-up their base hangars but clearly, the excitement of the moment brought out the boyish impulses of youth.

Several days before, Captain Guynan had travelled to the *De Havilland* Hatfield-works to take possession of his new charge, the first of three modern aeroplanes ordered by his newly-formed company. Growing excitement rippled through the throng of office staff, club pilots and technical bods as they gathered on the grassed area close to the peri-track to watch the plane make its long approach over the river, for a landing on the main runway. It was a moment of grand anticipation for all of them. The arrival of the new aircraft marked a turning point in their fortunes with the company.

The factory-painted, *Dragon Airways* colours were not clearly distinguishable at that distance but the morning sun glinted for a fleeting moment across the shine of the white-topped fuselage as the plane turned onto finals, lining-up with the runway. The *Heron*'s smoothly contoured shape and four humming engines had already captured the hearts of all, as they watched and listened. They heard the engine-note change as the propellers shifted into fine pitch for landing. They saw the broad wing-flaps lowering in stages. Moments later their beautiful new machine crossed the threshold, nose high as the Boss fished with the main-wheels for the runway, to make the smoothest of landings.

Euphoric comments and banter circulated freely. Everyone moved forward eagerly as the Boss turned the *Heron* at the intersection, taxiing quickly to the perimeter-track, and towards expectant crowd. They couldn't wait to examine the plane at close quarters, as it came to rest on the grassy parking area. The Boss, grinning from ear to ear, took off his head-phones and opened the cockpit's tiny quarter-light window. "She's a little Brahma, isn't she boys?" he shouted. "Goes like a bird."

Almost before the propellers had stopped turning, Nick enthusiastically opened up an engine-cowling to investigate the mysteries of the latest *Gypsy-Queen* engines. The rest walked around, stooping and ducking and craning to explore the beautifully shaped aircraft. Finally they all agreed; she was indeed, a little Brahma.

During the following days, enthusiasm mounted. Pilots came back from their type-conversion training, the luckier engineers returned from their technical courses at the factory and the Chief put on some technical lectures in the evenings, to give the others some idea of the systems employed on the *Heron*. Morale was at an all-time high. A week later, after proving flights, the first daily schedules opened between Liverpool and Newcastle and with the delivery of the second and third aeroplanes, in quick succession, the newly installed operations officer threw them immediately into a program of schedules for other routes. Suddenly they were extremely busy. The Northwest's newest operator of feeder-line services was in business.

The *Heron* operations didn't affect Edward for the most part since his duties remained with the club's activities. Club-flying and small charter-work were in full swing. With summer's arrival, the company expected to remain very busy on that side of the business, too. Membership had increased and the little *Austers* and *Hawks* began to put in many more hours of flying. Post-war austerity having diminished, people seemed more affluent, with more money in their pockets and a refreshing new desire to fly; an appetite for aerial sport had returned. For this reason the Club's fortunes began to pick up again. Edward felt sorry that the old Boss, who originated the flying club, hadn't been able to weather the storm of declining business in preceding months. He would have enjoyed the new times and better luck that had returned to the Club.

Along with the increased fleet size and the company's consistent operational activity, it became clear that the small hangar hadn't the capacity to house the support necessary to *Heron* maintenance. As the new fleet's flying hours mounted, heavier maintenance checks began to fall due. It was impossible for the club's small hangar to accommodate the larger aeroplanes. The company needed larger facilities.

It came as no surprise that the RAF unit, in the next hangar, was about to close down. It must have been on the cards even before the formation of *Dragon Airways*, otherwise the company would have had to find more adequate facilities for the *Heron*s elsewhere. Some of the airmen, with whom the club's staff had been so friendly over the passing of time, popped over sadly, to pass the word of the event, after receiving notification of their new postings to other units. Their equipment would go too, and the big hangar would therefore become vacant. For *Dragon Airways*, the opportunity was particularly fortuitous. The Boss negotiated a lease successfully and before long the *Heron* and *Rapide* maintenance teams moved across to the larger facility.

New storage facilities for materials were also necessary, and since the bigger hangar had ample space in rooms alongside, the Chief engineer delegated Cliff and Edward to move all the spares across and reorganise, for greater productivity. During earlier times, Edward had already had some experience with stores work, but he wasn't very keen on the idea of becoming a storekeeper. Edward was not being disrespectful to those who perform this type of work – he merely affirmed that his ambition was to become an engineer. He felt uncomfortable about the motives for the Chief's decision to place Cliff and Edward in the stores. Such a move would, most certainly, take Edward away from aircraft work and his future hopes and dreams for a career in aircraft maintenance, as a licensed engineer.

He indulged in moments of self analysis. Was this to be a repeat of his failure to achieve in art? Had he failed in some way to meet the standards? Had he become, like others, just a number on the pay-roll, to be used wherever and whenever for whatever? Edward felt thoroughly depressed and he talked to Cliff on the subject. His companion's response was sympathetic but he thought Edward was over-reacting. After all, Cliff was in the same boat too, wasn't he?

They pressed on with the task anyway, and found themselves up to the ears with documentation and paperwork. The task seemed never-ending: identifying spares, checking them physically against the packing lists and invoices, placing the parts in shelf locations and raising stock cards. A large package-deal of spare parts came with the *Heron* aircraft, designed to support the three aeroplanes for two years. One could therefore see that the larger storage area was essential to house it all. Proper organisation was a prerequisite, and they designed stock control procedures to ensure positive materials accounting. The company accountant had advised them that an annual audit would not look kindly on shoddy record keeping.

It was fortunate for Edward that weekends continued to provide contact with the Club aircraft and as the season drew on, pleasure flying with the *Rapide* began again. The terraces of the airport drew the usual crowds of people, out for a Saturday or Sunday afternoon's diversion. It pleased Edward to be away from the Stores environment and to be, instead, out in the sunshine and fresh air. He enjoyed the social contact, too; talking to the people when they came to ask questions about the aeroplanes.

One Saturday afternoon Edward set up his desk and sign, ready for the customers. It was early. There weren't many people about, so he went for a stroll along the curving terrace. Dicky, the pilot, agreed to hold the fort. On his rounds, Edward stopped a few of times to drum up custom. Several people bought tickets, making their way to wait close to his table and sign, in readiness for the first flight of the day. As Edward moved on, he spotted two girls who'd just arrived. They were struggling to open their metal deck-chairs. He recognised one of them and paused for a moment to recollect where he'd seen her before. To his astonishment and delight, he realised that she was the girl whom he'd met last autumn, on his visit to *Lockheed*'s. Wasn't she the one who had served him with tea and shown him to the sales office so pleasantly? Wasn't it she who had left a question mark over a promise to come to the airport one Saturday?

His heart bounding but without feeling the usual and annoying, adolescent affliction of bashfulness or colour rising to his cheeks, he stopped beside the two girls. "Good afternoon, can I help you with those chairs. There's a bit of a trick to getting them open," he ventured. "Are you going to fly this afternoon?" The girls turned with some surprise, because they were laughing and joking about the awkwardness of the chairs, and hadn't heard Edward's approach.

"Oh, it's you. Hello," said the one whom he knew, displaying the lovely smile that Edward, quite suddenly, remembered so well. "There you are, I told you I'd come and have a look, one Saturday." Then, in an undertone to her friend, as if winning a bet, but still smiling and gazing straight into Edward's eyes, she added. "This is the boy I told you about. You see, I was right, I told you he'd be here today."

After a long, gazing, smiling moment, the girl returned her eyes to the chair she'd half-opened. "These chairs are a bit silly. Do you know how they work? Can you help us?" she asked, and stepped back a little to let Edward get at the offending contraption.

"Sure," he said, and snapped the seat down to spread the metal legs, dusting it off with his hand, "they are a bit tricky sometimes." The girl smiled again and sat down daintily to observe the panoramic view, the airfield, the river, Speke Hall and the Wirral's hazy outline in the distance, under a cloudless blue sky.

The girl's companion continued to wage war on her seat until with a clank she was victorious, and she placed the chair close to the railings to face the airfield, next to her friend. The girl he knew looking up at him again asked, "The weather's super. Will there be any flying today?" Edward nodded, telling her that the first flight was almost ready to go and judging by the scores of people crowding onto the terraces then, they would likely fill quite a few more flights during the afternoon.

The conversation had lasted only for a mere five minutes, yet he felt as if he'd known the girl forever. He reflected that it was a whole winter and spring past, since he'd made that singular visit to *Lockheed*'s. Edward felt very strange. It was surprising that she remembered him after all that time. Then, she spoke again as if reading Edward's innermost thoughts. "I did come to the airport, the weekend after we first met – you know, at *Lockheed*'s, but there were no Pleasure-Flights on. They said the season had finished," she explained, pouting her lips and looking sidelong at Edward, "I was very disappointed."

He responded quickly, for he could see the pilot over at the desk, waving to catch his attention. "Yes, we only have a short season, even in a good summer. We have to fill the plane to make it worthwhile, and if there aren't enough customers it doesn't work. We usually fly until September, depending on the weather." Then a thought struck him. "Do you want to fly today?"

"That depends," she answered, lowering her head so that her impishly smiling eyes scrutinised him from beneath the mock frown of her dark eye-brows, "are you going to find us the best seats then?"

Her question caught him unawares, unlocking his memory to remind him of the conversation they'd had as she'd taken Edward to the sales office at *Lockheed*'s all those months ago and the suggestion he'd made about seating, if she came to fly.

"Absolutely," he said, hoping that his smile was as winning as hers. "Look, I have to dash off right now, my pilot is trying to tell me that we've a few more people interested in the first flight. My desk is over there," he added, pointing down the terrace to where Dicky was still waving his arms about. "Will you come over in a minute, there's plenty of seats on the second flight."

"OK, see you in a minute then," she said.

Edward hurried away, but not without over-hearing the girl speak again, almost excitedly, in a loud whisper to her friend.

"You see, you see, didn't I tell you he'd be here. I knew it. I just knew it!"

For the life of him, Edward couldn't think what was so special about him to warrant such an emphatic statement. Could he have really made such an impression all those months ago? He began to remember his visit to *Lockheed*'s very clearly. She had certainly made an impression upon him, though he hadn't realised at the time.

With the first load of excited passengers seated and strapped in, Edward closed the cabin door, wheeled away the steps and returned to stand in sight of the cockpit, whilst the pilot prepared to start the engines. The first engine started without difficulty but the second just didn't want to; the propeller rotated, jerking impotently through a dozen revolutions under the electric starter's whining impetus. It was over-primed. Dicky shouted from his cockpit window, asking Edward to 'blow-out' the cylinders by turning the propeller backwards half-a-dozen times. Edward obliged and the pilot tried the starter again. This time the engine fired at last, with a gout of blue-grey smoke belching from the exhaust, to be quickly whisked away by the prop-wash.

Edward returned to his desk on the terrace to find the two girls sitting next to it. They watched intently as the *Rapide* taxied around the perimeter to the end of the shorter runway, for take-off. Edward decided to introduce himself at this point, since they seemed to be on such good terms; and because, suddenly, he felt very strongly attracted to the girl.

"My name is Ted, by the way." He ventured, offering his hand.

"Oh, yes, I know; I'm Valerie, Val to my friends, and this is Patricia, everyone calls her Trish." Val said, clasping Edward's outstretched hand with soft cool fingers, and fluttering her other hand in her friend's direction. The plane still had Patricia's attention; but she glanced at him, bobbing her head with a bright smile as she said Hello.

"You can go on the next trip if you like," Edward suggested, "I've got almost enough people for the second flight, there's three seats left." He wondered how she already knew his name; a curious mystery.

Valerie still held on to Edward's hand until his comment seemed to remind her of the fact. She released his fingers slowly, almost reluctantly, and with a brief smile she opened her purse to find some money; handing Edward enough for two seats, in exchange for the tickets. The two girls turned to watch as the *Rapide* took off and droned away on its first aerial sight-seeing trip of the day. Meanwhile Edward gathered all the other passengers in readiness for the next.

Animated with unhidden excitement, Val and Trish asked a barrage of questions about flying and aeroplanes. "Why does the plane have two sets of wings?"

"The pilot won't fly it up-side-down, will he?"

"How many people can the plane carry?"

"Do we have to wear parachutes?"

"Do you fly, I mean, are you a pilot, too?"

Edward answered the questions as simply as he could, and there were many more; until Valerie touched his arm and said, "Did I tell you I have a brother in the Air Force. He's a pilot, you know."

"Yes, I think you did mention it; that day at the factory," Edward responded, "how is he getting on, I mean, where is he stationed?"

"He's in Germany at the moment," Valerie answered with a perplexed expression, "flying Meat-boxes, I think he called them: strange name for an aeroplane."

Edward chuckled. "That's a name the RAF chaps use for Gloster Meteors; they're twin engine jets," he corrected, "there's a couple of squadrons of them, across the river, at Hooton."

Valerie had changed little over the months since they had exchanged those furtive glances across the hatch-way counter at *Lockheed*'s. She was a bit older than himself; about eighteen he guessed. She looked fresh and elegant in her sleeveless, summer dress that modestly revealed a slender figure with soft curves. She clearly took pleasure in dressing-up. Her dark hair was long enough the catch the light breeze that wafted across the airport terraces. She had expressive, brown eyes set wide beneath refined, dark eyebrows and an oval face that needed little make-up. When she smiled, two little dimples appeared beside her perfect mouth. She was quite, quite – stunning!

Edward couldn't take his eyes off her and, to his own wonderment, he didn't retreat into the timidity that so often plagued his association with girls in those times. Perhaps it was her friendly and easy manner or perhaps it was because Edward was on his own turf, so to speak, there at the airport.

Valerie's white, summer dress, with tiny printed roses at the neckline, caught the breeze and threatened to billow out of control. She laughed, a tinkling, rippling, joyful laugh, as she hastened to restrain the light material.

Edward asked her where she lived and how she liked her job at *Lockheed*'s; did she prefer the country-side rather than town-life. He asked her anything that came into his head, to make conversation.

Val said, "I live in Woolton, not far from the big golf links; d'you know Woolton?" She smiled again and mused, "Its more country than town where we live; I prefer the countryside."

"Yes, I sometimes go to the pictures there, you know, at the Woolton Cinema." Edward commented, as another thought came to him. "Do you like the cinema?"

"Oh yes, we go there sometimes too, my family and me, I mean." She hesitated for a moment before continuing. "There's a good film on this week you know. It's showing all week."

Trish, Valerie's friend, interrupted, pointing to the West. "The plane's coming back, look."

The gentle whistling of air though bracing wires and over wings and the sound of throttled-back engines announced the quiet return of the *Rapide*. The plane swooped in over the perimeter fence and, levelling off, flared, to touch down neatly with hardly a trace of bounce, as its wheel's spun and rolled on the smooth tarmac.

Edward said, "I must pop down to see the aircraft in, you two can head the line for the next trip if you like; that way I can find you the best seats, OK?" Then, raising his voice above the babble and chatter, he called for the next group, "Passengers for the second flight, please be ready to go down stairs now," before leading them down to the tarmac where he asked them to wait until he'd disembarked the returned group.

As the *Rapide* quickly rolled into position on the apron, he wheeled out the low steps and positioned them close to the trailing edge of the wing, next to the cabin door. The people where smiling and talking excitedly about their fifteen minute experience as Edward opened the cabin door and handed them out, down the steps, to the hard-standing.

Amid subdued squeals of alarm and surprise, the passengers alighted to have their hair blown and dresses flapped and fluttered by the slip stream of the propeller, idling then, on the left-hand engine.

"Thanks ever-so-much," said one lady, "that was my very first flight and it was lovely."

"I never realised that Liverpool was so big," said another, "and the river – what a sight; thank you, young man." Edward smiled and helped the last of the passengers through the narrow door. They all seemed pleased with the experience.

As the first group of passengers made their way back up the stairs to the terraces, Edward boarded the next group, seating the two girls where they could get the best view. It's difficult in a *Rapide* because of the position of the wings and engines. Also, being a biplane, there are struts and bracing wires between the wings, obscuring the view somewhat. However, Val and her friend seemed content.

Trish managed her seat-belt without any help. Val had sat on hers but after some struggling, Edward and she managed to retrieve the straps and place them correctly across her lap.

Edward could smell the perfume she wore, as he leaned across to untwist the outboard belt-half, and their hands touched again as he showed her how to buckle-up. Valerie looked up at him, as he straightened. Her eyes met his as, with complete simplicity she said just, “Thanks.” Her smile was precious.

Edward seated the other passengers, and checked briefly with the pilot, up-front. Valerie and her friend were joking a little nervously now and nudging each other across the narrow aisle, as he passed back through the cabin to the door.

“See you later,” he said with a wave – imitating Val's.

28

As the aircraft taxied away, Edward sat by his desk feeling very strange and extremely happy. He felt as if he were standing outside his body, looking on at the delightful encounter of that afternoon. He felt butterflies in his stomach and yet he felt totally calm and in control of himself: no shy, retreating, bashful feelings, just sheer joy and excitement. He watched as the *Rapide* took off and climbed away into the distance over Speke Hall, before turning over the river to pass downstream, towards the great shipping terminal and dock-land of Liverpool. More people came to ask about the flights and Edward quickly sold another full, plane-load of tickets. His friend, Tommy, the *BEA* station-engineer, came by. He sat for a moment to pass the time of day.

Tommy, a short man with a limping, arthritic walk, must have been close to retirement age. Edward knew him quite well and they chatted for a while: idle gossip mostly. Tommy had worked at Roots during the war, so he knew Edward's father and asked after him. His square, weather-lined face bore the usual grin and eye-glasses that he always perched on the end of his small nose. His smile exhibited teeth that were too even and too impressively white to be his own, but he was a good man and Edward liked him. He'd worked with *BEA* for many years, ever since the war and always at Speke. Edward was sure that Tommy really loved his work.

He was forever on the job, rushing here and dashing there in his white overalls, with his uniform beret displaying the company badge, pulled jauntily down on one side. It often occurred to Edward, in a mischievous way, that perhaps Tommy slept with his beret on too, for he'd never seen the little man without it. He was a frequent visitor to the Club hangar, to borrow this or that, or just to have a chat and a cup of tea with his mates.

“This is your first day of the season then,” said Tommy, “you should do well today; plenty of folks on the terraces and a few who'll want to fly I expect."

"You're not wrong Tommy," agreed Edward, "the second trip's away and I've two loads more who're already booked; I'll be busy until sun-down at this rate."

Tommy pushed his glasses up to the bridge of his small nose to see his young friend better. It was a mannerism he needed to apply often, for his spectacles were always slipping down. His nose was far too small to support such heavy, horn-rimmed lenses. "That reminds me lad... met a young lass last year, in the cafeteria. She asked about you. Aye, a pretty girl she was. She came to go on one of your Pleasure Flights but the season had finished; it was a rough old day anyway – no flying at all that afternoon as I remember. The weather clouded up and it was clamped down until next day." Tommy grinned as he went on, "Any road, she asked me if I knew you; o'course I told her I did, and mentioned your name. She seemed to know you; p'raps she'll come back again this year."

"Oh, that's interesting," Edward responded, suddenly intrigued, "was she with a friend, a blond girl?"

"Aye lad, that's them; girl friends of yours are they?" Tommy's eyes sparkled behind his glasses as he grinned again and gave Edward a knowing nudge. "Better be careful son, pretty girls like that."

Edward coloured up for the first time that day but he realised that this was how Valerie already knew his name. The fact had him wondering all afternoon, but he hadn't the courage to ask Val directly how she'd come by it. Then Tommy spoke again, breaking into Edward's thoughts. "Ehup lad, here comes your plane again," he said, pointing towards the gas-works. The *Rapide* was on final approach. Edward imagined the passenger's exclamations of surprise at the momentary humps and bumps, as the aircraft flew through the turbulent air over the gasometers. Tommy got up to go, flinging a broad smile and parting words over his shoulder as he limped away to the stairs. "Keep em flying, Lad, I'll be seeing you."

"Yes Tommy, see you around."

Edward realised that his thoughts had been elsewhere and he hadn't paid too much attention to the rest of Tommy's gossip over the last ten minutes. He hoped that his friend hadn't noticed. Edward hoped too, that Val had enjoyed her first flight, so long postponed.

The two girls had indeed enjoyed the flight. They laughed girlishly at the indignity of emerging from the enclosed cabin, into the propeller slip-stream that threatened to lift their fluttering dresses. Edward helped each of them to step down from the doorway onto the steps, Trish first and then Valerie. She was the last out and gripped Edward's arm as she jumped down to the ground.

Her touch was like an electric shock to his nerve ends, sending strange and wonderful messages to his heart. Her breath against his cheek, as she bent close to support herself on his arm, was like a summer breeze. Something was happening to Edward's emotions and he felt ten feet tall.

After boarding the third trip of the day, Edward decided to buy some ice-cream for the girls, to round off their experience. It took him only a moment to pop into the cafeteria on the ground floor to make the purchase, and a moment more to climb the stairs to the terraces. Meanwhile, Val and Trish had seated themselves beside his table again, chatting excitedly about the flight. Edward didn't realise it, but they had already bought ice creams, including one for him. How they laughed. Great minds think alike! Well, they all liked ice cream anyway.

Later on, after several more flights, Trish suggested that they should have some tea, sandwiches and cakes or something. That sounded like a good idea, so off she went to arrange it, in spite of Edward's protesting that he should go. “No, no, it's my treat,” she said, “you're quite busy and the plane will be back again in a minute.”

Valerie stretched back in her chair; with her hands clasped behind her head, her eyes closed, and a tiny smile playing about her lips.

“What a beautiful day. This is better that sitting in an office all day long,” she mused, “I wish I could work out in the fresh air.”

“You're right,” said Edward, “they've just moved me to an indoor job too, and I don't much care for it; that's why I enjoy the summer season weekends, up here on the terraces.” He felt mightily attracted to this vision of youthful girlhood seated beside him. He knew that if it hadn't been for his weekend pleasure-flight duty up there on the terraces, at the terminal, they would probably never have met again. Was it fate that had brought them together in this way? He had to take the opportunity. He had to ask Val if she would go out with him. 'Now is the moment, Ted,' he thought, casting his eyes about him and noting the crowds. 'Trish is away getting the tea. We're alone, except for the fifty-odd people sitting all around us, but to heck with that.' He leaned closer to address Valerie directly.

“I don't suppose you would consider going to the pictures with me, would you?” He asked, tentatively, unable to avoid seeing the reaction of an older couple, sitting just beyond Valerie, close by.

They smiled knowingly at Edward's overheard petition and nudged each other without looking directly at the two youngsters.

"What a good idea Teddy, yes, why not," answered Val, with delighted enthusiasm. She used Edward's intimate given-name. This astonished and startled him, but it pleased him intensely.

"When shall we go then? Tonight? To the Woolton? You said they're showing a good film this week," he blurted out, somewhat hurriedly. The near-by couple nudged each other again, quite openly showing their clear approval of Valerie's assent to Edward's intense solicitation for a date with such an attractive girl.

"Yes, tonight, if you like; where shall we meet? Or will you come to my house? We could walk from there, it's not far from the town."

"All right, that's great. Tell me where you live and I'll come at about... What time does the movie start, around seven? I'll come at about six-thirty then."

"Yes, that would be perfect."

Edward leaned back with a smile and inwardly sighed with relief. Again, he caught a glimpse of the older lady, behind Valerie, nodded her approval and patting the hand of her spouse as if she'd known all along just what Valerie's answer would be.

At that moment the aircraft came back again and Edward excused himself to rush off and supervise the unloading. There were no more customers, so the pilot decided to call it a day and take the *Rapide* back to the hangar. Edward declined the offer of a ride in the back of the plane, saying that he'd stay on to tidy up around their slot on the apron and he'd walk or hitch a ride with one of the airport's MTCA cars later.

Tea was on the table and the girls were setting out plates as he got back to the terrace, and what a splendid tea it was, too, with cucumber sandwiches, sticky cakes and a real tea-pot. Since the aircraft had gone, Edward took down the sign for Pleasure Flights, so that no one would disturb them. There were still quite a few people about, enjoying the fresh air and taking tea, like them, on the terraces, as the late afternoon shadows lengthened.

Their conversation drifted along with references to the two girls and their work. Trish worked at *Lockheed*'s too. They both enjoyed their work with the company. They felt that it was secure and gave them independence. However, Valerie wanted to get into something else with travel as a prerequisite, perhaps with an airline as an air-hostess or what they had already begun to call cabin-crew or flight attendant. Edward suggested that she visit *Dragon Airways*. Perhaps the company would have something she might prefer – after all the company was in the process of expansion.

Much too soon for Edward, the sun began to sink towards the close of day and the girls decided it was time to go. Val gave Edward a scrap of paper with her address written on it. "We have to go Teddy, time's getting on and we don't want to be late and I have to change for this evening. I'll see you tonight then. Don't forget, will you?" She said with a sidelong glance and a smile that charmed Edward to distraction, along with that famous little wave of hers. Edward saw them to the stairs before clearing up his pitch on the terrace. Today has been a beautiful day he said to himself. He wanted to jump for joy, kick his heels in the air, shout, laugh and sing a silly song; he was so happy.

The couple, who had unavoidably overheard much of the young peoples' conversation at the pleasure-flight table, got up to go themselves and, as they passed Edward, the lady stopped for a moment to smile kindly and say, "Forgive me if it sounds like an intrusion, but we couldn't help overhearing some of your conversation earlier. I do hope you have a nice time tonight. Treat your new friend well, mind – she deserves it you know; such a lovely girl. Goodbye young-man." There was a look of happy reminiscence upon her face as she spoke: perhaps remembering her own young days of courtship.

Edward thanked her with a bright smile that must have shown just how he felt – wonderful. The lady's comment added to his euphoria at what had transpired that afternoon on the airport terraces. He tidied up the remainder of his gear and scrounged a lift with Tommy Rise, in his old Standard car, to carry him down the peri-track, back to the Flying club.

Edward didn't have his usual tea at home. His mother was a little surprised when he explained that he was going out. Her concern was that he hadn't eaten since lunch time, but of course he had.

"Oh well, if you're all right then, son. You have such a good tea as a rule. Are you going far? Out, I mean?" Edith was fishing but her Pisces son wasn't biting.

"No, only to Woolton, to the pictures. I should be back by eleven," he reassured her.

He washed and changed quickly. He even put on a tie with his clean white shirt. He wanted to look a bit smarter than the person in overalls that Valerie had agreed to go out with. A final check in front of the mirror showed that he'd forgotten nothing. The mirror reflected his image but, apart from the clothes, the visage failed to impress him. With a couple of crooked teeth, hair thinning a little, eye-glasses, and facial hair that grew so sparsely that it only faced the razor once a week, he'd never thought of himself as having film-star attributes, although he supposed that he wasn't exactly ugly! No, his reflected aspect didn't appeal to him much. He even began to wonder what it was that Valerie saw in him.

He pictured her lovely face in his mind's eye and thought of Beauty and the Beast. Nevertheless he felt extremely happy, and changed the tie for a cravat. It seemed to go better with the blazer.

He checked his watch. The bus was about due and in his haste he almost forgot to pick up his wallet from the chest of drawers. He grinned to himself. That would have been embarrassing; asking a girl to go out with him and having to ask her to pay for the cinema tickets. At the front door Edward called out Cheerio to his parents.

He needn't have bothered, for they both appeared at the living room door to see him off. "My word, you do look smart," his mother said, casting an approving eye over her middle son's attire, "someone we know?" Prompting him again, with a charming smile.

"No Mum, no one you know. See you later," he answered with a laugh. He gave her a kiss and grinned at his Dad who stood beaming all over his face. Edward could hear the bus coming, and had to make a dash for it, to the top bus stop.

Finding Valerie's house wasn't difficult. As she had said, it lay close to the top-end of the golf links, in a small Close of five or six semi-detached houses set out in a circle, off the road out of Woolton, towards Huyton. It took Edward only five minutes or so to walk the short distance from the bus stop to the corner of the Close. Valerie sat on a low wall in front of the house, her shapely ankles crossed and her hands resting on the wall at her sides. She was talking to a middle-aged man, who busily fussed with roses in the small garden behind her. Valerie introduced Edward to her father who, by then, had drawn off a glove to shake hands. He seemed pleased to see Edward and called towards the house, "Mother, Valerie's just off, are you coming out to say goodbye? Come and meet Ted."

A bustling little lady trotted out of the open front door, wiping her hands on an apron. She was smiling widely and proffered her hand, as she reached the group. She was out of breath and explained that she had come from the kitchen; baking cakes for tomorrow. "How nice to meet you; Ted isn't it?" She said, "Val'rie's not stopped talking about her trip in the air today. It's her first time you know." The abridgment of Valerie's name gave away the Northern upbringing. "We've a son in the Air Force. He's a pilot." Valerie's mother fired at Edward as she pumped his hand vigorously. "You're too young to be a pilot aren't you?"

Edward shook his head and told her that he was training to be an engineer, but that he enjoyed flying too, when he could get the opportunity. Everyone smiled and Edward noticed a couple of elfin faces at the curtains of the front room window. He waved at them; it seemed to be the right thing to do, and they waved back, laughing silently at him through the glass, before letting the curtain fall again. Edward glanced at his watch, more as a signal than to note the time, then catching Valerie's eye, he suggested they make a move, if they were to catch the movie.

"Well enjoy yourselves," Val's Mum said. "Have a lovely

time."

"Have her home by Ten-thirty, Lad, will you; shouldn't be out too late y'know." Val's Dad lectured with a smile. Edward nodded his response and set out for the town with Valerie.

Valerie giggled and told him how she had described the afternoon to her parents, including the double-order ice cream and the tea and stickies. "Do you like my parents?" She asked.

"Yes. They seemed very happy and sympathetic," said Edward, "but I hardly know them yet."

"They're a pair of old softies really. Did you tell your people about me?" The question startled Edward. Valerie had a way of doing that: startling him.

"No," he answered, "but they wanted to know where I was going and who I was going with. They were a bit inquisitive so I left them to wonder."

Valerie giggled again, "Yes, I know, parents are always a bit possessive and inquisitive, like that." They crossed the main road at the tram terminus. The cinema was up the hill a little and then to the left, up a very steep gradient to the stained glass, canopied entrance doors.

Edward explained to Val how he, as a kid, used to go to the pictures on Saturday mornings to the Saturday Matinee, at sixpence to get in. The serialised, cinema versions of Batman, Flash Gordon, and Superman were as popular then, as they are now on TV.

29

The Woolton wasn't a very large cinema. It had a silly reputation for being a bit of a flea-pit, but it wasn't. It was really quite pleasant inside, and seemed to get all the latest movies quite quickly after first showings at the larger cinemas in the city. Seats in the circle and stall seats were on the same sloping floor, with only an aisle across the auditorium to separate them.

Edward bought circle tickets and a box of *Cadbury's Roses* Chocolates for Valerie. The house lights were still up as they passed beyond the plush-velvet curtain. An usherette took the tickets and tore them before pointing the way to the circle's centre aisle. The seats had better upholstery in the circle. Edward supposed that was why they were more expensive. However, the screen seemed miles away; no Cinemascope or wide screens in those days. He tripped on a carpet corner as they started up the slope and ended on his hands and knees. He cursed roundly under his breath as he went down with a heavy thud, audible throughout the theatre.

Valerie giggled behind her hand as she helped him to stumble to back to his feet. I'm a clumsy oaf, he thought, and hoped his whispered oaths had gone unheard. Still, in the half-light of the cinema no one would see his blushing cheeks. How he wished that his face wouldn't do that. They found good seats and settled down. The lights lowered and the screen lit up as the curtains in front withdrew to the sides. The sound track of Pathe-News boomed into the theatre and the evening's entertainment began.

Valerie leaned close and whispered, “Are you all right, did you hurt yourself?”

“Yes, I'm fine, I mean no, I must have tripped over the carpet edge or something.”

“Sorry I laughed, Ted.”

"I felt a bit foolish, sorry about the language."

"I didn't hear a thing," Val said loudly, with dignity and in a very politic voice. They both laughed out loud, only to be rebuked instantly from behind, with indignant shushing. The news narrator's clipped, staccato voice informed them of problems in the Middle East, polar bear, Brumas's progress at London Zoo and the latest from the Test match at Lords. A quick cartoon followed and some trailers for forthcoming films.

Probably choking over the bag of nuts he'd been delving into, the chief shusher behind them, burst into a fit of coughing and snorting, accompanied by loud trumpeting through a handkerchief. Valerie and Edward collapsed in silent and not so silent mirth and hilarity. They shrank into their seats, trying to subdue their laughter. With tears of mirth streaming down her face, Val leaned towards Edward again to whisper, "Serve him right, for shushing us. He's been scrunching at that cellophane-bag of nuts ever since."

Edward gave her his spare, clean handkerchief to wipe her eyes and immediately she had him shaking with laughter again, as she imitated the shusher's Trumpet Voluntary. Edward fell instantly in love with Valerie's display of impish humour. The main feature titles flickered onto the screen but for the life of him, later, Edward couldn't remember what film it was; he only vaguely recollected that Virginia Mayo was in it and possibly Gregory Peck. He didn't care much anyway. His heart and senses were afloat on an ocean of distracted thoughts and new emotions that crowded into his being. Just sitting there so close to Valerie was enough.

At the interval he suggested ice cream and stepped, carefully this time, down the aisle to where the sales girl stood with her tray of torch-lit refreshments. On the way back to his seat, and just for fun, he imitated another carpet-trip, without falling completely down. Someone in the row beside him ventured to say, "What's up with you then, got two left feet or something?"

Edward told him no; he hadn't, the carpet was up. The lights lowered again as he regained his seat next to Val. She asked him if he'd really tripped again, on the same piece of carpet but Edward told her it was only a joke that had backfired, serving him right for being so silly. They laughed again over it, but the shusher behind them was too busy, noisily scooping large dollops of ice cream into his mouth, to consider another reprimand for silence.

As they watched the remainder of the film, Edward sensed that Valerie had moved closer. Their shoulders touched and she rested her arm on his arm, and her cool hand upon his. Edward was in heaven. This happy encounter on such a splendid day was indeed the most exciting thing that had ever happened to him. Although the recollections of those adolescent emotions remained indelible, they were fragile. They could never to be repeated, nor suffered, nor enjoyed, ever again, in the same way: likewise, never erased.

It was almost dark outside, after the show, as they walked slowly away from the town. Valerie took Edward's hand as they strolled, an innocent yet intimate link that seemed so natural. The summer night air filled with the aroma of the fields and the trees, now in full summer regalia, borne on a soft breeze: warm and scented. They didn't say much. They just breathed deeply. They just enjoyed.

As they turned into the Close where Val lived, she stopped for a moment, turning to face Edward. Even in high-heels she was a little shorter than him. Val smiled into his eyes as she spoke very distinctly. "Thank you, Teddy, for a perfectly wonderful day. I've flown in an aeroplane, I've had three ice creams and a box of chocolates. I've had tea on the airport terrace, I've been to the pictures but, most of all, I've spent much of the day with someone I first met, and began to like, more than six months ago. I lost him for a while, but today I found him again. Thank you and thank you, Teddy."

The sentiment and repetition startled Edward. He felt abashed and yet honoured; for indeed his own enjoyment of the day, equally matched Valerie's. Someone found him that day and stirrings in his heart told him that he had found someone, too.

Valerie's father came to the front door as Edward opened the garden gate for her. She turned to face him again as he closed it, she on one side and he on the other. Edward searched for words of his own, to express how marvellously that day had affected him, too. "Valerie; thank you for making my day such a very special and splendid one, too. Can I see you again? Will you go out with me again?"

She nodded vigorously with a joyful smile, then, leaning over the gate, she kissed him very quickly and lightly, with soft, soft lips. A wisp of her hair, caught in a faint gust brought by the night breeze, stroked across Edward's cheek. He thought his heart would burst as his emotions turned this way and that and he raised a hand to touch the lovely face, so close to his. Val's Dad coughed discreetly and Valerie turned to go in, with her famous wave and that brilliant, brilliant smile.

"Can you come tomorrow Teddy? Its Sunday and we have a picnic planned for the afternoon. Can you – will you come?"

"Yes, I'll come. After lunch then."

"Yes, perfect. Good-night and sleep tight." Val slipped past her father who raised his hand in a formal wave and said, resignedly, with a grin, "See you tomorrow then lad, good night."

Edward decided to walk home to Hunt's Cross. The night was warm and clear and he felt very, very happy. Had all this happened to him today? Was he dreaming? Did he deserve such rapturous joy? Sleep came creeping to him that night, stepping gently and silently, with sweet dreams of today and high hopes for tomorrow.

Sunday's sunrise promised another beautiful day. Edward told his parents that he had arranged to go out in the afternoon, after lunch. They asked him if he'd enjoyed his evening out. He answered that he had, but didn't say more, and they didn't press him. How could they have possibly known that his affirmation was an unqualified understatement, but they glanced at one another with knowing smiles.

The afternoon came and the bus retraced Edward's journey of yesterday, to Woolton. He reached the Close and as he turned the corner, he saw two young children swinging on the garden gate at Val's house; the same two he'd seen at the window, the day before. They rushed away into the house as he approached, calling out in gleeful, piping voices, "He's here, Valerie, your boy-friend's here." Valerie came to the front door, scolding the children good-naturedly.

She called to Edward, "Will you come in for a minute, we're almost ready. We're not going far, just through the back and across the fields. Come on." Her parents greeted him as he followed Valerie through to the kitchen, where picnic preparations were in full swing. The two youngsters eyed the visitor with inquisitive stares, whispering together and gazing at Edward with unhidden curiosity.

Finally, all was ready and they set off through a little gate in the garden fence, out into the summer-decked fields. They walked a while, with the picnic things shared between them, the two youngsters hopping and skipping ahead. Kevin, had made a point of telling Edward with pride, "I'm nine," and, referring to Wendy, his younger sister, "she's seven." Their curiosity satisfied, the two youngsters paid little more heed to Edward's presence among their family. He seemed to have their stamp of approval.

They walked on through the fields, chatting about the things to be seen about them. Edward found himself observing Val's parents. Her Dad was not a tall man. He had dark hair that was thinning at the crown, and dark eyes beneath heavy eyebrows. He worked at a bank and had served in North Africa and Egypt with army, during the war. He smoked a pipe and wore grey flannels with an open-neck shirt. Edward liked him. He had a sense of humour not unlike that of his own father.

Valerie's mother was plump and small. Her movements were brisk and birdlike and very precise. Clearly she was a very capable person who enjoyed being a wife and mother. In her anxiety to have the picnic things ready, she had almost forgotten to take off her apron before setting out from the house. She had whooped with laughter at her own forgetfulness – an infectious laugh that no one could ignore.

They reached the edge of a field where a shaded brook ran chuckling and gurgling through a stony bed and in the dappled shadows of the trees, next to the stream, they found an area of fine, short-cropped grass. From there they could see the golf-links on the other side, with a view across gentle slopes, running down towards Hunt's Cross and Halewood. It was the perfect picnic spot.

Valerie and Edward went to the bank of the brook with the kids. Among the water-grasses and reeds they could see tiny fishes, Sticklebacks, darting here and there, startled by the passing of their shadows across the running waters. They bent close from where they knelt and Kevin asked if they could catch one. Edward told him that the creatures were probably too quick to catch by hand; they would need a small net. Wendy picked wild flowers close by and Val's parents sat in the shade of trees, relaxing on a plaid car rug they'd laid out, resting upon their elbows, as they watched their children getting to know Edward.

It was indeed a magnificent day with warm, summer sunshine and the song of a lark high above the fields, its rippling notes drifting down through the gentle breezes. Other birds called and answered each other from the hedgerows dividing the fields all around. The rustling leaves trembled and shifted as the hay-scented breeze moved up the gentle slopes from the languid river that lay distantly, beyond Hale. Edward breathed deeply the sweet air as he stood back for a moment, to watched the smaller children and Valerie as they walked farther up-stream hoping to see more fish. He stepped back mentally and emotionally to take stock: to indulge in self-analysis.

In two days he had come to know a person in the simplest terms. Here he was with a family of people he hadn't known until then. He still didn't know them, yet he felt as comfortable with them as he did with his own people. Their behaviour with him was like being at home among family. He didn't feel bashful with them. He seemed to have conquered his habitual facial flushes. He felt grown up and responsible. The surprising thing was that these good people seemed to feel comfortable with him. As with his own parents and brothers, Edward felt accepted by Valerie's family.

And then there was Valerie. Since yesterday she was in his every thought. How could this be? What chemistry was at play here. Am I in love? He asked himself. Is Valerie in love with me? Did it show? Could it be, at his seventeen years of age, that such emotion, such joy in her presence, were symptoms of love? Edward mentally shook himself, returning to the living world from his inner-most thoughts. Valerie was looking at him intently over her shoulder, her summer-dress set about and around her on the grass where she had seated herself near the brook-side. The children squabbled over fishing rights not far away. She caught and held Edward's gaze for a long, long moment: smiling.

He walked across the grassy space that separated them and sat beside her. Valerie leaned her shoulder against his, still gazing at him. Perhaps she too had explored her own thoughts. She reached over and gently took off Edward's glasses, as if to see more clearly what lay behind the reflective curve of the lenses. “Did you know,” she said, “that you have the gentlest and most thoughtful eyes I've ever seen?”

Valerie had startled him afresh, with words and sentiments that he suddenly felt unworthy of, and again he was without words of his own. “Thank you,” sounded so lame but it was all he could say.

After a moment, she replaced the glasses, which pleased Edward because he was next to blind without them and he wanted only to look at her. He said, repeating Valerie's own preface, “Did you know, Val, that since yesterday I think about you all the time, and that you have a gift for making me feel, I don't know, that I am not who I am. It's like being in a dream.”

“Teddy, you are who you are. I am who I am,” she said reproachfully. Then she thought for a moment as her eyes moved across his face in a tender visual examination, ending again at his eyes, with a direct and frank gaze. “Are we in love, Teddy?”

Another startling question that left him almost speechless.

“I think we must be Val. What else could it be?” he ventured, adding, to qualify his statement. “In twenty-four hours my life has changed; I think only of you.” Edward tried to focus. “I know we met once, months and months ago and I must say that I liked you then, and if I had not been so shy, maybe I would have done something about it. How can someone feel as I do now, after only twenty-four hours? Why do I feel so ...so ...extraordinarily comfortable? Why do I feel this way with you? I'm confused Val.”

This outflow of sentiment burst forth like a river through a broken dam. Edward couldn't hold it back and then he began to feel embarrassed. He felt his face flush. Valerie touched his face with the back of her fingers, in a tender, stroking motion, as if to drive away the hated redness. He had to go on. “I mean – we still hardly know one another. Have you had other boy friends? Have you been in love before? I think I'm younger than you – does that matter?” The outpouring wouldn't stop. “I've wanted to – but I haven't even kissed you yet, and here we are talking about love. You kissed me last night, over the gate. If kissing means you love someone, then I'd better kiss you now because I do believe it’s love.”

Valerie hadn't taken her eyes from Edward's face. Her smile remained and she nodded with raised eyebrows, seemingly to hang on every word. “Its love then!” She affirmed with complete conviction, "That's what it is." Her tinkling, rippling laughter filled the air about them. Valerie turned to kneel, facing Edward, and took his shoulders with both hands. She shook him gently, "Love is what it is. Can't be anything else." Then she frowned, pouted her lips in mock seriousness, she nodded vigorously. "It is love," she announced.

Again, Valerie astounded Edward; this time with her sincerity and unabashed, frank, open acceptance of an emotion that now he knew they both shared, teenagers or not. Valerie moved closer. Her lovely face, close to his, framed by her soft dark hair, took on a more serious expression. She put her hands in his and spoke softly.

"Yes, I'm older than you, Ted; but only by a year. I'm eighteen. You're seventeen." She said, with analytic precision and a tiny shrug, "Yes, I've been out with other boys; but usually in a group, girls and boys together, or along with my parents." Valerie sat back on her heels before going on. "I've never had a real boy-friend; I haven't met a boy, until now, that I've really wanted to be with. Boys are boys are boys."

She shrugged. "You know Teddy, I'm quite shy too. When I ...we, met, that time when you came to *Lockheed*'s, something happened to me. I've thought about you ever since that day. I didn't know how to find you, except for the Pleasure Flight place. I wanted you to come to the office again. I wanted to telephone the Flying Club, but I thought you might think that I was too pushy or flighty or something. I didn't even know your name, nor you mine." Valerie frowned over eyes that glistened now. "I thought I was too sensible and practical to believe in love at first sight, but I think it's happened to me. Is that what this is Teddy? Is it? Could that be what it is?"

"Something happened to me too," said Edward, nodding as he looked into the brook, seeing everything in bright relief, "I didn't know what it was. I was confused and I was fantasising over another girl at the time – you know, when I came to *Lockheed*'s that day?"

He glanced back to Val's face again. "When I come to think about it, that day at your office, I do remember it all so clearly; I must have been stupid not to have realised what was happening to me. Now, somehow, I feel that I don't deserve to be loved by someone like you."

Valerie put a finger on his lips. "Sh!" she said, smiling then.

Their discussion ceased abruptly as Val's father hailed them to the picnic. The youngsters at the brook scampered off, shouting, as Valerie and Edward stood up, turning to walk back.

"Race you Teddy. Last one there is a water-rat." Val shouted as she ran behind the children. Edward followed but he couldn't make up the head start they had.

"Teddy's a water-rat. Teddy's a water-rat," the children sang, as he brought up the rear. Everyone laughed and they had a splendid tea. Afterwards the water-rat felt so happy that he showed the voles, Wendy and Kevin, how to play French cricket with a ball and an improvised bat.

As the sun went down that afternoon, and they all strolled back across the fields to the house, Edward imagined himself cut loose from conventions. He took Val's hand in his. It seemed so natural and he felt a freedom that he had never known before. He felt grown-up and yet still a child, among children.

Val's mother put the young ones to bed. They went unhappily and grumbling. Valerie asked her parents if she might walk up to the bus stop with Edward. Her father looked at his watch and agreed with a smile. He took Edward to the front garden whilst Val went to put on a cardigan. As he showed Edward the roses he said, "You must come again lad, the youngsters seem to like you and they do miss their big brother you know. I think Val likes you too. I've never seen her so thoughtful and animated." He glanced to the front door as Val's mother appeared with their daughter, "Valerie was a bit of a tom-boy when she was younger you know, but she's always been a good girl, all the same." He added in a confidential manner. Valerie came to Edward's side and he said goodbye to Val's parents and thanked them for a wonderful afternoon.

When the bus for Hunt's Cross came, Edward looked long and hard at Valerie. He cupped her face in his hands and bent to kiss her lips. Valerie responded and after the moment she drew back and said with an impish grin, "Its love you know. That's what it is."

"Can I come to see you next weekend Valerie?"

"Yes, please; Saturday evening."

The bus conductor rang the bell for the bus to go. He wasn't going to wait. Valerie pushed Edward away with a laugh and he jumped aboard as the bus moved off. She waved her famous wave as he looked back from the platform, watching her solitary figure at the bus stop recede into the distance.

After this happy, happy encounter and in the astonishing knowledge of what they had discovered in their hearts, the future took on new meaning for Valerie and Edward. Yet it was only Valerie's perception and intuition that had kept alive an emotion that he had failed to recognise, all those months ago. The seed of love had lain dormant through the winter months, until that glorious Summer of 1955.

Part 11

A Parting and New Pastures

30

When Edward's father received notification of his posting to a new job in the South West of England, he didn't know for sure if it was to be permanent or temporary. To Edward, anyhow, the news was perplexing and unexpected. Certainly, one thing was very clear; the assignment was too far away to provide any opportunity for the family to be together at weekends.

Until then, Fred's work as an AID examiner at English Electric in Preston, allowed him to spend time with his family for two days at week-ends. Working at Preston was bad enough. He commuted; travelling by bus and train to and from Warton each day. He came home every day of course, but because of the distance and train schedules, it was eight o'clock in the evening before he got into the house. It seemed that the new posting would take even that small weekend pleasure away. For Edward, the idea of more disruption in the family's routines was not something to reflect on happily.

In the mean-time, Freddy wrote frequently from Germany, where he remained stationed with B.A.O.R. He kept the family well informed of his achievements and promotional advancement in the Army. By then he'd earned the stripes of Sergeant, but by the next time he took home-leave, perhaps it would not be to the familiar surroundings of the home he knew so well in Halewood. That would be an unhappy wrench for him too.

Young David moved on at Primary school with steady progress. However, a change of school, childhood friends and the only home environment he'd ever known would be hard for him also, especially at the age of ten.

Far from being selfish concerning his own misgivings, Edward realised that his parents were tied unequivocally to a decision for the move away from Liverpool. They were certainly not to blame. Probably they were far more worried about the family's future and collective happiness, rather than concerns over individual changes that the prospect incurred to their children's lives. There was little or no choice in the matter.

Edward continued with his apprenticeship, in somewhat inauspicious circumstances at the work-face, along with his studies at evening classes. His friendship with Valerie was a unique source of diversion and profound happiness for him at the weekends. The love they had discovered gave them hope for the future until his father's posting added another dimension to the family's fortunes.

Fred took up his new position in the southwest. He kept his family advised by letter, writing to them regularly from Somerset. Eventually, the inevitable news came, confirming the posting as permanent and that he had begun to search for a suitable house to buy in Yeovil. Clearly, the family's removal to that part of England was then a certainty. This wasn't an appealing prospect to Edward. At seventeen and a half, still a teenager, he felt that such a drastic change to his prevailing lifestyle was an unwelcome intervention. On the other hand, it did occur to him that maybe a move out of Lancashire might present better prospects for his career. Whatever the case, his future still seemed unclear, confusing and difficult to contemplate.

In his letters from Somerset, Fred described the beauty of the West Country in detail, extolling its rolling hills, green vales and rural countenance. The Weinels all knew that one day, the green fields behind their home at Halewood would be the venue for Liverpool's post-war, industrial expansion plans. Edward didn't really want to witness the urbanisation of the rural locality that had been their home for so long. Therefore, the move away from Lancashire, possibly, would not be without a good side, from that point of view.

More important and frustrating to him, however, was the effect of a move upon his friendship and emotional involvement with Valerie. How was he going to break the news to her? What would become of their relationship? Come what may, the imminent movement to Somerset was a disagreeable necessity that tied Edward to compliance – at least, until he reached the age of twenty-one. It occurred to him that he might seek to remain in the district; continuing his apprenticeship in spite of the unhappy diversion from aircraft to store-keeping work.

The thought was worth considering, but the expense, both to him and to his parents, seemed to put it out of the question. His meagre wages certainly wouldn't stretch to paying rent or even to living in digs with a family. He could hardly expect his Dad to sponsor that either; after all, a mortgage on a new home in Somerset must clearly burden family resources to the limit.

Moreover, the idea was a non-starter unless someone would be responsible for him. Edward couldn't ignore that he was still under-age at Seventeen: still a minor. The family had no relatives in the North and it seemed to Edward that only relatives could provide the necessary umbrella of accountability under which he might remain in Liverpool.

Until that time, Edward had not mentioned his association with Valerie to his family, foolish though his silence in that direction might seem. On that score, he wasn't ashamed of himself, nor of Val, most certainly, but they hadn't known each other for very long and it was just that he felt crowded. He was still trying to sort his feelings out. It had only been a matter of weeks since they began to share their few hours a week together, on Saturday afternoons or on Sundays. Private telephones were not commonplace in those days, and neither of their families had one anyway. They grasped at whatever time they could share in each other's company.

Neither of them had enough spare cash left over from the small wages they earned at their jobs, and so they spent their moments together, mostly walking or just sitting in her parents' garden, talking. They were never at a loss for conversation; discussing their hopes and dreams, daily occurrences in their lives and all those subjects upon which they shared common ground.

Valerie studied in the evenings, like Edward, but at home with a correspondence course. Their snatched time at weekends, therefore, gave them much happiness in each other's company. They were in love; but more, they had become good friends. Their teenage relationship was utterly innocent and wholesome; it was pristine.

Nevertheless there they were, two young people presented with uncertainty over their future. Edward had no idea how to find a solution over his family's impending move away from Liverpool. He wrestled with the dilemma of split loyalties and the fear of losing something, someone, he so wanted to hold on to. What was he to do? He resolved to explain everything to Valerie at the first opportunity.

When the weekend came, Edward spent Saturday morning at work as usual and he had to cover the Club Flying programme again, during the afternoon. However, this suited him very well. He planned to go over to Woolton that evening, to see Valerie and, importantly, to confront their situation in an attempt to find an acceptable solution to what lay ahead.

That evening, they sat together in the garden at the back of Val's house. Edward was silent and Valerie sensed his uneasiness. She questioned his solemnity. “Teddy, you're very quiet. Is something wrong?” she asked, touching his hand.

"Yes Val, there is something wrong, and I don't know what to do about it." Edward turned his eyes to meet her gaze, not knowing where to begin.

"Tell me Teddy. What's wrong?" Val prompted with a frown of concern.

"My Dad's been posted to Somerset, and it's confirmed as permanent. That means the family will move away, and me with it," he said. "I'm going away Valerie. I don't want to but I have to." His words seemed so brutal, so inadequate.

"Are you going for good? I mean, won't you be coming back?" Valerie's eyes widened and her face was expressionless. Edward could see she was close to tears. Her eyes brimmed. He felt helpless and, through his wretchedness, he tried to explain further.

"Valerie, I don't want to go, I want to stay, but it's so complicated, the expense, our age, we're still regarded as minors, not responsible adults. I don't have any relatives here who might be willing to help me with accommodation and things."

Valerie looked away, pushing her shoulders up, with her hands at her sides, palms pressing on the garden bench where they sat. She was very quiet. She stared into the distance for a long moment until, at length, she returned her eyes to Edward, again, with a thin smile. "If you need somewhere to stay, you could stay here, with us. I could ask my Dad. You could use my brother Ralph's room."

"The thought is fine," said Edward, shaking his head, "but someone has to act as sponsor, guardian or whatever and be responsible, accountable, not only to my parents, but to the authorities. It would be an imposition and there would have to be some legal stuff too, I suppose. If only I had relatives here." Edward took Valerie's hand and tried to smile, but the prospect of what was to be, was too painful to contemplate. Silence hung heavily upon them both, until Val's eyes lit up again. Edward could see that another thought had come to her.

"We could write to each other. Every day we could write to each other. That way, there would be a letter on the mat every day of the week! It would be like now, only we wouldn't see each other," said Val. "You could come here for holidays sometimes and I could go to Somerset sometimes too. Ralph comes home on leave from the Air Force all the time; you and I could take 'leaves' as well, like him." Her eyes were bright now and she giggled. "It won't be so hard, keeping in touch between holidays, will it?"

Her enthusiasm to meet the problem head on, was astonishing. Edward found himself grinning and nodding with his own renewed excitement. "Yes Val, it's not the end of the world is it?" He squeezed her hand in his and looked around him as if new vistas had opened up. "That's it Val, letters and holidays; hundreds of letters and lots of holidays together."

Valerie said, "Let's walk for a little while, Teddy and let's not talk about you going away until its time, OK?" She stood up before going on. "Perhaps we'll be able to find another way in the time that's left, before you have to go, but if you can't stay because of all the reasons you spoke of then it's settled. We'll write to each other and have holidays together, until we're old enough to make our own decisions about how we can be together, for always, OK?"

Edward marvelled at the objectivity of which Valerie was capable. In many ways she was so much stronger than him, and resignedly he answered with a smile. "OK, agreed."

They continued to meet at weekends, their only opportunities. Sometimes they went to the pictures or bussed into the city to see the sights, or they walked in the Woolton countryside; always talking, exploring each other's minds. Sometimes Val went to the airport to help Edward sell Pleasure-Flight tickets. He managed to get her a few free trips whenever there was a spare seat available. How she enjoyed flying. They would have tea on the terraces afterwards, when the flying finished for the day. The airport terrace was their 'special' place.

The summer's end approached and although it seemed as if they had known each other forever, in reality, their friendship amounted to little more than three months, not counting the dormant winter and spring that separated their first meeting from the revival of their acquaintance and the recognition of their love for each other. Valerie never questioned Edward's reluctance to take her to his home in Hunt's Cross, although she said to him once that if he did, maybe the two families could agree about his staying on in Liverpool, perhaps living at Woolton.

In retrospect, he supposed that is what he should have done. In other circumstances, Edward knew that his parents would have welcomed Valerie into the family circle, with love and generosity but with the worries, anxieties and preparations for the coming move, he didn't think it was appropriate then. To burden them with his need to resolve his own emotional situation, seemed unfair; he thought it would be too upsetting for everyone concerned. Moreover, he didn't think that his parents would take kindly to the idea of leaving him in Liverpool, especially with a company that had no real track record. There were no guarantees that *Dragon Airways*, as a new airline, would continue and prosper. The future was too unclear in that direction, too.

Edward's father wrote to reveal that he had found a suitable house in Yeovil and that the family should start final preparations for the move. He had arranged temporary accommodation for them in the town, at his own 'digs', though hopefully they wouldn't have to be in 'rooms' for too long. His idea was to show them the house that he wanted to buy, before committing to a mortgage.

In due course he came home on special leave, for the family's removal to Somerset. They set about packing up. The furniture and household goods went into storage, together with many of their personal belongings. After some discussion they decided that they would travel south by car. It worked out to be as cheap as by train, and avoided the several train-changes a rail journey would entail; with possibly two night-stops on the way. The car-hire company confidently advised that they could make the trip, very comfortably, in a day.

Then it was time to talk to Valerie. The moment Edward had been dreading had come. He went over to Woolton one evening with a heavy heart and a miserable feeling in his stomach. It surprised Valerie that Edward should visit on a midweek evening. Her eyes lit up and she laughed with un-hidden joy at the unexpected arrival, but then her face dropped as she realised why. The late Autumn sun lay low in the West and a cold breeze chilled the air. Valerie put on a coat and suggested they go into the garden at the back, where they could talk undisturbed. “Its time then Teddy,” she whispered, “are you going to tell me that its time?”

“Yes, Valerie,” said Edward, “its time. My Dad came home on special leave and we'll be moving within the week; Saturday, probably.”

“Oh,” Valerie smiled a thin, thin smile, “I've bought a whole lot of writing paper and ink, and stuff; you know, so that I can write to you every day, like we said.

How bravely she wrestled with her feelings.

“Me too, I bought a whole lot of stuff myself,” Edward rejoined, feeling close to tears himself, “if we write enough, maybe we could publish it all, one day.” He added, trying to joke about it. It was a bad joke. It was no joke at all. Valerie giggled and glanced at Edward with that impish look in her eye.

She stood up from the bench and held out her hands to him. He took them and she pulled him upright, to face her. “You know Teddy, we don't have to cry or be sad or upset. We don't have to wish for better things or to wish that we were dead. We're young and we're strong. We have our whole lives in front of us and if we can weather this storm, the world is ours.” Her small oration startled him, in the way that only she could do it. How Edward loved her.

Her words captured his heart and he realised, not for the first time, just what an extraordinary girl Valerie was. “How right you are,” he said, “these few weeks have changed my whole outlook on life. Not a moment can be forgotten, not a moment can be erased. These moments can never be repeated either. New times we may have but these belong to yesterday. They can never, ever, be repeated and we can look back at them and enjoy the memory of them, even when yesterday is gone."

Val laughed again and verging on the poetic she exclaimed. "Then they're ours and ours alone; no one else can share them. Whatever happens – this time, these moments, what we share – before, now and forever, they belong only to you, and to me." Then, she added in a low voice, "Its love you know, that's what it is!" They laughed at their wisdom. Two teenage people who shared an emotion called love. They had the world at their feet. A parting, however long or however short, could not impede them.

Edward made his sad farewells to Valerie's parents and the youngsters. Nobody was very happy. Then Val and he walked a little way towards Woolton. They halted beneath the broad branches of chestnut trees for a moment and she put her arms around Edward's neck, pressing her body close to his with a spontaneity that demonstrated her moment of weakness. Edward held her in his arms tightly. They kissed and he felt her tears against his cheek. His own mingled with hers, as they kissed again. The wretchedness of that parting was unbearable but through the painful instant, emerged the certainty of the love that Valerie had for Edward, and his for her. She leaned back, smiling, still with her outstretched arms about his neck. She whispered, "It's love you know. Teddy, please, always remember that I love you. Please don't forget me. You won't forget me, will you?"

"Love, that's what it is Valerie and I can't ever forget you; I love you so much," echoed Edward. They were able to laugh through the tears, and they did.

Edward didn't want Val to come all the way to the bus stop with him. He couldn't bear the thought of speeding away, leaving her to stand and watch him go. Nor could he bear the thought of watching her lonely figure at the bus stop, diminish into the distance as the bus carried him off, unwillingly, to his new life and destiny, far away from Liverpool. "Valerie," he said, "let's part here, under the trees. It's like a halfway mark. We each turn and walk away and we don't look back. That way no one gets left behind. Does that make sense?"

"That's a good idea Teddy. I love you dearly for such a tender thought." Valerie agreed. They kissed again and whispered good-bye. They turned and walked, each in their own direction, their footfalls muffled by the wind-blown leaves of gold and brown. Not turning to look, not turning for that last glimpse. They both walked until they each turned corners that hid them from the sight of their witnesses, the trees.

31

Edward served out the remainder of the week's notice to his employers and it was a sad day for him when he made his farewells to everyone at the Club and *Dragon Airways*. He really had enjoyed the two years with them and although winds of change were indeed blowing, not only within the company, but in Liverpool at large, he was extremely sad. Valerie was in his mind all the time and he felt as if he were deserting her.

The journey to the South West was uneventful until the Weinel family got to Tewkesbury. The early Autumn weather had been closing in and finally they ran into thick fog. The driver tried hard, but in the end he said that he could go no farther. An unscheduled night stop was inevitable. He had prepared for this eventuality, and soon found them overnight accommodation. He also arranged another car, for the onward journey, the following day. This turned out to be very agreeable. Had they gone on, arrival in Yeovil would have been in the early hours of the morning. As it was, they arrived nicely in time for lunch.

One hitch was Edith's cat. They had transported the tabby along with the family in the car. She slept most of the way to Tewkesbury, curled up on the shelf behind the back seat of the car. The landlady at the over-night accommodation was very kind, and allowed them to bed the cat down in an outhouse – compensation for the fact that she couldn't allow animals in the rooms. Unfortunately the cat had found an unlatched window ajar, and had taken advantage of the opportunity for a night on the tiles. Edith was frantic and really thought that would be the last she ever saw of the crafty puss. Happily, in the morning, Edith discovered that the cat had returned. There she was, curled up contentedly, asleep in the shed.

Mrs. Lemon was Fred's landlady. She had rearranged rooms

in her house for his family to use while they settled into the Yeovil scenery. Edith hated the idea of living in 'rooms' but she remained very tolerant and stoically put up with the inconveniences. There was much to do. Fred had to get back to work.

Edward had interviews to attend, for a continuation of his apprenticeship, at the helicopter factory where his father worked and David was about to start in his new school, settling back into his primary education. Edith had to find her way around the town for shopping, and organise the thousand-and-one things that mothers need to do, in support of moving a home and family. Edward sat down to write to Valerie on the first night in Yeovil. There was so much to say ...things left unsaid at their parting and detail about the journey. The promise to write every day had begun.

The following Saturday, after arrival in the West country, Fred took the family up to the northern end of Yeovil, where he'd arranged for them to look over the house he had in mind for their new home. Several new housing estates were in construction then, and his choice was a small semi-detached close to the highest point in the district. The house was one of three at the end of a row of bungalows.

There was a good primary school at the back. David would therefore have only a three-minute walk around the corner. Coupled with this, since the school's playing fields were immediately behind the garden, there was little chance of fresh construction on that side of the house. The view across Yeovil, through a line of trees, would remain unblemished. Not far away, there was a public park with good playing fields for children and the local cricket and football clubs. A very convenient local bus service, already operating on this newest area of housing in Yeovil, passed the house every half hour, with its out-of-town terminus just a few steps up the road.

The house was not as big as the old place in Hunts Cross, but small though it was, it was ample enough. It stood back from the road, having a front garden with a low brick wall separating the plot from the footpath. Sufficient distance between the houses allowed space for the building of separate garages, and a good area at the back provided for a garden of manageable proportions, all bounded by wire-and-hickory-stave fencing. After viewing the place and discussing the finer points, the decision to buy came quickly, leaving Fred to set about the arrangements for a mortgage with the Bank. They would move in after Christmas.

Meanwhile, *Westlands* agreed to take Edward on as an apprentice and arranged for him start work, on the following Monday, in their satellite-works at the Royal Naval Air Station, Yeovilton. Much to Edward's delight they told him that he'd be working with aeroplanes: not least, jets. His father's office was also at Yeovilton, so he and Edward would travel together each morning and evening, by bus. This was another unexpected bonus.

Yeovilton lies beyond Ilchester, some five or six miles out of Yeovil. In those times, the bus normally took about half an hour to make the journey, and since the only service that could get Yeovil dwellers to work on time, passed around 06:45 a.m., an early rise was necessary.

By now it was autumn and on his first day as an apprentice with *Westlands*, Edward and his father walked together to the bus stop. It was only a short walk, around the corner from Mrs. Lemon's house but it was a poignant experience. It reminded Edward of his childhood days and, again, the Saturday morning walks that they used to take into Hunt's Cross. Edward had grown up a bit since those days, of course, and at nearly six-feet, he was indeed taller than his stockily-built father. The morning sky was black at that hour, street lights were still on and there were precious few people about. When they arrived, several men and women already stood waiting for the bus at the kerbside, in front of the Pickett Witch pub. Some had broad, Somerset accents that Edward found hard to follow, as they greeted their fellow travellers. The new experience dramatically brought home the radical changes to Edward's new lifestyle in the West country.

They boarded the bus and joined the smokers upstairs. Edward could see nothing of the countryside passing in the darkness beyond the glass. The steamed-up windows mirrored only the vague images of the passengers, illuminated by the bus's dim interior lights. Much of the journey was downhill, of that he was sure. The bus stopped several times on the way to Yeovilton to pick up more people. No one alighted. Eventually, they swung into RNAS Yeovilton and parked next to the guardroom. There were strong, bright lights all about, shining down from tall posts around the perimeter of a broad, tarmac area.

To Edward, it looked like a parade ground, and indeed, he was right. Also, he learned later, a small area, roped off with thick, white ship's cable, in front of the guardhouse, was the called the Quarter Deck, in true naval tradition. Edward and his father alighted, trudging off with the other passengers, through the parade ground gate, and out on to the approach-road. It was a lane really; quite narrow, with high hedgerows on either flank. Half the people from the bus were stepping that way too, so Edward assumed that they too, were going to *Westlands* satellite works.

The black sky was by then beginning to give way to pale shafts of sunlight in the East but there were heavy clouds almost to the horizon. The rays of watery sunlight would not persist for long. The trudging, straggling, group from the bus, followed the road through a bend and beyond. Edward could see the lights of their destination not far away. It began to rain – a light drizzle and the people turned up their coat-collars. Vehicles passed on the road and voices called out to some of the walkers, trudging, bent-headed, up the lane in a strung out line. "Marn'n Billy, marn'n Gearge."

"Ow bist boy?" Another strange new phrase to Edward's ears.

They reached the main door of the works and entered the building through a lobby, where a uniformed security man sat behind a counter, eyeing everyone who passed with unconcealed suspicion. Edward and his father bypassed the crowding, queuing workers, lining up to clock-in, and Fred took his son to a row of offices at the head of a long, brightly lit hangar. As far as the eye could see there were aeroplanes: jet aeroplanes: hundreds of them. Edward stood in awe for a moment until, at his Dad's nudge, he turned to face a tall, round-shouldered man, in shirt-sleeves. They shook hands as Fred introduced his son to the Production Foreman responsible for the shop-floor work in this enormous place. Fred withdrew as the foreman gave Edward a toothy grin and took him into one of the small offices.

"When everyone has finished clocking in," said the foreman, offering his new recruit a chair in the cramped space between two desks piled with paperwork, "I'll get your clock cards organised and set you up with someone to show you around." There seemed to be hundreds of workers and Edward began to get the jitters. This was his first encounter with such an enormous facility. "When you get back from looking around, I'll put you with someone, a skilled man, to work with," continued the foreman, "for the moment, just have a seat and hang on here, I'll be back in a minute or two."

With that, he hurried out of the office, to remonstrate with a couple of idlers he'd been watching, as they loafed and chatted in the aisle between the aeroplanes. Edward stared in disbelief at the long, long line of aircraft. Stripped of paintwork, cleaned to bare metal, the planes stood formally in two herringbone lines, separated by a walkway between two white lines; front fuselages to the left and rear fuselages to the right. Except for the first couple of aircraft in the line, standing upon their own undercarriage wheels, immediately in front of the office window, all the front fuselages rested on yellow-painted, wheeled trestles. The rear fuselages were the same, standing on similar trestles in their long line, with bits and pieces, pipes, tubes and wiring harnesses hanging out of the empty shells. Dive-brake panels hung obliquely on their hinges from the sides of the structures, with little red flags, limp, but visible, clipped to the sharp corners.

Hundreds of bright lights hung from the roof of the hangar, brightly illuminating the interior of the hangar. Workers switched on lead-lamps and small spot lamps, to give additional light beneath and inside the fuselage shells, where they had begun their day's work. Edward could hear the whine of 'windy' drills, driven by compressed air taken from distribution pipes around the hangar. Rivet guns rattled and tapped, metal on metal. A hydraulic-rig came to life close by – groaning and squeaking as it gave power to the systems of the lead-aircraft.

The *F86 Sabre* fighters, there at Ilchester, were undergoing a refurbishment and modification program for NATO. Edward was already aware of that. He could see perhaps a hundred in the long hangar and as the workplace woke up, swinging into action, he began to recognise the elements of a production line. This was something he'd never seen before. The immensity scared him; it was a far cry from the flying club in Liverpool.

The foreman returned from his mission, accompanied by a young fellow sporting a James Dean haircut. The lad wasn't much older than Edward and he was introduced as Dave, a third-year apprentice who had been at the Ilchester satellite facility for about eight months. They referred to the place as Ilchester, rather than Yeovilton. Edward never was quite sure why. The foreman said, "Dave will show you around and where everything is, then he'll bring you back here. I should have someone ready for you to work with, by then, OK?"

Edward nodded, and followed Dave out into the enormity of the hangar, with its noisy and busy workforce. Nobody took much notice of them as they passed; the work went on, and so did they. Dave showed Edward where the toilet and washrooms were, before they made the long walk through the length of the hangar, to its farthest end. Edward had a million questions and Dave tried to answer them without going into too much detail. Then, returning to the Time Office, where the workers clocked on and off their jobs, Dave began to describe the work system. "We work piece-work here," he explained, "each job is timed, so you clock on as you start and clock off again when you've finished."

He picked up a typical job card and its attachments, to demonstrate. The attachments he referred to were plastic bags containing spare parts needed to complete the task.

"Usually, if the spare parts are small, like nuts and bolts or suchlike, the job card comes with these attached," Edward's guide continued, "but if it's a big part or a component, time is allowed on the job card for the operative to draw the item from the Embodiment Loan store." He indicated a storeroom hatch, not far from the foreman's office. "That's the place, next door there, for smaller stuff, or at the other end of the hangar for the very big stuff. You have to take the old bit with you and sort of exchange it for the new bit, as called up on the card."

"Sounds a bit complicated to me." Edward ventured.

Dave grinned, "Oh, it grows on you after a while, it's just a way of accounting for everything. It's a Government contract you see. Everything has to be accounted for unless it's an expendable item."

Next, he took Edward to the Rate-Fixers office, where a couple of harassed looking fellows sat, busily writing and calculating. A girl at the back of the office looked up briefly from her work; typing swiftly on multi-page job cards. "In here," continued Dave, "are the rate-fixers. Their job is to calculate a time for the specific job, based on the Planners needs after Inspection have surveyed the aircraft."

By now, Edward's mind reeled. There's so much to remember, not like before ...so much to learn: and much, much more. He verged on the lyrical in his silent thoughts. Dave brought him back to earth again. "You don't have to commit all this to memory you know, because you'll be with a skilled chap. He'll deal with this stuff, especially if he has to negotiate with the rate-fixers for a better time on the job, but it's handy to know about."

"Thanks, Dave," bantered Edward, "very much." They laughed and moved on to the next area that Dave thought his charge should know about.

Eventually, they ended up at the Production Office again, where they found the foreman talking to a dark-haired, youngish man wearing a blue coverall boiler-suit and a dark-green corduroy cap.

"Ah, you're back," said the foreman, "this is Brian, you'll be working with him from today." Brian and Edward shook hands.

"Brian, this is Mr. Weinel's son ...mm, what's your first name son?"

"Ted." Edward prompted.

"Yes, Ted, your father told us you'd be starting today." The foreman reached for an insistent telephone, concluding the conversation and dismissing the two newly acquainted fellows.

Brian and his new apprentice moved off, out into the hangar again. Edward glanced at Dave and thanked him for his time and the guided tour. Dave grinned and pushed off into the depths of the production line. "Come on then, Ted," said Brian, "follow me, the job's down here, let's get on with it," and he strode off down the line to where he said 'the job' was.

Brian's sharp-featured face was more round than long, and he wore his black hair slicked back and parted. His blue eyes showed humour at the corners, enhancing a mischievous countenance.

As they walked down the line, Brian halted now and then to greet a fellow worker. "Ow bist boy?" That strange phrase rang out again, "Ow be get'n-naan then?"

"Dang'd if oi do know son, oi reckon oi had a few too many o'th scrumpy laas noight, boy,"

The Somerset accent intrigued Edward but on that first day, immersed in it, he hadn't a clue what the men were saying in those brief conversations. It was like a foreign language to him. The job turned out to be inside a rear fuselage, resting in its wheeled trestle, with straps over the top to secure it into the padded, contours. Brian took out his wristwatch; he always kept it in his pocket when working, and consulting the timepiece he made a mental calculation. “Right,” he said, “we've got two hours to fit half-a-dozen thumb-clips on these pipes, and torque 'em up. That means we'll have to do it in one hour, if we're to make anything on it.” What on earth did he mean, thought Edward. He still had his coat on and here was Brian talking about the job already.

Brian, as if reading Edward's thoughts, told him to take off his coat, and hang it somewhere. Then, he climbed into the interior of the structure to weigh up 'the job.' He asked for a lead-lamp, telling his new assistant that he'd find one on the bench next to the tool box. Edward found it and handed it up to Brian, coiled lead and all. “Plug it in then!” Brian urged testily, and Edward looked around for an appropriate socket, feeling a little foolish. “Up there, Ted,” growled Brian, with impatience. Edward looked up and found the hanging lead, complete with adapter socket, and plugged in his lamp. He didn't like Brian's attitude much; he seemed a bit too brusque and intolerant. “Right, get yourself up here then, and I'll show you what to do,” commanded Brian.

Inside the rear fuselage Brian's lamp brightly illuminated the structure. Yellow paintwork was everywhere, except for the flexible pipes and wiring looms. There were many ‘hard’ pipes clamped and clipped to the structure. Each bore a coloured tag wrapped around the circumference. These were not only coloured, but identified by little graphic markings and a printed word or two. Brian explained to Edward, noting his questioning look, that they identified specific systems such as Main Hydraulic, Alternate Hydraulic, Emergency Hydraulic or Fuel, or De-icing, etc. easing maintenance routines and ensuring that fuel systems didn't get connected to hydraulic systems by mistake, for example.

Constricted space inside the structure made working uncomfortable. They crouched on wooden duckboards, resting on the fuselage frames. Brian opened a bag of spare parts and counted the contents. The clock-card told them that they were to replace a number of pipe-clips, thumb-clips, as Brian called them – those marked with blue paint by the surveyor, during his inspection of the structure. After replacing these and tightening them, the time-card instructions called for the torque-loading of each clip, using a small, preset torque-spanner, with a slot instead of a hexagonal socket. The wrench came with the packet of bits. Edward's part in all this needed a tiny jar of red paint and a small brush with which to mark each clip's thumb-screw, after torque-loading.

How interesting, he thought, sarcasm getting the better of him, now I really am an engineer, painting red lines on pipe-clips. Within the hour, they'd completed the task. Brian went back to the office to clock-off the job and clock-on to a new one. This time the job was on a front-fuselage, up the line a way. They gathered up all the bits and pieces and moved Brian's wheeled bench to the assigned aircraft. Between them, they accomplished a number of relatively simple, similar jobs, over the next couple of hours, and each time Edward was the one to paint the red marks on the torque-loaded components that they fitted.

Eventually, Brian took out his watch again and announced that it was time for tea break. The rest of the factory clearly had the same idea as the noise of work diminished and a loud hooter announced the temporary cessation of work. Brian found two cups on the shelf under his bench and told Edward to wait for the tea cart to pass. Edward looked down the aisle between the aircraft and saw the trolley, pushed by a woman who filled cups from a large tea-urn, whilst making jolly remarks to her customers as she went. When Edward popped over to fill the two cups, he noticed that the tea-lady's trolley also had things to eat, set out on trays, next to the urn. He realised that he was hungry.

The food tray held large slices of rich, brown, bread-pudding on crusts of pastry. At sixpence each, he thought a slice might fill the gap until lunch time, so he bought one. The pudding was fruity and it was delicious. This became his standard tea-break snack each day, both morning and afternoon. The tea, however, was shocking. The tea-lady mentioned to Edward that, at lunchtime, she would have Cornish pasties on sale, at a mere eight-pence. He ordered two, and this became his standard lunch-snack for some time to come. Thereafter, Edward was always ready for a hot meal when he got home each evening.

Brian moved across the aisle to chat with a couple of friends during tea-break, leaving Edward to sit, sipping at the hot, tasteless tea and analysing the first couple of hours on the new job. It seemed to him that the work followed an inconsistent pattern. The jobs they carried out were unrelated. That worried him, but since it was his first day, he thought it best to keep his silence. He said nothing. Brian came back and they talked awhile, about Edward's background, and about Brian's. He too, was new at Ilchester. He'd recently come out of the Air Force and married a Yeovil girl. His aim, he told Edward, was to get into the engine side of things, at Ilchester, because he was engine-trained. He'd spent some time in the mob, as he put it, mostly on the RAF engine maintenance unit at St.Athan.

The call of the hooter sent them back to work again.

Lunchtime came and went and the afternoon slipped past quickly as they scurried from job to job, aircraft to aircraft and production-office to clock-office. When five-o'clock came, and Brian had consulted his watch for the umpteenth time, Edward realised that for the first time in his working life, he'd spent all day long, cooped up, inside. He'd seen nothing of the outside world: no sunshine nor rain: no blue sky nor grey. He didn't even know what the weather was doing out there.

At the blaring of the hooter, sounding long and loud throughout the hangar, all work-noise ceased. The workers made their way towards the time-office for the daily ritual of clocking-out. Wearing overcoats and raincoats, hats, caps, gloves, crash helmets and bicycle clips, they all stood in line, shuffling towards the clock. What a sight to behold. What a ghastly sight: Thermos flasks, satchels, ex-army bags slung over shoulders: plastic boxes, empty now, to be refilled with egg and tomato sandwiches tomorrow. Edward shuddered at the automaton march.

He joined the queue with Brian and shuffled along with the wretched, until his turn came to face the clock, and its two slotted boards, holding the cards. He felt intimidated. He found his card near the bottom of the first board. Copying Brian, he picked it out and dropped it into the clock's waiting slot. He pressed the handle, like the others had done, punching the time onto the face of the card to the sound of a bell: a death-knell, perhaps. Punching a clock-card was something Edward had never had to do until then, and he always, thereafter, despised having to do it, but it was a revelation that he would have to tolerate for years to come.

Outside, it was as if there had been no day. There was only darkness and rain, bucketing down. Edward found his Dad, smiling genially as he waited in the lobby and they took their leave together, heads bowed into the driving rain. They found the bus standing near the Quarter Deck, under the lights and in the rain. Its windows were opaque, steamed-up inside with the breath of tired workers, smiling and laughing now: homeward-bound. Edward felt depressed. His father detected the mood and they sat silently for some moments while the bus filled up before drawing away. Somehow, Edward felt cheated. He told his father that it really hadn't been a good day; rather, it hadn't been what he'd expected. The bus swung out onto the main road and as it ground along at snail's pace through the pelting rain, Fred consoled his son, suggesting that it was early-days yet, and not to get too upset with the newness of things or the discomforts of factory work. Of course, he was right.

That night Edward sat as usual to write his daily letter to Valerie. He felt deflated and although his day had been full, he had thought of her often. His letters were always long and he found that he could write about almost anything; not that his written English was good, nor the spelling, but he seemed able to find plenty of words to describe his new life.

Under indentures, Edward's apprenticeship charged him to attend night-school classes every evening, and one whole day each week (on day-release) at the local Arts and Technical College. The new day-school term was about to begin, obliging him to present himself at the College on a Wednesday morning, to sign on and to begin the course of studies. The following Monday, night-school term started, scheduling him for five nights a week.

At interview, the apprentice supervisor had explained to Edward that the only tech school possibilities open, would be the City & Guilds course in Workshop and Machine shop Engineering, in spite of previous studies in Liverpool related to aircraft maintenance. This bound Edward to start at the beginning, in company of class companions who had just begun their apprenticeships at the age fifteen to sixteen – more than a whole year or two younger, than himself. This was something else Edward didn't much like, but there was no other choice. His career, or rather his technical schooling, took on a different aspect that wasn't in line with his ambitions in aircraft maintenance. Would it eventually steer him away from direct contact with aeroplanes?

The unique consolation in this new life, was Edward's initial induction into the Ilchester workplace. There, at least, his hands were still on aircraft. What was much more important to him though, was the retention of contact with the life he'd left behind, through Valerie's letters. That was his salvation in the midst of utter confusion.

Part 12

If a Man Should Wish to Die

32

That first Christmas for the Weinels in their new hometown passed quietly. The winter moved on. And as the sun rose earlier each day with the approach of Spring, Edward's daily trek to Ilchester became more interesting. At least, he could see more of the countryside through which the bus carried him each morning and evening. Once out of Yeovil, the route to his place of work descended between the embankments and high hedgerows of a winding, downhill gradient towards the vales and flat, green fields of Ilchester and surrounding districts. Straightening out with perfect precision at Chilthorne Domer, the highway followed what was once part of a Roman tributary to the famous Fosse Way, crossing a low-lying flood plain, a marshy region flanked by the rolling hills of South Somerset and Dorset. The river Yeo and the brooks of Bearley and Hornsey that drained the rich pastures and paddocks, meandered close to the tiny town of Ilchester, through the open flatlands, near the county line boundary.

When Edward's father had first taken-up the new posting in Somerset, his letters had often described those settings more eloquently than Edward could ever hope to emulate. But, like his father, he tried to relate the beauty of that south-western county in his own letters to Valerie. He guessed he was lucky to have new and interesting things to write about. Valerie's letters began to reach him only three days after he'd arrived in Yeovil. They were full of interesting news and bubbling gossip – so well written that Edward blushed a little at his own crude attempts at literary continuity. Valerie had done well at school, gaining several School Certificate subjects, with Distinctions in English and Drama. Why she hadn't tried for a place at university remained a mystery. But perhaps, in spite of that, she would still realise her aspirations towards a career in journalism, one day.

How he missed Valerie. He missed her frank and open manner, those startling remarks, her hand in his as they walked, the sweet fragrance of her presence beside him and her joyful laughter. Such things couldn't ever find translation into written words upon a page. He prayed that their promised exchange of letters would be sufficient to keep the flame of their love alive and strong.

The days passed quickly at his work, in spite of the irritating discontinuity between tasks. Brian and Edward darted from aircraft to aircraft in compliance with the job-cards handed out by the Production foreman. Edward soon came to the conclusion that piece-work did not inspire him. Brian's only interest was in making good times against the clock, to make better money at the end of the week. It was a common central topic in his conversation and he frequently talked about getting on to something more positive and interesting; something that they could get their teeth into, instead of the constant flow of seemingly unrelated jobs of short duration that required little in the way of real skill.

One morning, a message came for Brian, asking him to present himself at the Production office. He returned with a broad smile on his face; he fairly beamed with undiluted delight. It seemed that he and Edward finally had been given a new task, one that Brian had been angling for since joining the company at Ilchester. The time had come to form a Flight Line.

As the first of the production-line aircraft came to the point of completion, it was to become Brian's responsibility to fit engines, test and adjust them, bringing each aircraft to a state of flight serviceability and readiness for dispatch. Edward would continue to work with him. This was more like it, the moment had arrived. Brian took out his watch as he told Edward that he'd clocked them onto a special card, providing time to set up the new area of responsibilities. There was much to organise: with tooling, equipment and facilities to arrange.

They went outside into the fresh air. It was morning time, the sun was well up and birds were singing in nearby trees as the two surveyed their new working environment. They stood for a moment to take stock of their surroundings. The hangar, which had entombed them over passing weeks, was long and narrow. Its huge, sliding doors on the airfield side formed the closure of its length, parallel with, and facing, the approach and threshold of the airfield's main runway, some five-hundred yards distant. Between the hangar and the airfield's perimeter fence, a wide expanse of concrete hard-standing extended for the whole length of the building. Beyond, at the farthest end, lay another smaller hangar, used as a paint shop.

In the other direction, towards the approach road and the works entrance, lay the main gate, cycle-sheds and car-parking areas, alongside which, a low, black-painted building stood. This contained the very secretive avionics centre and the Management's canteen, with kitchen facilities. Halfway along the concrete hard-standing, on the airfield side, lay a small 'blister' hangar, its doors facing the main factory building. They were to use this as their engine installation shed.

There were many more *Sabres*, parked on a grassy area, between the shed and the entrance gate. Most were still in RAF colours and markings. With engines, armaments and avionics equipment already removed, the planes waited to pass through the paint shop for stripping, although some of this work had already begun outside, where the aircraft stood.

Brian and Edward walked to the farthest end of the hard-standing, to a point opposite the doors of the paint shop hangar. There, they found the long disused, engine-ground-running base, with its special tie-down points and a few pieces of ground-equipment for use during the test program, such things as air-intake cages, fire extinguishers, a *Houchin* ground-power unit, an old *Fordson* tractor and an equally antiquated, wartime fuel-bowser.

The latter machine stood as if abandoned. Its steel cupboard-doors swung on rusty hinges to reveal the big, pot-bellied, single-cylinder, *Petter* diesel engine that powered the delivery pump. Their first job was to get this going and then to tidy up the area in preparation for the first engine ground runs, scheduled for the following week.

The bowser proved to be a good solid work-horse of a machine, once the diesel engine had responded to Brian's first attempt to start it. After filling the fuel tank with diesel-fuel, and the pot-bellied jacket around the cylinder with water, Brian inserted the crank handle into the end of the shaft at the centre of the heavy, spoked, outboard flywheel. Edward held down the decompression lever, whilst Brian turned up a few revolutions on the shaft. Once he got the revs up, Brian shouted for Edward to release the lever. The inertia of the double flywheels took the piston through several cycles, until, with a wheeze and a deep-throated cough, the engine struggled into life. Beating slowly and relentlessly, with puffs of black smoke emitting from the exhaust pipe, the rotational velocity increased, until the engine reached its governed speed of perhaps a hundred and fifty revolutions per minute.

As the *Petter* settled into its rhythmic beat, Brian suggested that they kept it running for a while to warm up, before testing the pumping gear. They left the bowser pounding away, gently bouncing on its thick balloon tyres, in time to the heavy beat of the engine. "We'll have to check-out the tractor and the *Houchin* ground-power unit too," said Brian, "although the tractor should be OK, because its often used to tow in the newly arriving aircraft from the Navy side."

The tractor was an elderly, pre-war *Fordson*, the type with two fuel tanks forming the 'bonnet' above the engine; one containing petrol for starting and the other containing paraffin for running once the engine warmed up. Brian stepped up on to the driving-platform and stood astride the metal seat with its cantilevered leaf-spring. He selected petrol for starting, pulled the ring-and-ratchet throttle-rod out a little and waggled the long gear-lever to find neutral. Stepping down and moving around to the engine, he 'tickled' the ancient carburettor until it flooded and then moved quickly to the front to pull the cranking handle through several compressions. The tractor's engine started easily. Brian returned to the platform with a grin and a wink at Edward. Selecting first gear with noisy grinding sounds as the old gear-teeth meshed, Brian took the rusting, yellow *Fordson* for a spin on the hard standing. He seemed satisfied and brought it quickly back to the ground running area. Brian was in his element; totally enchanted with his new responsibilities.

The ground-power-unit was something else, however. Brian explained that this machine had a *Ford-Vee-Eight*, side-valve engine to drive a generator that would provide external electrical power to the aircraft. Although they were to use the machine mainly for engine starting, it would also provide power for the aircraft systems, when the aircraft's engine wasn't running. Edward remembered the trolley-acc (accumulators or batteries mounted on a two-wheeled truck) he'd helped to build for the *Heron*s, at *Dragon Airways*. That device had performed a similar function but the power came from heavy-duty batteries that needed recharging regularly.

Brian further explained that this *Houchin* model was also quite old: probably one of the first that *Houchin* ever put on the market for the aircraft industry. They checked all the vitals – petrol, oil, radiator water and battery levels, etc., before preparing for a start. A hinged panel on one side of the blue painted metal, box-shaped body-work of the trolley, revealed the operating controls. Brian turned the ignition key and pressed the starter button, but there wasn't enough urge in the battery to turn the *Vee-Eight* over. "Get a grip on the crank handle, Ted." Brian commanded.

The handle rested in a couple of clips, under the control panel. Edward took it out and inserted it in the hole at the engine end of the machine, beneath the radiator grill. "Right, Ted, when I press the starter again, give it a good old crank on the handle. Don't forget to keep your thumb on the same side as your fingers though. If she backfires, she'll break your arm."

"Right," said Edward, thinking, now he tells me.

After some energetic cranking, the engine came to life, but

like all old *Vee-Eight*'s, it ran erratically until the temperature came up. Meanwhile the old bowser, throbbed away happily, nearby. By then, however, steam had begun to drift from the cylinder's water jacket. "That's normal," Brian commented, without too much concern, "you just have to keep an eye on it and top it up occasionally."

They stopped the diesel after a while anyway, and using the tractor they moved things around to restore the area to some sort of orderliness. Later, Brian requisitioned a couple of yard-brushes and between them, they swept the whole patch until it was clean to Brian's satisfaction. Quite like old times it was, marching up and down on the end of a brush, until the hooter sounded to announce the end of another day. It was Friday and another weekend to look forward to.

What an interesting day it had been, too ...a great change from the artificial lighting of the hangar. At least, outside, Brian and Edward had the fresh air, and sunshine when it came, not to mention the additional interest of seeing the Naval aircraft movements, as they practiced circuits and landings on the main runway, close by.

Beyond the lane at the works entrance, many of the Navy's green-painted hangars, the control tower and the distant hills were visible, complementing the tiny, rural village of Yeovilton; its square-towered church was set amongst trees, on the other side of the airfield's main runway approach. Edward concluded (as he'd often done before) that working outside was altogether much better than being cooped-up, working inside. With this thought in mind, Edward boarded the bus for Yeovil, along with his father, to chat excitedly about the day's challenging new enterprise.

It was normal for the Weinel family to make a Saturday trip into the town for weekend shopping. However, that weekend, they expected to be moving out of their digs, and into their new house. The removal van from Liverpool was due around midday, with all the stuff that had been in storage over the intervening weeks. A taxi delivered the family and all their suitcases to the new house and Edith set about detailing jobs for all to help, getting the place ready for the carpets and furnishings when they arrived. The house was clean and in good order. It didn't take long to make the preparations. By eleven o'clock they were able to sit down for a cup of tea from the thermos and wait for the arrival of the van. They pottered about in the garden area and discussed the best approach to cleaning it up, and where best to put the lawn and flower beds. The construction company had used it as a builder's yard and they'd left it full of half-buried, broken bricks, roof tiles, weathered platform-timber and a huge lump of concrete where the cement mixer had stood for many months, whilst the houses were under construction.

Fred wasn't a great lover of gardening but he and Edward put some ideas together for later, after settling in. It was all very exciting. Until now, they had always lived in a rented house but this was their new home, their own house.

One of the most important items to come with the family's household stuff was Edward's bike, of course. Old and well used though it was, his trusty two-wheeled steed was a welcome sight. The open road would be his again, after so many weeks of reliance upon the buses. He wrote at length to Valerie that night, to give her the new address in Yeovil. He included details of the interesting things about the move and of the pleasure at rediscovering personal things, long in storage. The bike was a major feature. They had often laughed and joked together many times about it. Val rode it a few times and fell off once; so, it too, had become part of what they shared.

He began to use the bike for the journey to and from work. The distance to Ilchester and Yeovilton wasn't too far, perhaps six miles, and all through open countryside. Going to work was fine because most of the way was downhill ...no chance of being late. Coming back, naturally enough, the journey took much longer and it was far more strenuous, but it didn't take long for Edward to run-in his dormant cycling-muscles and his respiration again.

However, after a few days, Brian suggested that he could pick Edward up each morning, and bring him back home again in the evenings. He'd just finished restoring an old, *BSA*, three-wheeler car; not unlike a *J.A.P* powered *Aero-Morgan*. One didn't need an alarm clock in the mornings, for Brian's car was noisy enough to wake the dead. Edward could hear its approach to the house soon after Brian left his, a mile away at the foot of the hill, near Pen Mill. The car's big, *BSA* 500cc *Vee-Twin* had a good, throaty, if staccato, roar and not a few rattles, detonations and reports to accompany the climb to the top of Yeovil's highest point.

Brain loved his three-wheeler, with its flapping, canvas enclosure. He drove it with gusto, reaching between spokes of the enormous steering-wheel to shift through the three gears. This was the only way by which he could manage the gear lever, in the confines of the tiny cockpit, for it was a narrow vehicle, having only two seats mounted forward of the single back wheel. Brian joked about one having to be careful when turning left at speed, because of the torque. With no passenger to add weight to that side, the car would roll and tend to lift a front wheel on sharp-left hand turns.

The new job at Ilchester began to develop. Brian and Edward had, by now, installed the first engine into the leading aircraft's front fuselage. They completed the connection and functional testing of all systems, and they'd taken the plane outside, on to the ground running base. They tied it down using cables attached to fittings at the main landing gear, with the other ends secured to the strong-point ring set in the concrete base. They set in place a safety cage for the air intake, at the nose of the aircraft, and positioned the *Houchin* ground power unit behind the left hand wing, ready to be plugged in, when needed. They inserted the non-slip, wheel chocks and, apart from putting some fuel into the aircraft's tanks, the aircraft was almost ready for the first of its engine ground-runs.

Westlands were unable to conduct engine overhaul and refurbishment on-site since they had no aero-engine capability or the facilities for such work. The power-plants therefore went to the engine division of *Bristols*, who provided a technical man based at Ilchester, to oversee reinstallation and ground testing.

Mr. Cook, the representative on detachment from *Bristol*'s, came out to inspect the facilities. He seemed suitably impressed with the ground running area and made only one or two minor suggestions to bring it into line with his needs. These didn't take long to accomplish, and excitement mounted as the first ground runs drew closer to being reality. Brian was ecstatic, and as for Edward, at last he was back with aeroplanes again; with more experience to gain and more knowledge to tuck away for the future.

It began to rain. It was only light to start with but became heavier as tea-break time approached. Brian and Edward hurriedly covered the tied-down aircraft with a large tarpaulin to keep the rain off the exposed engine and the un-canopied cockpit. It was impossible to consider engine-running in such weather. Brian suggested that they retired to the works entrance lobby, where at least it was dry and they could grab a cup of tea from the tea-lady as she passed through on her way to the production line.

The lobby was full of men in oilskins: handlers, expecting the arrival of more aircraft that day. They loafed about waiting for the weather to clear up, in spite of that doubtful possibility. At least, the lobby was warm, but there wasn't much space inside, crowded as it was, and noisy with the sound of joking, bantering voices. The room was long but it was narrow, half the width being taken up by the security man's counter.

Movement on the edge of the gathering turned all eyes from the windows to the hangar door as it opened and a girl squeezed through. The men stood aside to make space for her and the banter ceased for a moment, and it was then that Brian's mischievous streak became evident. The girl worked upstairs in the Chief inspector's office. Her name was Lorna. Edward had seen her many times, walking through the hangar in company with other office girls, their arms folded and their eyes averted, running the gauntlet of wolf-whistles and tiger growls, as they made their way to the wash-rooms.

Lorna was certainly good looking, short, well shaped, dark-haired and altogether, quite desirable. Some of the gathering spoke with her, not impolitely, but with somewhat unguarded, amorous humour. She blushed a bit and tried to smile but she looked as if she would rather be somewhere else. No one asked her why she had come into a lobby full of men clad in oil-skins, and a boy, Edward.

After a moment or two she wiped away a small circle of condensation from one of the long windows to see what the weather was doing outside. The rain was pelting down, still. Then, casting her eyes around the group of loafing men, she asked if anyone would like to help her get some papers and documents from her boss's car. Her eyes came to rest on Edward, much to the amusement of the gathering, lounging about the lobby. The situation reminded Edward of a Giles cartoon. It was all there: the gawking faces, popping eyes, stupid grins, long, tall, short and fat: all craning for a glimpse of a comely girl and straining to hear the reaction of an embarrassed lad to her plea for assistance.

Suddenly, in that moment of relative silence, Brian spoke in a loud voice, with a wicked grin and a dramatic wink at the others. "Go on, Ted, now's your chance. Get in there, boy." He gave Edward a shove towards the outside door, amid peals of laughter from the others – their voices lifting with unrestrained encouragement.

"Go on then, boy, what ye be waiting for!"

"Daang'd if he ain't scared o'the gal!"

"Don't ye hang about lad, she won't wait forever, thee knows!"

Brian was having a whale of a time just watching Edward squirm and colour up as he tried to conceal his embarrassment. The girl too, looked uncomfortable, but she managed to smile through it all, and continued to gaze at Edward with some sympathy for his situation amongst the misplaced frivolity. As a local girl from Tintinhull, Lorna also had the Somerset accent. "Will you help me please?" She asked again in an attractive, low-pitched voice, and in spite of all the noisy racket in the lobby, Edward said that he would. This raised a ragged cheer, lead by Brian, who by now was almost wetting himself with mirth at his apprentice's discomfort amid the gibing.

Edward asked the doorman for one of the spare oilskins that hung behind his desk, and draped it over Lorna's shoulders. She had the key to the car ready and they lurched out into the rain, followed by another rough and ragged cheer. They both got very wet, but somehow they managed to keep the documents and files dry beneath the oil-skins. Yet another ragged cheer greeted them as they re-entered the lobby to face more ribald comments.

"Did yer get yer 'and on't, lad?"

"Twern't lang enough fer more'an 'at, boy!"

"Could've 'aad er away in the back o the car if ee'd troid!"

"Aar, 'es still green be'ind the ears, baint 'e; wad ee espect!"

The place was more like the public bar of a pub on a Saturday night, than a works lobby, as bawdy laughter and licentious remarks continued to flow freely. Edward took the oilskin from Lorna's shoulders and helped her to arrange the cardboard boxes into a manageable pile that she could carry. "Whoi don ee tec em up for 'er, boy; 'er moight give ee somethin fer it aafter!"

Edward shrank and coloured-up again as laughter drowned out further lewd commentary, filling the lobby with noise.

At this point Lorna exploded. She looked angrily around the grinning men and spoke a few words in their own accent.

"Why don't ee wrap up, all of ee; bide ee quiet won't eye; yuum a bunch o sniggering idiots."

The men began to shut-up.

"Caan't ee see 'ow wet it be out there an 'ow wet we be, too?" Lorna's sharp response only served to draw wicked nudges and winks, and more course laughter from the mischievous group.

Brian's little tease upset Edward and the obvious pleasure at someone else's embarrassment, was a clear demonstration of Brian's other self.

Anyway, for him and the others the fun was over. Lorna gathered up the documents, and thanking Edward with a thin smile, she left the lobby to the sniggering idiots, and a red-faced boy.

Still smarting from the uncomfortable situation that Brian had instigated during the day, Edward arrived home to find a letter waiting for him.

It had arrived in the morning post after he'd left the house for work.

The handwriting was not familiar to him but the letter came with a Woolton postmark. Edward had his meal as usual with the family and took the letter to his bedroom, to read.

There had been no letters from Valerie over the past several days and he was anxious.

He wondered who else from Liverpool would write to him. Perhaps one of the chaps at *Dragon Airways* had finally decided to drop a line, since Edward had written to Cliff and Nick several times, without response, but so far as he knew, none of the fellows he'd known at the Club lived at Woolton. Valerie and her family were the only people in Woolton who had Edward's new Yeovil address. So from whom could this be, he wondered again. With some misgiving he slit the envelope and unfolded the letter with growing concern:

Woolton,
20th April 1956

My dear Ted,

You of all people, should not be the one to suffer at the contents of this letter, but I have taken it upon myself to write to you, because of your association with our daughter Valerie over the two or three months before you moved to Somerset and, since then, by the many letters you have exchanged.

It is not easy for me to tell you this, but I feel that it is only right that you should know, if only because you are young and enjoyed a friendship that I now know was sincere and profound. I am so sorry for the pain that this letter must give you.

Our daughter, Valerie, died last night. Five days ago, a hit and run driver knocked her down in the street. Valerie regained consciousness briefly yesterday, in the hospital. I was with her and so was her mother.

You should know, Ted, that before she died, Valerie spoke of you and she made me promise to tell you her exact words. She clearly knew that she could resist no longer.

She said: "I love you dearly, Teddy, more than I could ever express in words. I've missed you so terribly but I've tried to be strong. You are the one Teddy. Only you. It's love you know. That's what it is. Live your life for both of us and if it's with someone else, please be sure that she is, like me, in love with you. Love is all that counts."

Ted, my lad, Valerie died with those words on her lips. I think you must have given her so much happiness in the short time you had together. Her mother and I thank you for that.

We can't begin to tell you how sorry we are. We know the desperate feelings you both must have felt when you had to move away from Liverpool. Real love is a rare, rare thing. We are sure that Valerie truly loved you, and we are proud of you both.

Please visit us if you have the opportunity, but then perhaps that might open old wounds. Maybe it's best just to remember Val for who she was. Have no doubts, Ted, that we will always remember you.

May God bless you and watch over you, son.

Valerie's father had signed the letter.
Edward couldn't believe what he was reading. This isn't true. It's all a lie. No, this can't be. My Valerie lives. She writes to me every day. I write to her. We, who set aside our misery in parting, with the idea of our hundreds of letters and lots, and lots of holidays. Valerie's voice echoed in his ears: "It won't be so difficult, will it? It's love you know; that's what it is!"

Edward curled up on his bed and truly wept. He wept like a child. In the silence of his room, he wept. He thought his heart would burst and his brain would burn.

He wished to die.

What more was there to live for?

Anger and sorrow welled up inside him, consuming all other thought. He wanted to find the killer and kill him too. The bastard, the cowardly bastard; the one who killed his dearest, dearest friend, the girl who knew Edward's love for her, and who gave her love to him, even as she died.

The words of that letter wrenched at Edward's heart. He thought he would go mad, as all their moments together coursed through his mind, from the beginning to the end. He remembered his own words, that their moments together, for those few months, could never be repeated, enjoyed or suffered ever again, and Valerie's; "They are ours Teddy, they belong only to us. No one else can share them." His misery left him exhausted and he slept. He had no dreams.

Edward woke next morning with immediate recollections of the letter. He sat on the edge of his bed for a while before dressing, to gather his thoughts. He resolved to go on just as Valerie had said, living his life for them both. Life had to go on, like it or not. Edward determined to write a letter to Val's Dad, to console him and his family in their loss, and to thank him for his kindness in writing.

First he would grieve, but he didn't know how. Were there to be only tears? Was there only to be a feeling of total emptiness? Should he confide, after all, in his parents? He had never mentioned Valerie to them in any direct sense. They had never met her. They didn't know how Edward felt about her. What could they say that might help their son in this moment of such terrible distress and sadness?

His only consolation was that it was a weekend. He wouldn't have to face the outside world and the fellows at work. The two days would serve him well, to be alone, quiet, in thought: to collect himself and to remember with all his broken heart, his dearest, dearest Valerie.

Edward decided to say nothing: to bury himself in his work and study. He felt that Valerie would want that. She had such zest for life, she hated sadness and always looked for the solution to make things better. For that, Edward had to respond, and find the solution to his grief. The only way he knew, was through his work. He would tell no one of his joy, shared with Valerie yesterday, nor of the deep hurt he suffered now. No one would know. He would seek no sympathy nor pity. Valerie's memory would remain with him for always.

Part 13

Picking up the Pieces

33

Edward had little in the way of conversation when Brian picked him up for the Monday morning run into work. Indeed, he had little to say to anyone that morning. It took all his moral strength to face the world at all, as the loss of Valerie continued to dominate and consume his thoughts. Concentration on anything else was almost beyond him. It probably showed. But he fought to focus his mind on the day's work ahead.

The rain had stopped overnight. The first aircraft from the final assembly line stood amongst large puddles on the ground-run base, raindrops, still clinging to the tarpaulin cover, glistening in the early morning sunlight that struggled to shine through less ominous clouds. All was ready for preliminary engine tests that Brian had re-planned to begin on the following morning, Tuesday, weather permitting. Some days earlier, he and Edward had prepared the small, round-roofed blister-hanger in front of the main building, to use as a covered work-place for engine-installation.

Fortunately the *Sabre* is very ‘field’-maintainable and requires very little additional ancillary equipment to fit or remove engines and rear fuselages. Such work can all be done with ‘bolt-on’ lifting tackle for the engines and adjustable, wheeled trolleys for the rear fuselages. They'd shifted the second aircraft and its rear-fuselage over from the main hangar, in readiness for engine pre-installation preparations, and this was their work for that day.

Before joining s, Edward hadn't any experience with jet aircraft or gas turbine engines. For him, a new learning curve had begun. It should have been an exciting time for him but the dreadful, unbelievable news cast its awful shadow to exclude all thought, save that of the terrible loss he felt. Edward couldn't raise enthusiasm. He moved listlessly to follow Brian's instructions like an automaton. There was little conversation between them, although, sensing his apprentice's preoccupation, Brian did try to make cheerful comment over their progress, but he soon gave up. Perhaps he thought his apprentice had taken umbrage over the teasing in the works-lobby, before the weekend, but it wasn't that. How could he possibly know how Edward really felt inside?

In other circumstances, Edward would have been over the moon with excitement, like Brian, who bubbled with exuberance and enthusiasm. Having trained and practiced as an engine-man with the Royal Air Force, it was to Brian's great satisfaction that he had secured a similar job in civilian life. His ebullience carried the job along in spite of Edward's apparent lethargy and preoccupation.

Several refurbished engines had arrived from *Bristols*, delivered to the Ilchester works' central stores over the passing weeks. They'd arrived in special, steel canisters designed to float in case of accidents at sea when being transported by ship. Each contained one engine wrapped in a thick, rubber bag, sealed for added protection against moisture. Brian and Edward had already transferred several to the cradles of wheeled handling trolleys, and moved two over to their work area in readiness for the second and third airframes as they came off the production line.

An interruption halted their work during the morning, when the Production Foreman and the *Bristol* Rep came over to the shed to view progress and to discuss further planning details with Brian. The Works Chief Inspector and Edward's father, in his capacity as AID examiner at Ilchester, joined them and the gathering developed into a full-scale meeting. It was all high-level stuff and Brian left his apprentice to continue with the work in hand. Edward preferred to remain distant anyway, pressing on in silence with the tasks that Brian had set him to, out of sight inside the engine compartment of the second aircraft: engrossed with his jobs and with his own thoughts.

After the meeting, Brian appeared to be even happier than before and he joined Edward to finish the airframe preparations. By lunchtime, when the hooter announced the midday break, they were ready to lift the second engine to begin installation.

Brian decided to go home at midday. His wife being pregnant, and the baby due at any time, he thought it prudent to make the trip and have lunch with her. This left Edward to his own devices for an hour ...time in which to relax and be alone. He needed that, and stayed in the shed, sitting on an upturned box among the engines and the second aircraft to eat his lunch, but he had no appetite, and couldn't manage more than half his pasty. It cloyed his dry mouth; his mind was beset with memories, so painful and poignant in their clarity. Yet those months shared with Valerie, before leaving Liverpool, and then after, writing to each other, seemed like a dream.

Had he been writing only to a dream?

If that were true, why had he woken up?

Why had the dream ended with such awful finality?

Now, all those moments were his. Those happy, happy moments that he could never share with anyone else, were his and his alone.

What was he to do? Inside a tiny hangar, in a far-away place, the solitary figure of an eighteen-year-old boy sat among the tools and machinery, next to engines and a half-finished aeroplane. He felt very isolated. He was close to tears again: thwarted, desolate and desperately unhappy.

Brian returned from Yeovil, full of bubbling joy. By then, Edward had pulled himself together and tidied up the workplace, even before the hooter blared again for the start of the afternoon's work. They slung the engine and offered it up to the waiting airframe, working together like a well-oiled machine, as they carefully slid the power plant into place on its front support-runner, until they could drop the main trunnion-mountings into place and cap them home. By late afternoon they were ready to fit the tail-pipe, and they felt more than satisfied with a good day's work. The second aircraft was almost ready for its engine testing.

That evening, at dinner, Edward's father asked him how he then felt about the apprenticeship and the new involvement with engines at Ilchester.

Clearly, having regained interest and enthusiasm over the new work-content, Edward told him that he'd begun to feel much better about prospects, especially now that he had his hands on whole aircraft again, rather than being stuck with unrelated jobs on the production line.

Edward had to admit, of course, that the experience gained on the line, over the passing weeks in the main hangar, was essential to familiarisation with the airframe. His father had been right about sticking with it and learning all he could.

Theoretical and workshop study at Technical College in the evenings and on day-release once a week was another question, however. The course wasn't on the same track as Edward's work in the hangar, nor with his ambitions.

His Dad assured him that perseverance would be for the best. There were no other options but to follow the courses through, especially as Edward, sooner or later, would move to other departments at Ilchester, perhaps the Planning department, the Inspection department or even to the Production office. At the main works in Yeovil, Edward might find himself posted to the machine-shops for a while, or to the helicopter assembly hall, or maybe a spell in the tool-room.

Apprentices followed a pattern of pre-determined on-job

training designed to give them a broad experience in many aspects of aircraft production, before specialising in a particular trade. The work at Ilchester was just a part of the overall picture and the Apprentice Supervisor had placed Edward at Yeovilton only because of his aeroplane background thus far.

Neither Edward's father nor his mother knew anything of his involvement with Valerie, nor of the devastating news that he'd received only several days before, though they seemed to sense his preoccupation and unhappiness. They didn't press their son to explain but they surely knew, instinctively, that something was wrong. Their middle son had lost his usual bounce and sense of fun. Perhaps they put it down to adolescent disquiet, and that rebellious feeling that frequently troubles lads in their teens.

Later, Edward sat in his room and wrote to Valerie's father. The task was not an easy one, and he fought valiantly to stem the tears that threatened to flow again. Words of comfort to Val's parents and family gave him an outlet for his feelings, denied until then by his own dereliction in not confiding in his parents at the outset. He regretted that, but foolishly, he'd cast the di himself.

In his letter, he tried to explain the reasons why he hadn't mentioned any part of his relationship with Valerie to his own family. Perhaps those reasons had something to do with the adolescent insecurity that he felt, or an inherent shyness of the limelight that surely his parents would have poured upon him once they were aware of his true feelings for such a lovely girl. Perhaps it was that fear of the unknown view that his own people might have taken, or simply anxiety, born of a very real desire not to lose touch with what he felt to be so uniquely special to him.

Valerie and he had known each other for only such a short time, when his family's (and unwillingly his own) movement away from Liverpool occurred. The move had only served to compound and complicate his thoughts over their association, making it more difficult to see a clear path. He'd tried to steer a grownup course through emotional confusion, frustration and split loyalties.

He ended the letter by suggesting that his loss must be as nothing compared to theirs – a daughter and sister, taken so cruelly before her time – a girl so full of life and with such unlimited capacity to give happiness and love. Their loss must certainly be harder to bear than his. He promised to write from time to time.

When he'd finished, he slipped out to post his letter, walking slowly to the mail box not very far from the house, next to the post office at the top of the hill of Hundred Stones. He felt sick at heart and weary. He felt like a ship without a rudder. Utter misery descended upon him as he dropped the letter into the post. It was like closing a book before reading the last chapter, never to know how the story may have ended.

Next morning, the rainclouds had given way to a clear blue sky as Brian and Edward clocked-in at the works. On the ground running base, they removed the tarpaulin cover from the first of their two aircraft. The *Sabre* looked bizarre with only the front fuselage and wings standing on its landing gear. With the engine tail-pipe protruding at the back, its metal-foil, insulation padding wrapped and laced around, glinting in the early sun-light, the aeroplane looked like some strange, silver-tailed insect, drying its wings in the warmth after emerging from the chrysalis.

Brian towed the bowser over with the tractor, and they pumped *AVTUR* into the aircraft's fuel tanks until the undercarriage struts sank with a jerk, as the weight increased. A quick look around underneath showed that there were no leaks. After checking the restraint cables from the main landing gear to the base ring, they made some final inspections around the aircraft and its engine. Brian told Edward to check the ammo boxes below the gun panels, to ensure that the ballast was there, having forgotten this when they tied the plane down some days before.

A quick look inside confirmed that all six boxes contained the necessary bags of lead-shot needed to compensate for the weight of the removed guns and belts of ammo. It wasn't so important at that stage, since the aircraft was without a rear fuselage for the moment – the aircraft being nose-heavy anyway, in this configuration. But it was best to be sure before they put the tail on after successful ground-runs.

Everything was ready. A slave-seat fitted in the cockpit, the safety-cage positioned in front of the nose at the engine's air-intake, the *Houchin* warmed up and plugged in, double anti-skid chocks in place, tie down cables attached and warning notices posted at various points, to advise personnel of hazard.

"Nip over to the Rep's office, Ted, and tell Cookie (the resident Rep) we're ready," said Brian, rubbing his hands together in anticipation. The *Bristol* Rep wanted to do the first run, to establish a ground-run schedule and, what was more important, to instruct Brian on the procedures for starting, running and shut-down. Brian would watch the whole process, standing on the wing root next to the cockpit; while Edward handled the *Houchin*. But first Cookie gave them the procedure for ground-power from the *Houchin*.

"When I give you the thumbs up," he said, "give it full throttle. Adjust the volts to the indicator mark here, with this knob, then on the second thumbs up, I'll initiate the start sequence but watch the other indicator. Be ready to adjust the needle quickly to this mark."

He continued, pointing to a red line on the second dial. "If you don't, the volts'll drop off and the start will abort. The starter motor will demand all the juice it can get. We'll do a dry run first to get things circulating and give you a chance to understand the *Houchin* procedure."

Cookie settled himself into the slave-seat in the cockpit and took Brian through the whole of the pre-start cockpit procedure. As Edward watched with the *Houchin* running, waiting for the sign, the Managers and Inspection people joined them again, to observe. The thumbs-up signal came and Edward punched-up the *Houchin* to full throttle. The *Vee Eight*'s exhausts bellowed, and its radiator cooling fan roared as he adjusted the volts to the mark, before looking over his shoulder for the second signal. He could see panel-lights twinkling in the cockpit from where he stood, confirming that they had power on the systems.

Then, thumbs-up a second time: Edward glanced back to the *Houchin*'s control panel, grasping the control knob for a fast reaction when the aircraft's engine starter engaged. Directly, the *Vee Eight* took the load, revs tried to drop, the exhaust roared even louder as the governor compensated, the needle dropped back, but Edward caught it with the controller and adjusted to keep the needle on-mark. Above all the din of the *Houchin*, he heard the rumble and whining of the *Sabre*'s engine, as it ran up the starting RPM. Thumbs up again from the cockpit told him to reduce the *Houchin*'s power and stand by.

With the roar of the Ground Power Unit reduced to a gentle mutter, he heard the *Sabre*'s engine, it's compressor and turbines spinning, singing with an ever-lowering note, as it ran down again. It was all a bit of an anticlimax, really.

After more discussion in the cockpit, Brian shouted for Edward to be ready for a full start. Then, Cookie and Brian climbed down to ground level for a moment, shone a torch up the intake and had a swift look around the engine, through the various open panels in the fuselage, enclosing the power-plant, and finally shone their torch up the tail pipe. They were looking for signs of oil leaks, Edward surmised, as they clambered top-side again to resume tests.

Cookie's thumbs-up sign came for a full start. Edward opened up the *Vee Eight* the maximum power again, adjusted to stabilise the voltage, before looking over his shoulder for the second sign. When it came he was ready and kept the needle on its mark. The *Sabre*'s engine wound up to starting revs, and above its whine Edward heard the intermittent cracking of HT igniters and the dull boom of light-up.

He glanced briefly at the tail pipe to see the rushing shimmer of heat. The jet efflux thundered heavily as the engine speed increased, the rushing, super-heated gasses distorting and blurring the images of the trees and airfield beyond the jet-pipe orifice.

After a brief moment Edward turned his eyes again to the

cockpit. Cookie nodded and gave the unplug sign – two fingers of one hand pulled from the closed fingers and thumb of the other. Brian grinned down at Edward from his position next to the cockpit. There was a very happy man!

For auditory protection, each of them had two tiny plugs inserted in their ears. These were supposed to reduce the effects of ear-damaging, high-frequency, sound emissions from gas turbines, but they weren't much good. Later they acquired proper ear-phone-type defenders shaped rather like two half-coconuts and held over the ears by a head strap. These were much more effective than the plugs. Edward also discovered that they kept one's ears very nicely warm in winter, too.

Cookie continued with the ground run checks and Brian became engrossed with his instruction. There was little for Edward to do then, so he busied himself with cleaning up an aircraft tow-bar that needed some repairs, whilst keeping a watchful eye open in case there was need for him on the *Houchin* again.

The *Sabre*'s engine ran at idle speed for some time until, eventually, Cookie opened the throttle slowly to near full power. The ground trembled with the thundering roar. The fierce crackling sound rent the air, reverberating from the hard surfaces of the base. Indeed it was like standing next to a huge blowtorch. Birds had scattered when the engine started. The saplings in a hedgerow at the airfield perimeter fence, some hundred yards distant behind the aircraft, swayed and bent under the hot, high-speed air flow.

Cookie shut down the engine after five minutes and he climbed down to ground level from the cockpit. The hugely smiling Brian suggested that it was time for a break and they retired to the little hangar for tea and discussion. One of the interesting things to come out of this was the subject of main-bearing lubrication, labyrinth-seal aspiration. Edward voiced his observation, "That's quite a mouthful, but what does it mean?"

Cookie explained. It seemed that the engine's main bearings contain oil seals, described as labyrinth-seals because of their design and function. These prevent oil, under pressure to lubricate the bearings, from migrating into the compressor air-flow and the turbine hot-gas-flow. Lubricant normally returns from the bearings for re-circulation through the pump, filters and reservoir-tank. The labyrinth seals capture the small amount of oil that may find its way past the plain seals at the bearing housings, retaining it in cavities. Drain-off is essential, since the lubrication system cannot retrieve this oil for recirculation. It must therefore drain to atmosphere, aided by suction.

With the aircraft's tail-unit fitted, the engine's jet-pipe orifice lies within an annulus (called the aspirator) that provides a local depression by venturi effect, as cited in Bernoulli's law, when the engine is running. The exhaust efflux effectively draws ambient air through the rear fuselage and its aspirator, not only as an aid to cooling the jet pipe and airframe structure around it, but also to create negative pressure at the annulus.

A pipe, connected to all the labyrinth seals in the engine and extending below the jet-pipe inside the rear fuselage, rests with its farthest end close to the aspirator-annulus, where the pressure is negative. Thus, induced suction acts upon the seal-drains and draws the captured oil out before it builds up. Without the rear fuselage, its aspirator and suction-pipe, the seals receive no assistance to aid the extraction or draining process.

Since the initial, post installation ground testing of engines were to be carried out without the aircraft's rear fuselage, they had to find a method of emulation to perform this function. Cookie came up with the answer, suggesting a very simple device. Brian and Edward made up a metal pipe of similar dimensions to that mounted in the tail unit. This they connected to the engine at the appropriate fitting, clipping it to the jet-pipe, with its open end in the jet-stream efflux zone at the end of the jet-pipe. Half way along the pipe, they arranged a glass U tube, or manometer, containing a quantity of red-coloured water, the idea being to get a measure of the suction that their jury-rigged pipe applied to the seals. They didn't want too much.

After fitting the equipment, they were ready for full-run checks. Brian climbed into the cockpit this time, with Cookie leaning over from the footholds in the side of the fuselage, next to the cockpit, to advise and guide his engine-running student. Edward was responsible for the *Houchin* again and between them they made a good, clean start. Cookie had asked Edward to keep an eye on the manometer as the engine power increased and, after shutting down the *Houchin*, he moved across to make his observations.

Again, the battered air reverberated and the ground shook under his feet, trembling, as Brian opened the throttle. Edward watched the manometer intently. The red-coloured water had climbed a little in the tube, showing that there was some depression in the pipe, as predicted. The noise increased around him as he continued to monitor the level, but then, as Brian slowly increased the power, the coloured water began to oscillate in the glass tube. It swung wildly this way and that, until, quite suddenly, it swept up the suction side of the U to disappear completely, sucked away into the jet-stream; indicating that there was too much depression.

Cookie glanced over his shoulder from his position next to the cockpit to look at Edward, standing behind the wing's trailing edge beside the jet-pipe and the empty manometer. Edward gave the thumbs-down sign, pointing to the empty U tube. Cookie nodded with a smile and returned his attention to the cockpit.

After shut-down they decided to fit a small gate-valve down-stream from the U tube, to provide a measure of control over the suction aggregate inside the bearing-aspirator tube.

The modification didn't take long to fit and after further engine-tests and fine adjustments to the depression at the manometer, they ended up with an effective piece of equipment that served for all subsequent engine initial-ground-runs they performed, before the fitting of rear fuselage units. Again, they were very content with another good day's work. Edward's learning curve advanced and he realised that he'd had little time for reflection upon his personal grief and unhappiness.

With night-school every evening and a whole day release once a week for tech-college studies, there wasn't much time for much else in the way of hobbies.

He missed the setting aside of that hour every night to write his daily letters to Valerie and set himself to reorganising his time, filling it with the ever-increasing amount of night-school homework.

He became deeply engrossed in the acquisition of knowledge, both at work and at school, but he was glad of it, for such intense occupation left little opportunity for him to brood over the terrible loss he still felt.

He'd reached the age of eighteen. His life had changed dramatically from that he had known in Liverpool. After Valerie's death he'd closed the book on that part of his existence, so full of joy and expectation, certainly until the emotional wound healed, at any rate.

One day, perhaps, he would be able to face the memories with renewed joy, remembering the wonderful encounter of his youth. It troubled him, however, that he was able to put aside his feelings in this way. It seemed so callous, so heartless – but what else was he to do? Somehow, he had to pick up the pieces, and carry on.

At home, the Weinels filled their Sundays with family activities, especially in the garden. The plans they'd discussed on the day they moved into the house were beginning to take shape. Edward and his father cleared the rubble left behind by the builders, and levelled-off the garden areas at front and back. With the arrival of Spring, they seeded the lawns and cut beds for plants and shrubs. They had neighbours by then and they, too, were hard at work to get their new gardens in shape.

Fred bought a do-it-yourself garden shed.

The kit came in prefabricated panels that only required screws and bolts to hold them together. The concrete base, they put down one weekend and they erected the wall panels and roof on the following Sunday. The shed wasn't very big, but there was room enough for the bikes, lawn-mower, deck-chairs and garden tools. Next door, their neighbours built a garage. The Weinels might have done the same, but since they had no car there seemed little point in such an expense.

34

The thing about a new house, Edward discovered, is that one has to start from scratch, pioneering, so to speak. Fred and his family enjoyed that, and along with the new environment into which fate had thrown them, came joy in working towards making their home into a haven. Edith named the house on their first day. A wooden plaque with the words 'The Haven' graced a space on the wall, next to the front door.

Sunday lunch in the Weinel household, as always, was very much a family gathering. They'd sit for ages afterwards, just talking and discussing the occurrences of the past week, plans for the future and the many topics over which they had little time to confer during the weekdays. Edith, of course, had spent a great deal of time and energy turning the house into a home. She spent many hours at her sewing machine, running-up new curtains and cushion covers. She had worked for the Singer company as a demonstrator and mechanic before she married Fred, and so she really was an expert. The loving care with which she prepared all the attributes of their home, never ceased to amaze her family.

Freddy wrote to tell the family that the Army wasn't for him after all. His endeavours towards the Commission he so badly wanted had failed to bear fruit. He decided to take the option of resigning after three years of service instead of going on for the twelve that he had planned. Everyone felt sorry about his decision for they all knew the extent of his enthusiasm and the effort he'd put into his plans for a career in the Army. Freddy wrote that he would be home in time for the summer.

Meanwhile, Edward's work at Ilchester continued to provide great interest to him. To improve matters, the company decided to move the flight-line to another hangar across the airfield. Hangar number fourteen had been empty for months. It belonged to the

Navy, but since they weren't using it, *Westlands* negotiated for its use to provide an airfield base from which to operate their *Sabre* flight testing and dispatching. This avoided the need to move aircraft to and from the Works each day, thus providing much more convenience.

Tired *Sabre*s flew into Yeovilton as RAF squadrons re-equipped with the new generation *Hawker Hunters* and *Supermarine Swifts* each week, and it was important therefore to keep finished aircraft output moving. Brian and Edward worked hard to keep pace with production, sometimes with tasks at the Works end, installing and initially ground-testing engines and sometimes at 'fourteen' hangars, preparing the flight-ready aircraft for air-tests and dispatch.

On a personal note, there were two things Edward learned about himself during that period of his training. The first confirmed that he had little stomach for confined or tightly enclosed spaces.

He had already experienced this phenomenon, years ago, at Bradda Head's watch-tower on the Isle of Man. On those frequent occasions when it was necessary to crawl into a *Sabre*'s nose air-intake to fit the engine accessory fairing, or into a soot-blackened jet-pipe, to check engine-turbine clearances, he found himself gripped with a dreadful fear: almost panic. The air-intake wasn't too bad. Apart from the curved contour beneath the cockpit and the upward incline when approaching the front end of the engine, the tubular, bare-metal enclosure was wide enough and light enough to complete the task in relative comfort.

The *Sabre*'s jet-pipe, however, was no more than shoulder-width and projecting thermocouples (temperature detectors) impeded the body-sliding progress towards the rear of the engine, where the turbine lay. Movement was slow, with little space to elbow-wriggle up the pipe, clasping a torch in one hand and the clearance checking tool in the other. The bulk of one's squirming body cut off most light from the opening at the tail end of the pipe and there was precious little reflected light from the sooty, matte-black surfaces of the interior. The confines of the narrow tube amplified the sound of one's own voice and the rustle and scraping of clothes, in a muffled and cheerless chamber of horror. Edward performed this task many times over, but always with the same foreboding and ill-at-ease feeling in the pit of his stomach, but he did it all the same.

The second characteristic, he recognised, was that such fear is surmountable, and by controlling and focusing upon the job in hand, and the reasons for performing it, panic induced by claustrophobic feelings, subsides. Nevertheless, he always abhorred enclosed spaces and over-crowded areas, avoiding such whenever choice presented itself.

Included in routine flight-line work at Yeovilton, was compass-swinging, for which Brian and Edward would tow an aircraft to the distant compass-base located on the farthest side of the airfield. Their job was to move the aircraft around its axis on the base, pointing it on the various headings so that the avionics chaps could make their adjustments and corrections against a Landing compass. This was grand in the summer, out in the sunshine all day long. Edward's preference and love for the outdoors persisted.

During that summer of his nineteenth year, two prominent incidents occurred at Yeovilton. The first was very serious – occurring when an aircraft oxygen system explosion almost resulted in the loss of life.

The *Sabre* has a low-pressure system that provides the pilot with breathing oxygen, needed for high altitude flight. There are four storage bottles mounted in the nose of the aircraft, two on either side of the engine air intake duct, coupled in parallel to the pilot's oxygen regulator in the cockpit. On the right hand side of the aircraft's nose, behind a small access panel, there is a replenishment connection.

Apparently, during a standard topping up operation, a faulty regulator on an oxygen replenishment-trolley allowed high pressure into the aircraft's low pressure system. Two of the storage-bottles in the aircraft burst, causing serious damage to both localised structures, skinning and engine air-intake. The explosion sent shrapnel-like fragments in all directions, some piercing the steel-skinned doors of the hangar, such was their velocity. Of the other two bottles, on the farthest side of the engine air-intake duct, each distended to the size and shape of huge vegetable marrows.

The operator was lucky only in that he was out of harm's way, standing on the wing-root, looking into the cockpit at the oxygen regulator. The contents-needle must have been off the clock! Brian and Edward were not present at the scene, but they saw the damage afterwards and later they had some involvement with the extensive repairs that were necessary.

In recalling the second incident Edward often told himself that amusing though it may seem now, there was little evidence of humour at the time, for obvious reasons. It happened in the lane to Yeovilton that passed the main gates to *Westlands* Ilchester works. Immediately opposite the works entrance, another gate provided access to the airfield, on the Navy side of the road. It was an access frequently used to move aircraft to and from the airfield and hangar number fourteen.

The incident arose when handlers couldn't tow a newly-arrived *Sabre* through to the works because the Navy gateman on the airfield side had misplaced the key to the padlock securing his gate. The tractor crew towing the aircraft, shouted across to the *Westland* side and quite soon a number of senior people gathered to view the situation. The naval CPO in charge of his side of things was distraught when someone suggested tearing the gate down with the tractor. Finally, the naval authorities decided that they could lift the aircraft over their gate, over the road and over the *Westland* gate (why open that too, if the aircraft was up in the air) using their big Coles crane.

A hurriedly organised Navy crane soon arrived on the scene, while a couple of *Westland* guys handed a *Sabre* lifting sling over the gate to the Navy people. The sling had four heavy steel cables, joined at the top with a lifting ring, and falling through two spreaders, to end with steel lugs through which heavy, threaded pins would secure them to the aircraft. The pins screwed into four strong points on the aircraft's fuselage, placing the lifting-ring, more-or-less, on the point of balance at the aircraft's centre of gravity.

The Navy people fitted the sling and hung the lift-ring on the crane's hook. The crane took the strain, raising the *Sabre* to a suitable height to clear the gates and swung the aircraft across until it was over the road. By then, some road traffic had appeared on the scene and was halted under the instructions of the Navy police who, by then, had also turned up. The Yeovilton bus-service was first in the line, the usual double-decker from Yeovil, carrying shoppers home to the villages. The passengers upstairs, had a grandstand view of the proceedings from their vantage point on a level with the swaying *Sabre*, dangling over the road from the crane's jib.

At this point, the worst imaginable happened. It was total disaster. One of the front screw-pins came adrift, leaving the aircraft hanging on only three. It hung there on the skew for a moment as the crane driver gaped, not daring to move the jib. The second front pin came out with an almost explosive bang and the aircraft's nose dropped, hitting the paved road with a tremendous crash.

The nose landing-gear buckled on impact, it's top end piercing through into the engine air intake. By now, the frantic crane-driver had begun to lower the hook. This probably prevented more serious damage to the remainder of the aircraft, as it dropped only a short distance, onto its main gear as the other pins popped almost simultaneously. Applause and cheers from the bus were not pleasing to hear at the time, for there was an awful mess to clear up.

Whoever put the screwed pins in place had not turned them fully home, as a result the threads tore away when the crane took up the weight. The whole episode was the subject of an inquiry of course, although the outcome remained obscure. After recovery of the aircraft, it ended up last in line for refurbishment and major repairs. It was the last *Sabre* to leave *Westlands*, out of the two-hundred refurbished at Ilchester.

On another occasion, Brian narrowly saved the life of a man. The circumstances were frightening. Cookie was running an engine on the base when Edward saw a labourer carrying two, five-gallon cans, one in each hand. They must have been full and heavy for he was making hard work of it as he plodded towards the ground-running base on his way from a paint and oil store located at the far side of the engine test area.

Edward was at his usual position, next to the ground power unit when he first noticed the labourer moving closer with his load. The man clearly wanted to pass across the base and he selected a route that would take him behind the aircraft. It turned out later that the poor fellow was a bit simple-minded, but in spite of that he followed, implicitly, the instructions about not walking in front of aircraft when the engine was running. Unfortunately those instructions originated from years before, and applied to aircraft with propellers. There were notices of danger zones posted all around the *Sabre* and the usual, yellow-painted safety-cage positioned in front of the engine air-intake at the nose but no-one had ever thought to update this fellow's instructions related to Jet aircraft.

Edward was frantic because when he noticed the man, he was within only a few yards of the running engine's high speed, high temperature jet efflux. There was no way that Edward could get to the fellow before he reached the danger zone, nor could he catch attention by shouting, there was far too much noise from the engine, by then at full bore. Edward ran around the wing tip waving his arms at the cockpit.

With Cookie in the seat, Brian was standing on the wing-root next to the fuselage, in preparation for making fuel control unit, high-speed-stop adjustments through the open panels. Pointing vigorously to the tail with one hand and throat-cutting with the other, Edward managed to catch their attention. Cookie closed the throttle instantly at Edward's protestations, just split-seconds after Brian had turned to see the fellow, head bent with his load, and who, by then, had reached an arm's length from the hot efflux gasses. Brian made a running leap from the trailing edge of the wing-root, hitting the concrete on the run and flinging himself to bring the man down in a flying rugby tackle; five-gallon cans, and all. Thank God for only badly bruised knees and elbows. The man most certainly would have died tragically had he walked into the jet-stream.

Such were the happenings and event of the times at *Westlands*

Ilchester works, during Edward's early months in Somerset. On the one hand his work was interesting enough, although his continued schooling at Technical College did not follow the direction he would have desired for his career in aviation. On the other hand he felt rootless in that beautiful but very different part of the country. Moreover, the emotional upset over Valerie's death only served to turn his thoughts inward and he became more introvert. In spite of this, somehow, he managed to develop a positive outlook.

Freddy came home in early summer. His excursions with the Army, (if that is an acceptable phrase), had not worked out favourably for him. His frustration and disappointment were obvious, but he quickly settled into home-life again.

He set about finding work, after a spot of leave, and quickly secured employment at Yeovilton, as a civilian electrician's assistant, with the Navy. At this time Edward had moved from the shop-floor to the Ilchester Works planning office as part of the ongoing apprentice training program. Brian took a move to the flight-shed at *Westlands* Main Works in Yeovil, to work with helicopters so, without the convenience of his lift each day, Edward started to use his old bike again.

Having found work at Yeovilton, Freddy cycled along with Edward each morning and evening. It was quite like old times. However the bikes, Freddy's *Rudge* and Edward's old *New Hudson*, were getting more than a little past-it. They decided to buy new ones. There was a good bike shop in Yeovil; a specialist place, where one could order customised machines. Freddy ordered a Raleigh and Edward a *Rudge*. They asked for the three-speed versions with caliper brakes, saddle-bags and dynamo lighting. The shop owner threw in some wet weather gear as part of the deal. The price of a fully equipped bike in those days was around twenty-eight pounds. However, even with a new push-bike, Edward's mind continued to dwell on having a motor cycle one day. He often visited the motorcycle shops in Yeovil, to cast his eyes over the second-hand machines. A brand new one was out of the question on his meagre wages.

In the Planning office, under the supervision of a Senior Planning Engineer, Edward partnered another apprentice named John, or Johnny to his intimates. He was a year or so older and carried the nickname 'The Dude' because of his snazzy apparel. The Dude owned a *500cc Manx Norton* motorcycle. With its huge racing tank, clip-on handlebars (half way down the front forks it seemed), its rear-set foot-rests, megaphone exhaust and half-circular fly screen, this was a bike-and-a-half. One lunchtime, Johnny took Edward for a spin, perched on the improvised pillion seat and hanging on for dear life. That cinched it for Edward. To have his own motorcycle became an insatiable ambition: an obsession. His aspirations didn't stretch to a racing bike like the Dude's, but simply to having any kind of motorcycle, big or small, slow or fast. He yearned for two-wheeled, motorised transport.

With a reputation in the locality as a ladies' man, the Dude was a benign womaniser. His exploits with the young female population of Sherborne and Yeovil were common knowledge, if undocumented. However, when Edward moved up to join him in the Planning office, he had settled down somewhat and indeed, he had a steady girlfriend with whom, he insisted, he shared the real thing, although he didn't think marriage was imminent.

Johnny was a dashing fellow in his way – good looking, if a little overweight. Always with good humour, he disclaimed that he was fat, contending instead that it was all muscle. Edward liked him. He had a sympathetic and gentle nature, in spite of his showmanship, flashy appearance and his devil-may-care attitude. It was the Dude who introduced Edward to smoking cigarettes. It was quite a spur-of-the-moment thing, for Edward had never seriously considered the cigarette habit as something he really wanted to take up.

Perhaps it was the suave way that The Dude proffered his pigskin-bound cigarette case, flicking it open to display Players in one side and Capstan in the other; or perhaps it was Edward's own, private desire to project a more grownup image that prompted him to accept his first taste of cigarette tobacco. He'd had already tried a pipe and quite liked it but the preparation before lighting up was plainly more for relaxed moments, when time wasn't important. Apart from childhood, parental warnings about stunting one's growth and shortening of one's wind at energetic sports, the more serious consequences and perils of smoking were relatively unknown in those times.

At close to six-feet tall by then, Edward had little care on that score anyway, and so far as physical exercise went – well, he cycled many miles each day and managed to negotiate the steepest hills and gradients of the district faster than most cyclists. Smoking therefore had small import, other than to prove a youthful (if unsound) point about being more adult. The smoking habit tended to put Edward on a par with others of his own age, although, unlike the Dude, clique-membership wasn't really a part of Edward's character makeup. He'd become a loner in many ways.

Their boss, a pipe-smoking, senior Planner-Engineer, also named John, was a pleasant enough chap. He too, was a keen motorcycling enthusiast and rode a treasured *Velocette MSS*. He was quite young and very enthusiastic about his work and under his guidance, the Dude and Edward processed spare part listings against each of John's sequentially planned jobs.

Edward's work in the planning office was interesting enough

to begin with but after a while it became routine and, to him anyway, somewhat boring. He often found himself day-dreaming, gazing from the upstairs window into the great outdoors that surrounded the works, or mentally calculating how long it would take for him to save enough to buy a motorbike.

The office had windows on two sides; one set looked into the hangar and the other provided a view of the airfield and the local countryside. It was a commotion in the hangar one day, that attracted the planning office lads to the windows on that side. One of a team of workers had trapped himself inside an aircraft cockpit, after the fitting of a new cockpit-canopy. He was shouting almost hysterically through an unglazed, quarter-light window of the windscreen. The closed and jammed canopy wouldn't move. Other team members on the outside, heaved and struggled to slide the heavy, glazed bubble open again, in a vain attempt at rescue.

It turned out that the man suffered with claustrophobia. The canopy's transparent Perspex had a covering of a brown-coloured, protective material that must have made the interior of the cockpit quite dark. When it became clear that the canopy was well and truly jammed, the poor fellow became distraught. In his panic, he began to squirm through the windscreen's unglazed quarter-light opening he'd been shouting through, despite his team mates' supplications to desist. He squeezed through like toothpaste from a tube but not without some bruises and grazes to his shoulders and hips. Someone took measurements afterwards which clearly showed, in theory anyway, that the man's escape by that route should have been quite impossible. They did get the canopy off again, eventually.

35

After only a couple of months in the planning office, another move took Edward into the Production Foreman's office. He knew the foreman quite well, and he took Edward under his wing with interest and good humour. He confided to his new assistant that the Navy had advised a closure of the Yeovilton air-station to accommodate further extension to the main runway, and that he, too, had a new task with new responsibilities.

To continue their flying activities, the Navy had reopened the long disused, satellite aerodrome of Merryfield, near Ilminster, to the West of Yeovil. At the new site, the Navy provided a hangar for *Westlands* use, but the difficulty lay in getting flight-ready aircraft, by road, to the place from where they would fly.

The long unused airfield of Merryfield was in the depths of the countryside, approachable only through very narrow lanes, and although Doug was responsible for the flight-line out there, he also had to coordinate the road transportation of the finished *Sabre*s from the Ilchester Works to the new flight-hangar facilities. The flight-line ground crew travelled each day by company van from Yeovil to Merryfield; the driver called at fixed pick-up points en-route. Most of the chaps lived in the Tintinhull area, outside Yeovil. The van delivered them to the hangar each morning and took them home to Yeovil in the evenings. They stayed all day, taking lunch in a small canteen on the new base at company expense, against off-base displacement. The foreman set up his office in the hangar, where he and Edward worked alongside, organising the various teams who worked to keep the flow of serviceable *Sabres* moving.

After a tenuous road-journey on the back of long 'Queen Mary' trucks, the aircraft arrived for reassembly by the ground crews, in preparation for flight tests and final dispatch to their new bases in Europe. Edward enjoyed his work, and the environment, at Merryfield. The team was small and independent.

Along with this and in spite of being office-bound for the most part, Edward still had many opportunities to assist in the hangar, often outside in the fresh air and sunshine that he so loved. However, after several very enjoyable months at Merryfield with the production foreman, Edward received formal notification from the Apprentice Supervisor's office indicating that he would move to another department. This time he found himself condemned to *Normalair,* a subsidiary of *Westland* at the Yeovil factory complex, specialising in aircraft pressurisation and oxygen equipment.

On the first day of his new assignment and according to instructions, Edward presented himself to the Superintendent of *Normalair*'s special-purpose machine-shop, for training in turning and milling work. As such, the work would follow his night-school and day-release courses more precisely. He couldn't say that he welcomed the idea very much, since it took him away from aircraft again but he had to be philosophic about it. Put it down to experience, his father had said. So he did.

After induction, the workshop charge-hand took Edward straight over to a lathe and began lecturing on the basic rudiments of its function. He was a kindly fellow and it pleased him that Edward already had some night-school experience in turning. However, this machine was a whole lot better and much more modern than those the students used at the Technical College. Edward began to work up some enthusiasm. Burt, the charge-hand, dug up some odd pieces of round bar material, off-cuts, to use as trial pieces and told him to spend the day experimenting with the machine. He popped over, periodically, to see how his new lad was getting on, giving advice where needed. "Don't ye worry, lad," he said, reassuring Edward in his strong, West-country accent, "it'll come to ye, bye and bye."

Burt proved to be a good friend and instructor. He told Edward that he had served with the company since the days of *Petter*, before the time of *Westlands* and yes, he was a Yeovil man, born and bred. The farthest he had been, away from the town, was Bristol, many years before. In due course, Edward had real work to do. The jobs were interesting because many of the parts he made on the lathe were experimental pieces, for use in the laboratories and test facilities next door. Mostly small items, the machining tolerances were minute. He learned much about how to read drawings correctly, and how to perform the various tool-cuts, in proper sequence. He learned to know the machine's capabilities, taper turning and thread cutting, and how to grind the cutting tools with the correct angles.

There were a number of apprentices in the shop. Edward was

the youngest at eighteen. One of the others operated a small universal milling machine, across the aisle from him. He told Edward that his next move was to operate a brand new jig-boring machine, that the company was installing in a closed room on the other side of the workshop. He suggested that Edward might like to apply to take over his current machine, to gain more experience. The idea appealed and Edward approached Burt with the notion.

Burt smiled, as he explained, "I've already put your name forward, son. You've spent a while on the lathe, and your work is excellent, so I think you deserve the chance to do some really technical stuff." This was an unexpected compliment, leaving Edward speechless for a moment. Burt continued to grin and sauntered away, whistling, with his hands thrust deep in the pockets of his smock. Clearly, he'd made someone very happy.

By mid-winter Edward had moved over to the universal mill. At once, he began to understand the milling and boring techniques upon which he'd theorised at night school. Much of the work was jig-boring for the tool-making department. Tolerances were close and margins for error were next to nothing. The machine, a Friedrich Decal, was beautiful; small and compact, it had all the attributes and more, of its bigger brothers farther down the shop. Its worktable equipment covered all the needs for accurate work. The dividing head, rotary-table and special clamps and vices were all parts of a wide variety of equipment supplied with the machine. Attachments converting it from horizontal to vertical functions added to its universality. Burt spent a great deal of time guiding his young prodigy along and advising best methods and techniques to produce quality work. He enjoyed the machine as much as Edward did. Edward missed the aircraft though, and often wondered if the apprenticeship would eventually return him again to the outdoors and aeroplanes. He found the training and the work interesting enough, but aviation was what he started out in, and suddenly it all seemed so very far away.

After his nineteenth birthday the chimes of change rang yet again when he responded to a call from the Apprentice Supervisors office for interview. The company was in a state of expansion and they had decided to extend their main tool-room to meet the need to build more aircraft. From a long list of apprentices, selections were in process to find suitable people to place in various sections of that workshop. Edward's reaction had to be positive when they offered him a place with the toolmakers.

They were supposedly the cream of the work-force, highly respected as skilled tradesmen and receiving better rates of pay. Additionally, Edward thought that perhaps this would be one way to get back, closer to aircraft. Within a week, his transfer from *Normalair* came through, and he moved across to the *Westland* main factory. Until then, Edward hadn't thought too much about National Service and the call-up for military service. Indentured apprentices were immune by deferment, until completion of their training and studies. A communication from the Government offices had come through the mail, however, reminding him of his obligations but confirming that unless war broke out, for which the army might immediately conscript eligible men, Edward would remain on the deferred listings. His Dad and he had already discussed the possibility of concluding the apprenticeship, in favour of volunteering for the Royal Air Force, if Edward's interest in working with aeroplanes was greater than the current training plan with *Westland*, but the idea didn't appeal to him, in spite of prospects for a good career in the service. In the meantime, David had moved to a brand new secondary school, not far away and his ambitions towards the Air Force had not diminished. He still wanted to be a pilot. He and Edward shared a bedroom again after Freddy came home from the Army and they had grown closer, especially because their kindred interest in aviation.

Freddy had turned his interest towards the church again, and he had some thoughts about entering the ministry but he found, not unlike the army, excessive bias existed in the selection process for religious theological training. His job, too, was not something that he wanted to continue, and in due course he landed a job with the BBC in London. He decided to take to bull by the horns and took off with ambitions for a career in the technical side of Television.

Situated at the back of the factory, *Westlands* Tool Room lay next to the single track railway connecting Yeovil with Taunton – one of those smaller lines that Lord Beeching's re-organisational plans for British Rail closed down in later years. However, in 1957 there were still several trains a day, back and forth. Being linked with the region's bigger towns and markets, the line was used mainly by country people. Taunton was the county town in those days. Edward found no difficulty in locating the Tool Room and after presenting myself to the Foreman in charge, a charge hand took him to a large room at the back of the shop, with windows overlooking the little railway. The room was full of large machines called jig-borers. There were vertical machines and horizontal machines, in two lines with an aisle between. Separated from the rest of the workshop by windows and wide swing doors, the jig boring room clearly had some independence. The charge hand took Edward to meet a couple of chaps who worked in the area. They in turn introduced him to a vertical boring machine, a very old *Newall*.

This was the smallest machine in the room but nevertheless it still towered above their heads. The fellow who was to instruct in its use assured Edward that, in spite of its age, the machine still retained its fine accuracy. The man showed Edward where to put his tools and coat before asking if he'd like some tea. His name was Arthur, a small, dapper man who always wore a clean collar and tie. Arthur had been with *Westland*, in the Tool Room, since he was a lad, and his trade was jig boring. His machine was next to Edward's. It was much larger and not so old, but the common factor between the two machines was the method of measurement for the precise location of the work tables. Arthur had started to instruct Edward over his first cup of tea as an apprentice in the Tool Room.

The vertical jig borer is a machine for boring holes with great accuracy and precision, not only in sizing but also in each hole's location or relationship to the next, or to a specific hypothetical tooling-point. Such machines are capable of the finest dimensional accuracy, down to tenths of thousandths of an inch, in the right hands. "Do you understand trigonometry, Ted, and can you use logarithms or a slide rule?" Arthur asked.

"Yes, I was doing small jig work at *Normalair*." Edward explained.

"Good, you'll need that knowledge to interpret the tool designer’s drawings, and calculate the positions of holes that you'll drill and bore," he went on. "The machine does the cutting-work, but it's the operator who controls the machine." Arthur tapped the side of his head with his forefinger. "It's all up here, lad."

Edward liked Arthur and thought that he would learn much from him as time passed. Edward spent much of the day looking over the machine and understanding its controls, not least of which was the precise method of positioning the worktable beneath the spindle and its rotating tool holder. Arthur showed him the two rows of cylindrical rollers that, mounted on the machine's base frame, provided the two axis' by which to make accurate measurement. One was set to the side of the worktable and the other in front. Ground and polished to precisely one inch in diameter the rollers lay fixed in their positions, side by side, in the two rows.

A movable micrometer, mounted on a V block, sat upon the rollers, one for each row. Since the micrometers provided adjustment from zero to exactly one inch, by means of a hand-rotated spindle, it was possible to make adjustments in tenths of thousands of an inch within the one. The worktable's range of positioning (relative to the tool spindle) was therefore infinite, within the confines of overall movement fore and aft and from side to side. The micrometer spindle-end-faces in contact with dial indicators mounted on the moveable work-table ensured minutely accurate positioning. Edward, having absorbed all this information, began to work with some trial pieces for practice, under the eagle-eye of his new tutor, Arthur.

At night-school he studied subjects very much related to the kind of work with which he was involved. The course included technical drawing, maths, sciences, materials, workshop technology, English and Practical work in the school's machine-shop. He absorbed the schoolwork like a sponge for the most part, in spite of the juvenile waggish behaviour of many students in his class. Most were a couple of years younger than he was because the continuation of his apprenticeship at *Westland* required that he studied a commensurate course at the Yeovil Technical and Art College, from the start. A City & Guilds Course normally started at the age of fifteen. He began at seventeen and a half. What went before didn't count with *Westland* and the Technical College. He had to start at the beginning. But he ignored the others for the most part and tried to concentrate on filling his mind with technical knowledge.

Of all the masters, three impressed him deeply and influenced his future career. Of those, it was Mr. Wilkins who teased out the wrinkles in Edward's understanding of materials, engineering techniques, technical drawing and trigonometry. Mr. Miller developed Edward's manual skills in the machine-shop and Mr. Morris nurtured his ability to write coherent English.

Dennis Wilkins was one of those special people who, Edward discovered, one rarely met in life. He commanded respect without the need for pompous rhetoric. He enjoyed a joke, but mastery of his subjects coupled with a down to earth approach drew forth absolute attention from his classes of would-be layabouts. A kindly attitude backed up by his stern appearance never failed to promote silence and learning. In later years, when Edward needed help, he rediscovered Mr. Wilkins' kindly manner and they struck a lasting friendship which, although it amounted to only the annual exchange of Christmas cards with brief family news, remained intact over the passing of time, together with happy memories of that time of learning.

Mr. Morris was an ex-Sherborne Grammar School prefect. His command of the English language and methods by which to plan the written word claimed long-standing results in Edward's ability to write. Mr. Morris was popular with the riffraff of the City & Guilds courses, in spite of his clearly more affluent upbringing and Grammar School background.

In time Edward's thoughts turned again to having a motorcycle. One Saturday morning, he made his usual visit to Pankhurst's, the motorbike dealer that he visited frequently. It was a big shop and the showroom had space enough for a large number of bikes. He browsed as usual until his eyes lit upon an old machine, priced at seventy-five pounds – an *Ariel Square Four* of 1949 vintage, a big motorcycle of 1000 cc, with four cylinders. He sat on it, walked around it, looked closely at the engine and felt himself falling in love with it. A passing salesman stopped to question Edward's interest.

"Yes, I'm very interested, but the price is a bit beyond my means." Edward said, with a shake of his head. The salesman then told him that he could allow hire-purchase on it, over six or twelve months, provided Edward could get a sponsor to sign the papers.

"Perhaps your Dad could do that for you."

Hire-purchase had not entered Edward's head and he leapt at the possibility. He rushed off home with the papers, to ask his father if he would be willing to stand as guarantor for him. Edward had enough money for a deposit and a couple of payments. To his absolute joy Fred agreed and Edward raced back to the town again to make the down payment. The salesman promised to deliver the bike after the weekend, perhaps on Monday or Tuesday, giving Edward time to buy the L plates, get a provisional licence, find a company that would provide insurance cover, buy a crash helmet and all the other things he needed to be able to take the bike on the road. Oh dear – his excitement was a little dampened. He hadn't thought too much about all the other stuff related to owning a motor vehicle.

They delivered the bike on the following Tuesday, as promised. Edward's mother came out to look at the monster as he wheeled it down the sloped pathway to the side of the house. "It's a bit big, isn't it, son," she said. "Will you be all right on it?"

"Yes I think so, although it really is a bit bigger than it looked in the shop." Edward answered with half-agreement and not a little apprehension on his part, too. He was eager to get some petrol for it and at least start the engine. He couldn't use it on the road for the time being because he still needed the road-fund disc, and the insurance cover note. Everything else, he had, including a provisional licence.

Since Edward had the day off, he walked up the road with an empty can, to the nearest petrol station at the top of the Hundred Stones hill. He bought a gallon of petrol and a can of oil. His father had arrived for lunch when Edward got back and cast critical eyes over the monstrous machine that graced the pathway next to the house. He'd had his own motorbike years ago, a *Dunelt*, so he knew something about the thrill of owning and driving two-wheeled machines. He greeted his son with a broad grin but his comment was clearly one of concern.

"It's a bit big isn't it, son," he said, "to start with, I mean."

Edward thought both his parents were a bit worried, but he resolved to get the beast running, anyway. He tipped some of the gasoline into the huge tank, checked the oil tank level and sat astride his new steed. The question then was how to start it. He knew about the gears, clutch and brakes but since there wasn't any ignition key he looked around for a switch of some sort. The only one he could find, that clearly didn't have any obvious use, was in the middle of the instrument panel, on top of the fuel tank. It was a turn-switch, so he turned it. Then, after opening the petrol supply-valve under the tank, he kicked the engine over with the kick-start lever. The engine didn't start, so he tried again. Success! The melodious note of two exhaust pipes burst upon the lunchtime air. What joy. What ecstasy. The deep tone of the *Ariel*'s engine was music to his ears. Should he try it up the garden path? Should he give it a whirl up to the front gate? Should he?

Edward rocked the bike off its rear-wheel stand, grabbed the clutch lever, eased the gear pedal into first gear with his toe – then, opening the throttle a little and letting the clutch out slowly, he was away up the path! What joy. What absolute joy! He let the bike roll backwards, guiding it down the path again before turning off the engine ignition, to park the bike again next to the gate at the side of the house. Could life get better than this? Perhaps it was the nearest he could get to total happiness, without Valerie.

It was over two years since Valerie died and although his thoughts often dwelt upon her and their time spent together, Edward no longer felt anguish at her loss. He pondered on what Valerie might have thought about his new acquisition. She knew of his interest in motorcycles and she had startled him in her inimitable way, when she expressed her enthusiasm, too. He remembered the day that they stopped at one of the showrooms in Liverpool, to look at the motorbikes in the window. Valerie got so excited about a shiny, new *Vincent Black Shadow*, that they had to go inside so that she could sit on it.

"How do I look Teddy? Isn't she a beauty? I think you should get one of these."

"You look splendid Val and the bike really is a beauty. Have you seen the price though?"

Valerie turned the price tag over, and with a laugh she patted the petrol tank and said to it, "You deserve to be expensive because you're so beautiful. One day I'll buy you for Teddy." Edward continued to preserve his silence and he never told anyone of the friendship, love and affection that he and Valerie had shared in Liverpool. There seemed little point. Time had passed by and time was a healer, so they said. In the probability that this was true, Edward had slowly come to terms with the fact that Valerie had gone. The pieces of his shattered life were gradually falling back into place.

Part 14

Fettered and Frustrated

36

Edward's long awaited, two-wheeled aspirations finally fulfilled, the motorcycle gave him a refreshing new source of distraction. Several days after its delivery and his first experimental jaunt up the garden path, he received the remaining documentation allowing him to take the machine on the road. The Saturday morning post brought the insurance cover and thus he was ready to take to the highway.

Between-times, he'd spent time preparing the bike for the big day and it gleamed in the morning sunlight as he wheeled it out of the shed. He kicked it over and the engine started readily. Slipping it into first gear, Edward headed up the path as before, stopping at the front gate to check for traffic. A clear road outside gave him the moment he'd been anticipating for so long, and he let out the clutch-lever to move off. For a moment, the bike didn't budge. Then, with a sudden lurch, causing him to twist the throttle wider, the *Ariel* leapt forward like a bolt from a crossbow. He was out of control with too much power. The machine took off across the road, exhausts blaring, and Edward hanging on for dear life.

Struggling to turn, he managed to get the throttle closed as the front wheel hit the footpath kerb at an oblique angle, on the opposite side of the road. He nearly dropped the bike there and then but somehow managed to hold it up as he twisted on some power again. Edward was all arms and legs for a moment until, with considerable effort and a pounding heart, he straightened up and resumed his first expedition in a more sedate manner. Later, he discovered that there was a kink in the clutch-cable, causing the inner wire to bind momentarily, delaying clutch engagement upon releasing the lever. A new cable was necessary.

Edward didn't know it at the time, but his mother had watched the whole episode from the front room window, aghast at his floundering departure. It frightened her rigid. Poor Edith, she didn't realise how frightened her son was, as he fought a machine that seemed to have a mind of its own.

His alarm subdued, he pressed on with his maiden trip, turning left at the top of the road to descend steeply from the brow towards the wide, main road out of Yeovil, leading to Mudford. To his dismay, halfway down the hill, the engine stopped abruptly. He pulled over close to the pavement and kicked the engine over, but it refused to start again.

He spent five fruitless minutes cranking on the kick-starter and checking all the obvious things to find the problem. In the end Edward decided to wheel the bike back to the house. How disappointed he felt and not a little embarrassed. There was nothing for it, but to turn the *Ariel* around and walk it up the hill, pushing the heavy machine by the handlebars. The effort completely exhausted him by the time he'd reached only a hundred yards or so. The bike was heavy and the gradient steep. He had to stop and pull the bike on to its rear-wheel-stand to rest for a moment. It took a few moments to catch his breath and then he decided to try the engine again. To his utter amazement and gratification the first kick started it. Why?

Then it struck him. The small amount of fuel that he'd bought earlier in the week was hardly enough to wet the bottom of the petrol tank. With the bike pointed downhill, what little fuel there was ran to the front of the tank, uncovering the outlets to the carburettor at the rear. Of course the engine stopped; it was starved of fuel. Pointed up-hill, the engine ran beautifully, the petrol having returned to the rear of the tank. He decided to motor on up to the brow again and the down the other side, coasting if necessary, on the downhill stretch, until he got to the other petrol station not far beyond.

There were two stations in the locality and it was six-of-one or half-a-dozen of the other which of them he used, they were both close by. Ten minutes later, without further event and the petrol tank filled, he was able to go on with his proving run. The two incidents overshadowed his initial joy, but he persevered and took the *Ariel* out onto the quiet country lanes towards Ilchester.

Gradually, he became accustomed to the bike's controls as he learned and practiced the coordination needed between hands and feet. Clutch in the left hand, throttle and front brake in the right. Rear brake under his left foot and gear-shift at the right. Handling the throttle needed care, since the engine developed substantial power with the four cylinders totalling a swept volume of close to a thousand cubic centimetres. Under the twist grip it wound up smoothly, like a turbine.

Although the bike's rigid frame didn't have springing at the rear wheel, the front-end had telescopic forks, and a steering damper to stiffen up the turning moment of the handlebars. In spite of the two incidents, Edward felt altogether very pleased with the machine, and upon his return to Yeovil he'd regained his confidence. It was only a matter of time after that, driving with L plates displayed, until he could take his driving test to obtain a full motorcycle licence. Edward had at last joined the ranks of two-wheel drivers and he was on a par with his peers. Most of the other apprentices of his entry, at *Westland*, already owned motor-cycles.

At work in the tool room, Management moved him onto another machine as he progressed in the complexities of jig boring. They gave him more responsibility and, as time passed, less supervision. The introduction of shift work gave him new working hours in alternating weeks of mornings or afternoons. This suited him very well since it gave him time in which to follow more of his own pursuits, during daylight hours. Sometimes, he'd join a group of tool-room apprentices who frequently motored down to Lyme Regis on the Dorset coast, to hire a boat and fish for mackerel. They'd make an early start, around six in the morning, run down to Lyme on their motorcycles, fish for a few hours from the boat, hired complete with its owner, the coxswain, and return to Yeovil with bags full of Mackerel in time to make the afternoon shift at two in the afternoon.

As summer approached again, plans for a family holiday came under discussion. The Weinels had vacationed at Paignton, near Torquay, the year before. They'd stayed at a small hotel close to the sea-front road, and since they had enjoyed it so much, they agreed to try it again. Edward decided to motorcycle down to Devon while the rest of the family travelled by train, and he prepared the bike with some care, together with a small kit of spares and a few maps. He'd never before undertaken a journey of that distance and consequently he was looking forward to the excursion as he set out very early on the first day of the holiday.

His trip took him through some of the most picturesque countryside in the southwest – Crewkerne, Chard, Honiton, by-passing Exeter and turning off to head for Dawlish and Teignmouth. Taking the coastal route was much slower, but it was far more interesting with twisting, winding roads that climbed and descended through the rocky, red cliffs of the seacoast. Torbay is a beautiful part of Devon's eastern coast-line and Edward stopped frequently to take in the views. He was first to arrive at the hotel, his parents arriving an hour or so later with David. They settled in and their holiday began, making for the beach almost immediately after lunch.

They spent their days at the beach, just relaxing and taking the sun. It was good to get away from work for a couple of weeks to bask in different surroundings, with bracing sea-air, the cry of seagulls and the sound of waves rolling in from Torbay to wash the Devonshire coastline. The red-sand beach stretched all along the bay, from the rocks that separated Torquay from Paignton, southwards, beyond the pier and almost as far as Brixham. Fred and Edith preferred the Torquay-end, since it was almost opposite their hotel. It was quietly elegant and not so commercialised.

Edward became restless after four or five days and one evening, pleading a headache from too much sun and declined to go out with the rest of his family, who had decided on a movie in town, or a variety show or something. Instead, he took a stroll to the empty sea front. When he reached the paved roadway overlooking the beach, close to the family's usual sunning spot, he stopped for a cigarette. Except for a distant couple, strolling hand in hand, the deserted esplanade was his.

The tide was in, the rollers covering the sands to churn and pound at the sea wall. He sat on the wall, swinging his legs over to the seaward side and looked out over the white-topped waves as they crashed in from Torbay. The turbulent waters beyond, stretched as far as the eye could see, to a darkening horizon. It was late evening, the reddening sun sinking Westward behind him after a hot day. The waves continued to roll in at full tide, breaking on the sea wall just a few feet below the road level. There were no people about and he felt very solitary. He sat there alone, thinking about his future.

Where was he going? What would he be doing in ten years from today? He felt uneasy as his thoughts focused. Was he likely to become one of ‘the wretched many’, tied to a factory job for the rest of his natural life? He could see little happiness in that. Was he to remain just one of the shuffling line that he had first met with when he joined *Westlands* at Ilchester, clocking in and out each day: every day? He shuddered. Was he to become another member of a Trades Union with whose methods of negotiation he didn't wholly agree? He rebelled at this thought. Would he be happy trudging to and from the same factory each day, working with same people each day, watching them grow older each day, as he would surely do? Could he face the sameness of every day and every night, in a man-made rut that eventually would become too deep to climb out of? What happened to all those hopes and dreams, all those adventures waiting for his participation? Was he, and would he be, too indecisive to try for better things? Was he so irresolute?

One thing was very clear to him. As an apprentice, his indentures to *Westland* tied him. Possibly conscription into the Forces would provide the adventure that somehow he craved for. When he reached twenty-one the Queen's Government would have him, if not for eighteen months, for three years if he volunteered.

He couldn't come to terms with this either. Convention and the Establishment seemed to control everyone and everything. Yet, he'd met men who were proud of their commitment to these idols of adulthood. Perhaps they preferred to have someone else think for them, and focus themselves only on the trade that they had learned. Do their job. Keep their mouths shut. Follow the rest of the sheep.

No, surely this was a negative thought. Such men had responsibilities to family and by whatever means, their objectives were clearly to earn enough to live, and perhaps to provide a better future for their children. More than that, such men were skilled tradesmen whose expertise and abilities contributed hugely to the development of the country.

As Edward sat thus engaged, smoking a second cigarette, he sensed movement behind him on the esplanade pathway. He glanced over his shoulder to see a young girl, not much younger than himself, walking alone towards him. She wore a short, flared, sky-blue dress with a white cardigan thrown about her shoulders. She looked as if she should be going to a dance or, as it occurred to Edward, (even though she wasn't wearing pinafore or those hooped stockings that girls wore in Victorian times) perhaps she was searching for a white rabbit, so Alice-like was her appearance.

As she came closer, she smiled and crossed the road close to where Edward sat. He recognised her as a girl he'd seen on the beach every day during the week. She and her family seemed to like the same spot on the sands, just as his family did with theirs. She stopped near the wall, two or three yards away, still smiling, but looking far out to sea in a detached manner, breathing deeply at the salty tang carried on the breeze.

Trying very hard not to smile as widely as Lewis Caroll's Cheshire cat, Edward volunteered a simple "Hullo," and alluding to the sea that this parody of Alice gazed upon with such obvious affection, "it's beautiful isn't it?" The girl turned her blue eyes towards him in mock-surprise, her long, dark hair blowing in the light sea breeze, and her sharp featured face lighting up at his comment.

"Hullo; yes, I like the sea when it's rough with surf and white-horses," answered Alice, adding a question. "I've seen you on the beach haven't I; are you on holiday?"

Amused by his own imaginary characterisation of the girl, Edward had half expected a reference to a Walrus and a Carpenter but kicked himself mentally to answer the paired questions sensibly.

"Yes, this is our first week. We're staying for oysters – I mean, we're staying for two," he said, hurriedly correcting his fantasy-inspired error and asking an obvious question to cover his self-inflicted, but equally apparent, discomposure. "What about you; are you on holiday?"

"Yes, we're here for two weeks as well," said Alice, disregarding Edward's confused faux pas, although a questioning frown flickered for a fleeting moment across her brow. "Staying for oysters?"

She moved a little closer to sit next to him on the wall with her feet on the landward side, before continuing. "not many people about tonight, are there?"

"I suppose they've all gone to the shows. My folks have gone to the variety show at the theatre in town, I think. You look as if you're going out somewhere too."

"Oh, yes, my parents are taking my sister and me to the show at the pier theatre. It starts a bit later than the others. I came down here for a breath of fresh air before we go," said Alice, flinging a sideways glance at Edward as she added, "I saw you walking across the green from my bedroom window at the hotel."

Edward caught her glance and held her blue eyes for a moment, smiling in spite of his surprise at the girl's comment. Had she followed him with the idea of a chance meeting? Was he her white rabbit? Before he could respond she asked, "What's the time?"

Edward looked at his watch. "It's nearly eight."

"I must go then, or I'll be late," Alice said, "will you be on the beach tomorrow?"

"I should think so." Edward answered, with a nod, inwardly thinking that surely it was the white rabbit who should have said the first part, about being late.

"You looked very brown on the beach today, you seem to take a tan very well," added the girl with another smile and a sidelong glance, as she stood up to leave, "I must go now, perhaps I'll see you on the beach again tomorrow. Bye."

"Yes, perhaps. Bye," Edward said absently, absorbing the evocative compliment and inwardly conceding that his Alice thoughts had perhaps been a trifle facetious. She walked away quickly, throwing a happy glance and a wave at Edward before turning off the road, passing between the bathing huts to take a short cut across the playing fields beyond where, just possibly, she might have passed the Queen of Hearts playing Croquet upon the green, knocking hedgehog-balls through the hoops with a flamingo-mallet!

That short conversation with an almost complete stranger, eminently trivial though it was, had pushed to one side his warren of anxieties over the future. Certainly the girl had stirred his imagination. It was as if a breath of fresh air had instilled a new feeling of hope into him. Perhaps that was what he needed. He felt like the Mad Hatter or the March Hare, or perhaps more like the reclusive mouse, who'd just climbed out of the teapot, for he couldn't help asking himself if this rendezvous with Alice was pure coincidence. He had seen her on the beach. Their eyes had met several times in exchanging mutual glances of youthful interest. Unquestionably, she had made an impression upon him. Anyway, he felt much better and swung his legs back over the wall to stand up, his mind refreshed.

He strolled back to the hotel as the last of the sun disappeared beyond the distant horizon behind the town and was that a white rabbit he saw, running across the green, in the twilight.

37

Next day it rained – a light, incessant drizzle. At breakfast, the family decided to take a walk into the town. Edward's parents wanted to make a few small purchases, films for the cameras, post cards, souvenirs and small gifts for people back home, in Yeovil. Edith suggested that they could take lunch in town afterwards, just for a change and because the weather wasn't good enough for beaching. It showed no sign of improvement.

There is nothing worse than a rain-wet, seaside, holiday resort; it's depressing. People mill about in plastic raincoats that always seem to get wetter on the inside than out, shopping for things they don't really want and waiting for the pubs and cafeterias to open. The Weinels' damp excursion into town was no exception to Edward's prediction. It was depressing and boring. They milled about with the rest, avoiding puddles and dodging umbrellas that threatened to gouge one's eyes out at every street corner. What a relief it was to find a small restaurant with space enough and a quiet table for the four of them to take lunch together, as a family, in unfamiliar surroundings. Edward had to admit, it pleased him to get back to the hotel, away from the madding crowds, kiss-me-quick hats and plastic Macs.

After the rain, two or three days later, they found themselves on the beach again, enjoying the sunshine. The girl, Alice, was there too, not far away, with her family. She waved and smiled. Edward waved back, pleased that she had noticed him. Quite by chance, (or was it) they met again that evening, as he took another walk on his own. This time she didn't look as if she was going to a dance. She wore a simple summer dress and walked with her parents and a younger girl, her sister. She smiled and lifted her hand a little, at her side, in a rather clandestine wave at hip height. Edward wished them, "Good evening" as they passed. The family responded with smiles and polite "Good evenings."

They were talking together and Edward overheard the girl's mother use her daughter's name. "Isn't that the boy you mentioned the other evening, Jane, the one you met before the show?" Edward walked on as their voices diminished until an exclamation from behind caught his attention. "Excuse me," the voice called, "young man." He turned and Jane's mother spoke again.

"Would you like to have a drink with us at the cafe?" She said, pointing to the building at the end of the esplanade. The unexpected invitation surprised Edward but they were all smiling and clearly the invitation was unanimous.

Caught unawares, and not a little tongue-tied, Edward nodded and said, "Thank you, yes."

"Come along then. We've seen you on the beach haven't we?" She continued conversationally, "Jane told us that she bumped into you the other evening."

"Yes," said Edward, rather lamely, "I sometimes walk here in the evenings, I like to walk by the sea." Jane moved across to walk with him, whilst her younger sister peered at him suspiciously from behind her Dad's legs. He was a silent man of burly build, about as tall as Edward but thickset and heavy.

"I hope you don't mind that I told them about the other evening." Jane said, whispering as if there were some intrigue and mystery attached to the innocence of their brief, trifling conversation of several days ago. It was clear to Edward that the girl, having noticed him from her hotel window as he walking alone to the seafront that evening, had precipitated their meeting out of curiosity. He also wondered if his solitary stroll had found inspiration in the hope that she might have had the same idea. The complexities of youthful attraction are sometimes hard to define.

"No," he replied, "I enjoyed talking to you, and you made me feel better. I wasn't in a very good mood that night."

"Yes, I thought you looked a bit fed up, sitting on the wall there, all by yourself."

"Well, I feel much better tonight," Edward said with a grin, "my name's Ted by the way. You're Jane aren't you?"

"Yes, and that's my sister Judy, and my Mum and Dad of course."

The cafe wasn't full but the tables were small. They pushed two together so that Edward and Jane sat together at one, and the rest sat at the other. To Edward, the gathering seemed like a mild, family inquisition, a sort of tribunal, established to see if he was the right kind of chap to know:

We're here to find out who this fellow is, he's clearly interested in our daughter and she in him it seems, the way she keeps talking about him and all those furtive glances on the beach, secret waves and whispered-asides. Are his motives decent and proper? He felt like Charles the first's, child-prince facing Cromwell's parliamentary committee: 'When did you last see your father?'

Edward shook himself mentally as his imagination got the better of his thoughts again, but he did feel a bit like an insect under a magnifying glass. To the melody of 'Summer Place' oozing gently from a radio behind the bar, they ordered tea and spent a stilted half hour of smiling, searching chat about the day's events. Later, to conform with what he sensed, and to satisfy well-disguised, though perfectly natural, parental curiosity, Edward told them a little about his family and himself until it was time to go.

He thought he'd acquitted himself quite well, presenting himself as an entirely innocent admirer rather than a furtive predator that he suspected they may have thought him to be and to bear this out, the conversation continued in a more jocular manner as they strolled away from the cafe.

However, the jocularity turned to disappointment when Jane's father explained that they were cutting their holiday short and leaving next day for home. Some business had suddenly come up, demanding a speedy return to Sussex. Edward walked with them as far as the turning that led to their hotel, where they stopped briefly to say good-bye with commiseration's suggesting that they might meet again next year.

Edward quite suddenly realised that he had developed a strong disposition for the blue-eyed 'Alice', albeit simply based on physical attraction, eye-contact on the beach and two short conversations that hardly spelled out knowing one another very well, but Jane seemed to be like-minded about Edward. It was sad that they would have to wait for a whole year to see if they really had discovered something special. Paignton had become Edward's 'Summer Place'.

It wasn't until the next day that he realised that they hadn't exchanged addresses. Except for somewhere in Sussex, he didn't have a clue where Jane and her family lived: a silly oversight. He retraced his steps to where their hotel was, on the off-chance that the proprietor would be willing to give him the home address he sought. The man was reluctant at first, but in the end he wrote it down, on the basis that Edward knew the family well enough. Several days later their own holiday ended too and the Weinels returned to Yeovil, tanned and relaxed. Then it was back to work and back to the daily routines of life at the factory.

On Edward's return, the offer of a new opportunity in the jig boring section was waiting for him. The installation of a new machine, some months before, had been admired by many old hands in the shop. Some were vying for selection to operate the new equipment. Finally, however, a younger fellow, having recently finishing his apprenticeship, had initial command of the machine and the Management asked Edward if he would like to work a turn-about shift with him.

This was quite a compliment and a positive lateral movement, enabling Edward to work with the latest technology. He accepted with pleasure and joined his new shift-partner next day, to begin learning the fundamentals of the machine's operation. It occurred to him that after all, if his future was already mapped-out for him by his supervisors, he might just as well take advantage of every opportunity that came his way.

He wrote a letter to Jane at the weekend, to which he hoped that she might respond without prejudice to the way in which he'd discovered her address. There had been a hesitant attraction between them, cut short just when it might have developed into a less immature acquaintance. He wanted to know her better, for she and her parents, had seemed so sympathetic. He wrote how sorry he was that she and her family had to break their holiday earlier than planned, and he included a few words about the remainder of his own. As an afterthought, he also mentioned the mental comparison he'd made about her, likening her to Alice, and his likening of himself to the teapot-mouse.

He didn't have long to wait for a reply. Within a few days, much to his delight, he received an astonished response to his letter. Jane couldn't understand how he'd managed it, because she too, had realised that they hadn't exchanged addresses. She wrote at great length about what a clever piece of detective work Edward had carried out, and how pleased she was that he had.

It amused her immensely, she wrote, that he should have compared her to Alice. She didn't mind a bit, and indeed, she thought of it as an unusual compliment, especially as she had so enjoyed the reading of Lewis Caroll's Alice in Wonderland and Through the Looking Glass, in her earlier years. Her letter was full of interesting observations and descriptions about where she lived in Sussex – notably that she was still at Grammar school, and enjoyed swimming and horse-riding as her principal pastimes. Edward answered her letter the same evening. And so they began a friendly exchange of views and opinion through their correspondence.

During the following months they exchanged many letters, but Edward found himself making mental comparisons between Jane and Val. Clearly, Jane was not the same as Valerie, and her sentiments, though sympathetic, did not carry the same qualities. However, their postal friendship developed in some ways, through common interests, and Edward found the activity of writing to be good therapy.

He'd taken up painting and drawing again, to re-kindle the skills he had once thought he could build a career upon and he sent some examples of his work to Jane, for her opinion. She returned them sufficiently impressed to encourage him to do more, especially illustrative work that might be suitable for books. Jane explained that she had an interest in writing, with perhaps a career in journalism, reminding Edward of Valerie's ambition to follow a similar course.

Edward's cherished motorcycle took up much of his spare time, and he often took it out to distant places that he'd never been to before. The machine gave him much joy, often reflected in his letters to Jane. The bike was his only escape from the ordinariness of his lifestyle of those times.

At work the jobs rolled in and their complexity brought added interest. He came to terms with the capabilities of the *Newall* and over the months Peter (his shift-mate) and he began to handle the most complex jig-boring work in the shop. Jobs came in to them from the many subsidiaries and satellite factories as far afield as London and Bristol.

Westland had bought into several other Helicopter manufacturers as the aviation industry in Britain became stagnant. Old established firms were under threat of takeover and slowly the British aircraft industry had fallen into the monopoly-trap of the times. Competition from Europe and America was too strong for the old established firms *like Vickers, Avro, De Havilland* and *Bristol,* etc. *Fairey, Hawker* and *Gloster* had succumbed, along with *Blackburns*, as they too, bowed to the pressures of big business and multi-national firms.

As the major helicopter manufacturer in Britain, *Westland* acquired much of the smaller, rotary-wing competition – subsidiaries of bigger aeroplane companies. Perhaps the fault lay in diversification when some of the old aircraft companies chose to enter the field of helicopters, instead of developing in their own specialties of fixed wing design: who knows. During those years, it was only the military who used helicopters. Development into the civil field was yet to come. In any event, *Westland* were building American helicopters under licence, while pushing forward their own plans to develop a number of wholly British designs.

Edward's driving test came up and he popped down to Yeovil on the specified day, to meet the examiner near the livestock market, next to the Odeon cinema. He directed Edward to drive around a set course in and around the town centre, while he walked to various vantage points from where he could observe progress, hand-signals and handling of the bike. The test didn't take long and eventually after a small disaster he handed Edward a ‘pass’ certificate.

This came as a surprise because Edward's bike skidded on loose gravel during the emergency-stop test, the last sequence on the list. In braking hard, he lost the bike and ended up in a heap right in front of the examiner. Edward picked himself up and the examiner helped him with the heavy motorcycle. He expressed concern at the incident, apologising profusely for initiating the braking test without regard for the poor road condition. Then he asked if Edward had precisely understood the situation. Edward answered the question in the knowledge that his front wheel had broken-away on the gravel and that he had dropped the bike intentionally to avoid mounting the pavement and to bring the bike to a positive halt, as if in a real emergency.

His answer seemed to please the examiner and he apologised again for his error of judgment, adding that he hoped the bike (and Edward) had not sustained serious damage. There wasn't any real damage, except for a little gravel rash on one of the exhaust silencers and the footrest rubber. They spent a few more minutes together because the examiner expressed interest in the bike, walking around it and asking many questions before returning to his office. Contented, and happily clutching the pass certificate, Edward drove home. With great ceremony, he took off the tattered L plates and threw them away.

There was little happening in his social life then. Shift work, though appearing to provide more spare time, night school and homework in preparation for the forthcoming City & Guilds exams, took priority over self indulgence. Edward had little desire to participate in the social activities that most of his fellow apprentices indulged in. He really had no heart for such excursions outside his working and technical college life. He had become somewhat introvert and his observation of those around him, at work, made him thoughtful of the future. He'd begun to see a fixed-line, career-trajectory that fell into the category of a typecast and repetitive daily routine. Conscription into the armed forces lurked at some point in the future, but he still saw himself sinking into the complacency of the establishment ...a life of indifference and dull insignificance.

The mundane loomed and he pictured himself as one of the unmotivated of the so-called working classes, clinging to a factory job, attended by a sameness that he so abhorred. He didn't want to be an automaton. He continued to write to Jane, but her responses had cooled somewhat with her own preoccupation for exams at school. This overshadowed their correspondence and friendship. They seemed to have run out of interesting things about which to write.

Edward pondered deeply again about National Service that Government regulations would oblige him to fulfil due course. This seemed to be the only possible route of escape from his portentous thoughts of obscurity, if not oblivion. Therefore, he had no real qualms about the eventuality of call-up, when it came. Perhaps he would volunteer, just as Freddy had done, as he had discussed with his father some years before. That way he might get back to the aeroplanes he so dearly loved, and perhaps even secure a little adventure and excitement to boot, unrestrained by family ties and predetermined conventions of civilian life.

His restlessness grew as time passed. Technical College exams came and went with pass results. Yet he felt no elation. Management accepted his applications for ability-pay on the job, as his skills at jig boring improved notably. He earned recognition for turning out good work on time, but still he felt unfulfilled.

Correspondence with Jane petered out over the months until one day she wrote a very short note, just to say that she wouldn't be writing anymore. It didn't surprise Edward, nor did it upset him greatly. The relationship was hardly what one might call, whole. It really didn't matter that much.

His twentieth birthday long since passed, he felt resigned to his future in the tool-room at *Westland.* On the one hand, his work was interesting and fortunately most of the jobs were not repetitive. On the other hand, although home life with his parents and younger brother wasn't so confining, the captive feeling refused to leave him. It was a combination of many things that impinged upon his desire for change.

He needed something more to stimulate his mind. Perhaps he needed an escape or, at least, a route that would offer more adventurous opportunity. In a year's time he would finish his apprenticeship with *Westland* and no doubt, before then, he would receive his call-up papers for National Service. The prospect of only another year heartened him. At any rate he could endure until he was twenty-one.

His thoughts in this regard had centred on joining the Royal Air Force when the time came but, over the years since Ilchester and Merryfield, he'd been out of touch with aircraft work. As an embryo Tool Maker, he could see no suitability for his newly-acquired skills in the Air Force. Many of his peers and fellow apprentices were thinking about the Merchant Navy as a strong contender to National Service. They too, felt much the same as Edward about the future with *Westland*, with its obscurities, uncertainties and stagnation. They discussed the subject often, over tea and sandwiches.

Finally, Edward concluded that he should grit his teeth, make the best of things for the time being, and try to get the most out of his current situation. He postulated over the possibility of promotion that might come his way. He would certainly have suitable qualifications, after the next round of exams at Tech College. However, just before his twenty-first birthday, the company pre-empted all these considerations with their own plan to offer senior apprentices a further two-year deferment from the armed forces. This surprising turn of events raised more discussion, but it became clear that the company needed to retain high levels of skill and experience, if they were to continue competitively in the industry.

Should the apprentices opt for the plan, continued attendance at tech school was an obligatory element of the proposal. This meant that they would have an opportunity to obtain additional qualification certificates. Their wages would be commensurate with those of skilled-trade rates, but they would remain classified as apprentices. On the face of it, the option was attractive.

Edward discussed the idea with his father. He needed learned advice before he could make a decision one way or the other. His father agreed that there were certain attributes about the scheme that might be beneficial. Promotional prospects, coupled with additional qualifications and better wages, fostered new enthusiasm in Edward's thoughts. Along with this, the strong possibility that conscription might end during the intervening two years was an additional factor for consideration. Edward therefore elected to take up the option, signing for a further two years of service with *Westland*. He settled down to endure the normalities of life for the duration.

38

It was winter-time one foggy morning when Edward set out for work as usual, at five-thirty. The fog was thick, its heavy moisture condensing on his goggles as he motorcycled down the hill to Yeovil. Once through Princes Street in the town, it was his habit to turn right into Huish road, passing the football ground and the old public baths building on his way. The street lighting was very poor at this point and the fog was thickening.

Presently, he detected the glow of a red light burning on the centre line of the road ahead of him. As he approached, more red lights appeared, glimmering in the murky mist and running in a straight line beyond the first. Edward remembered that when he'd passed this point the day before, on his way home, workmen had the road up on the side where he was now driving. He took a line to the right, intending to pass with the roadworks on his left. Quite suddenly he realised his mistake. A low mound of dirt loomed in front of him and before he knew it, the front wheel of his bike was over the top. Beyond the mound of diggings lay a ragged-edged hole in the tarmac paving. The men must have finished on the other side yesterday, filled the original hole and opened up a fresh one on this side. They had forgotten to place a light in front of the new hole and there wasn't even a barrier.

Somehow, Edward managed to throw himself sideways off the bike, as it fell into the pit with a crash. The lights stayed on, although the engine stopped as the machine hit the bottom. The hole wasn't very big around, but it was deep enough to receive the whole length of the bike, that now stood vertically on its front wheel, in the hole. After catching his breath, Edward reached into the hole to grasp the rear wheel in a vain attempt to pull the motorcycle out but it was far too heavy.

Sound and movement behind him attracted his attention as a

pedal-cyclist was about to follow the same route into the hole. The wobbling light hardly penetrated the thick fog. Edward lunged towards the cyclist, to warn him, but that only made matters worse and the startled fellow swung away from Edward's lurching figure, towards the road-works. He might have thought that the Martians had finally arrived as he faced a tentacle-waving leader clad in shiny black space-gear, with enormous, goggled eyes staring from beneath the round dome of its head – and the spaceship buried in the hole it made on landing.

At the last minute he saw the mound of dirt and hole with its ghostly glow from the lights of Edward's bike. He swerved wildly to avoid falling in himself, stopping just short of the pit to gape fearfully as Edward spoke though the scarf that covered his mouth.

"Can you spare a moment to help me get my motorbike out?"

"Not likely mate, I'm off," rasped the man, his staring eyes and chattering teeth belying a stout facade of bravery in the face of ghostly happenings, "bloody fog's gettin' thicker, I'll be late for work." With that, he pushed off and left Edward in darkness next to the roadwork where his bike languished silently, in a frozen parody of a vertical nose dive.

Since he couldn't move the machine by himself, he decided to walk to the factory. He could phone from the gateman's office and get the police or someone to come and retrieve his fallen steed. First though, he had to switch the lights off and make sure the petrol wasn't leaking into the pit, just in case some passerby threw a cigarette end into it, and set the bike afire. The humorous side of Edward's imagination conjured up the image of mini-headlines in the local newspaper:

"Fire in the hole at Yeovil, as motorcycle set ablaze in roadwork."

What if there was a gas leak and they were digging to find the pipe?

"Residents of Yeovil street evacuated as blazing motorcycle in roadwork hole causes fear of gas explosion."

Or worse: "Witness claims encounter with Martians in Huish street."

Later that day a message reached him from the factory's main gate, security office. A policeman had arrived with the recovered motorbike. Edward hurried to the gate-office and got a very hard time from the officer. He'd ridden the bike from the scene of that morning's incident and claimed that the vehicle wasn't in a roadworthy condition. Edward retorted that he (the policeman) wouldn't have been in a very roadworthy condition either, if he'd fallen on his nose down a six-foot hole. At least, the bike was ride-able. There wasn't that much damage, in spite of the policeman's comment. At any rate, Edward was happy to get the bike back in one piece.

The year drifted on into March. His twenty-first birthday passed with a small, family celebration. He received his apprenticeship completion certificate. His wage packet grew fatter with the promised coming-of-age pay award, to skilled rates, and an opportunity for promotion arose. One of the charge hands in the tool-room was about to retire. Edward applied for promotional consideration to fill the slot. He wasn't successful. Instead, the selectors gave the job to an older fellow whose track-record as a skilled toolmaker didn't meet the scrap-rate criterion. Everyone knew that, through poor workmanship, a high percentage of his work didn't pass inspection, ending up as scrap.

Edward questioned the unpopular selection with his Superintendent. It seemed to be inconsistent with the company's latest policy of placing strong emphasis on the need for the highest degree of quality. "It's better to move this guy up, if only to get him off the shop floor," the Superintendent explained, "we daren't sack him in view of the Company's published policy to retain workers because of the expansion program. The Union would have a field day. At least he can do little harm as a charge hand."

"Surely that means the workers and the other Supervisors will be forced to carry the fellow, making up for his shortcomings," Edward suggested.

The Super thought about this for a moment, before replying, "Yes, but it's better that, than having his high scrap rate. Anyway, he might make a good charge-hand."

To Edward the whole episode stank. He wasn't the only other contender for the post and it wasn't jealousy on his part, but it seemed unfair that the company should overlook so many recently qualified tradesmen in this way ...not an encouraging start to the new look, quality-first, plans. It seemed to Edward that striving for high-class work diminished ones promotional prospects. The incident annoyed him, but he tried to be philosophical about it. There would be other opportunities, of that there was no doubt.

In the meantime the 'time-served apprentices' had another ongoing battle to contend with. This showed them the reality of their situation as workplace outcasts. They had agreed to accept an option that gained them further deferment from National Service call-up. The accord charged them to continue with tech-school studies. Unfortunately this was not wholly compatible with the shift-work roster. Day-release lost them half a day's pay each week, while every other week, when working on the two-until-ten pm shifts, they lost three hour's pay for every evening that they attended night school, which was every night of the working week.

As apprentices, before completion of their service time, they received compensation against their weekly earnings for the hours lost at work when they were attending at Technical School. They asked for the reinstatement of this. Management showed little interest. Under the continuation of their apprenticeship contract to the Company, attendance at Tech-school was obligatory and binding. If this interfered with their earning capability as skilled men, it was just hard luck.

In spite of better wages (in theory) than before, given day-work instead of shift-work, they could expect to earn a near proper wage, only losing a full day's pay for the day release to school each week – but the company refused their request to revert to normal day-shift hours. Understandably, the company argued that this would have meant finding other work for half of the shift-working crews, and machine utilisation would reduce to eight hours a day instead of sixteen, adversely affecting output and productivity.

Equally, the Union wasn't sympathetic. They classified the ex-apprentices of the special deferment plan as Apprentice-members, even though they'd served their time and, in theory, received wages at skilled rates – in spite of paying Union dues at the fully-skilled-tradesman rate.

Under these circumstances, since the Company was immovable and the Union refused to represent their case, the ex-apprentices tore up their union membership cards in front of the Shop Steward. This lead to a serious situation because the company then appeared to be employing non-union workers in a closed shop environment. As a result, the unhappy lads were Black-legged.

The Union took the problem to Management, who had no immediate answers. The Works-Convenor called for a mass meeting outside the factory gates and union members took a vote, on a show of hands, to strike until the issue of non-Union labour in the tool room reached an acceptable resolution. The ex-apprentices (as they really were, in spite of their new contracts with the company) did not participate the strike, nor did they anticipate it.

The company had several choices open to them but, whatever decision they made, it was clear that blame for the entire episode rested firmly on the shoulders of the renegade ex-apprentices. They were in a whole lot of trouble from both sides.

Whilst the two parties negotiated over the strike, the ex-apprentices continued to work. They operated much of the key machinery in the shop but the strike curtailed overall output from the tool room. After a day or two, word came that decisions were imminent and negotiations were about to end. The major worry was that the Company might terminate the new contracts. An end to deferment would ensue and some Regimental Sergeant Major would have the lads at his mercy, on the parade ground of some obscure, dismal and lonely boot-camp. They lived in hope however, that an agreeable resolution to their wage shortcomings, related to the shift-work versus tech-school situation, might come about.

The company could put them on day-work or they could pay them a percentage against missed work time, whilst on release to studies. Either way, the lads had agreed amongst themselves that they would rejoin the Union if only to retain their jobs.

Finally and after all the strife, the state of limbo remained. The workforce got back to work and neither the Company nor the Union acceded anything to the ex-apprentices. The dispute found resolution through an additional clause in their contracts. In future, during any union dispute involving strike action, the Company agreed to place the contracted ex-apprentices on temporary, unpaid suspension. The Union had secured their closed-shop status. The Company retained their shift system and best machine utilisation. The lads were off the hook but they were no better off financially. On top of this they'd earned a reputation as trouble makers.

So, the manacles were on; the last link to their chains securely fastened. They earned less in wages than before they qualified as skilled men, and their action had only served to expose the imperfections of the established employment regimes of those times. It just went to show that they couldn't win against those odds. Harold Macmillan's proclamation to the country, that 'You've never had it so good', sounded very thin to the lads of Edward's ilk in 1959. The rut had gotten deeper and the future seemed to be even bleaker than ever.

Part 15

In Search of Freedom

39

The events of the passing year only served to reinforce Edward's resolve to climb out of what he perceived to be an ever-deepening rut. As predicted, military conscription ended and along with this event Edward felt sure that the way to fresh opportunities would prevail. The final key lay in his completion of the deferment-contract with *Westland*, by which he remained bound. He anticipated that freedom from these two obligations to the establishment would allow him to follow a clearer path towards the future, and his hopes to get back to a career in aeroplane maintenance.

Propelled by the need to have a plan of action as he entered his twenty-third year, Edward examined his attributes closely. The post-apprenticeship, forces-deferment-contract had twelve months to run, expiring on his next birthday. He had one year in which to prepare himself. He attached great importance to being ready for changes and for decisions he felt necessary to take when the time came. If all else failed, his safety-net lay in the strong probability that *Westland* would keep him on (ex trouble-maker or not) in the tool-room, as a full-blown, skilled, toolmaker.

With his mind thus focused, he began the search for other possibilities to make the changes he sought, at least anyway to begin with, a change of venue. So, what had he to offer? His attributes included several years in light-aircraft maintenance, a year or so with military jet-aircraft, some planning and production experience, machine-shop and tool-room training, a certified trade in the skills of a tool maker specialised in jig-boring and two City & Guilds certificates in technical subjects.

After his compelled separation from aircraft maintenance over the intervening years, the apprenticeship and technical school studies had concentrated Edward's training in machine shop aptitudes such as turning, milling, tool making and jig-boring. All these capabilities related to factory work and manufacturing. Specialisation in such activities was something of a contradiction to his aspirations. On the one hand, he had acquired useful skills by which he could earn a living, and upon which to fall back if his efforts to change his career-direction failed. On the other, such skills placed constraints and restrictions upon him, limiting his scope to re-enter aeroplane maintenance. He had a talent for drawing and painting, with a strong interest and preference for the technical facet of this capability, but he had no real paper-qualification in this kind of work.

The minor diploma out of art school really meant nothing concerning the possibility of using his artistic abilities in a gainful occupation. Nevertheless, it occurred to him that perhaps he could give his talent another try, to prove a point. He felt that his knowledge and understanding of mechanical engineering and manufacturing processes, gained over the passing years, placed him in a strong position to try his hand at Technical Illustration. In particular Edward's interest lay in the exploded and cut-away, three dimensional illustrations such as those that frequently appeared in aviation and motoring magazines, and in technical literature for machinery, cars and aircraft.

The more he thought about it, the more such a notion appealed to him, and he set himself the task to study and practice in this form of illustration. With less than a year to run on his contract with *Westland*, he could use his spare time productively in an attempt to acquire at least the basic skills needed for this proposition. Thus Edward formulated his plans.

Meanwhile, outside his working environment, he felt in need of fresh diversion. His motorbike had seen better days and he thought about changing it for a newer machine, though cost versus income prevented the acquisition of something new. A brand new bike was out of the question. The trade in motor cycles was then still very popular, and Edward decided to search for a suitable replacement on the second-hand market.

It was during one of his excursions into the town, to browse the many motorcycle showrooms in Yeovil, that a *Norton Dominator* took his fancy. Some years younger than his *Ariel*, the *Norton* was a 500cc, twin cylinder, sporting model – a touring bike with a large 'clubman' style fuel tank and swinging-arm rear suspension, unlike the *Ariel*'s, hard-ride, rigid back-end. The *Norton*'s 'featherbed' frame wasn't the latest duplex type, with twin down-tubes to counter flexing, but in all other respects the bike was truly of the *Dominator* mark. The price, based on a trade-in value of the *Ariel*, was certainly well within Edward's budget, and after a test run he decided to make the purchase.

Of course, he was sad to let the *Ariel* go. It had given him many hundreds of miles of pleasure over the years, but he put such feelings aside and delivered the *Ariel* to the showroom as agreed in part-exchange for the *Norton*. The salesman quickly arranged the details and documentation and within the hour Edward drove his new acquisition away to the open road. Pleased and satisfied with his purchase, Edward began the second of his projects aimed at a career change. He scanned the aviation magazines and newspapers for suitable job opportunities, and applied to those offering possibilities.

The *Bristol Aircraft Company* was among the first to respond, inviting him for interview at their engine division in Bristol. They'd advertised for technical illustrators to join their publication department, and offered six months limited training for less experienced applicants with potential aptitude and talent. This was a stroke of luck. Edward quickly arranged to take a day's leave of absence from *Westland*, to attend the interview at Bristol.

His visit began with a conducted tour of the less sensitive areas of the works, ending at the Publication Department. A senior illustrator showed Edward around and introduced him to sections of the art-work currently in hand within the department. Much of it was in full colour, and the quality impressed Edward so much that he inwardly questioned his own potential to produce such work.

Finally, the interviewers examined his curriculum vitae in great detail, asking many questions to qualify his background. Edward's anxiety, however, was ill-founded. After the discussion he had a strong impression that he was a likely candidate. He explained that while he was still under contract with *Westland*, he'd undertaken a self-imposed course to practice illustration during his spare time. This seemed to impress the personnel people and they made some suggestions about drawing equipment and books that would help in the studies. He also explained that he would not be immediately available because of his contract. His release wasn't until the following March. The interviewers accepted this in good humour, recognising that there were only a few months more to go. Although *Bristol*'s needed several illustrators and some trainees to meet future development plans, time scales to reach their recruitment goals were wide enough for them to agree that Edward's application should remain on the short-list of trainees, deferring his employment until he was available.

Edward's spirits soared as he rode home to Yeovil. He sang at the top of his voice, over the din of the *Norton*'s lusty, twin exhaust-pipes ...something he hadn't done in years. Perhaps things weren't so depressing after all. There was still much to do, however, for he wasn't planning to rest solely on this solitary stoke of good fortune that Lady Luck had sent his way. He would continue the search and hopefully attend more interviews during the remainder of the contracted time with *Westland.*

Although ultimately Edward was looking to get back into aircraft maintenance, he wanted to prove his artistic value beforehand. After all, he thought; there was no harm in having two or three strings to the proverbial bow. At home again in Yeovil, he visited a small shop he'd discovered years before, which specialised in art materials and drawing equipment. It was there that he'd bought his drawing board and drafting machine.

The shop stocked a wide selection of drawing pens and he bought two sets, some ink and various other materials with which to start his technical illustration exercises. It was pleasing that he'd kept himself in practice over the years, dabbling here and there with painting and drawing. Edward frequently used his drawing board for designing and drawing the plans for model aircraft, and other related projects that his elder brother and he had thought up from time to time. Edward's enthusiasm surged, and he soon began to develop his own techniques for producing presentable illustrations rendered in pen and ink. The work absorbed his spare time and gave him much satisfaction.

While he engrossed himself in the art work exercises, his mind turned to thoughts of working overseas at some point in the future. Several of his colleges, ex-apprentices like him, had already made the plunge. They hadn't taken up the two-year option to stay with *Westland* as Edward had done. Instead, they'd elected to join the merchant navy as cadet engineer officers. This exempted them from the Military National Service call-up. After initial training, they'd shipped out as fifth engineers, to sail the seven seas, visit far-flung places and learn the arts of naval engineering along the way. The notion of shipboard life didn't appeal to Edward that much, attractive though it seemed on the face of it. His thoughts and aspirations sought a life of living and working in foreign lands along with his beloved aeroplanes.

Africa and the countries of South America had a certain magnetism that appealed to the romantic in his character. He'd read several travel-books, written at the turn of the century. They had enchanted him with intense descriptions of those, then little known, continents – their rivers, mountains, deserts, jungles and primitive peoples of the regions, as seen then, by the authors in those distant days. Such vivid, descriptive pictures wetted Edward's appetite for adventure, reminding him of himself as a child, adventuring at play with his brothers and youthful friends.

Clearly, in 1961, neither of the two huge continents were any longer unknown places, waiting for the intrepid explorer to wrest fame from his exploits and lecture at the Royal Society. Indeed, the

Africa and South America of Edward's own youth were less than a couple of day's travel away by air but, true to say, the magic and mystery lingered. Plans fermented in Edward's mind and slowly he to put into perspective the route by which he might satisfy his compulsion for adventure and escape. To travel to work and live overseas was an enticing way of life that appealed to him.

Given the opportunity, he could, perhaps, give himself three to four years in which to taste the fruits of such adventure. After all, he had no real ties. The Establishment had released him from immediate obligation. He was young and not so much ambitious as inquisitive. He craved adventure and the freedom to seek anything that might prove different from his current way of life. How many times had it been said to him, "The world is your oyster, go, find the pearl", but to try his hand at technical illustration was also important to Edward, for two reasons ...the first to prove his artistic talent and the second to make a little more time in which to find a way to overseas employment.

Not long before his twenty-third birthday, *Bristol* wrote to him with disappointing news. The illustrator's job that Edward had thought might be waiting for him, was, after all, not to be. This turn of events, while unexpected, didn't upset him greatly. He had other irons in the fire and he had secured a number of other interviews in the meantime. Finally, a company in Basingstoke, not far from London, responded to his application and offered Edward a position as an illustrator. The interview was very positive and Mr. Stewart, the personnel manager, confirmed that Edward could start with them as soon as his release from *Westland* came through.

His extended apprentice-contract completed, Edward gave *Westland* the customary seven days notice, and made his withdrawal honourably in the affirmed knowledge that should he wish, they would have him back; a good second string for the bow that was about to launch his arrow of youthful desire, enterprise and personal development. Edward's new employers manufactured industrial tractors and fork-lift trucks, and his job was to assist with the production of technical manuals used by the sales people and product support department. It was a radical step to take, moving out of the world of aviation related work and out of the trade for which he had by then qualified, but Edward believed it was to be his first step out of the constraints that oppressed him.

He reported to the personnel department on a cold Monday morning, after a very early start from Yeovil and a long journey from Somerset by motorcycle. Mr. Stewart welcomed him and immediately took Edward to the Design offices to meet his new boss over a welcome cup of coffee. After a brief introductory conversation, the personnel manager withdrew, telling Edward to report to him at five that evening, to receive details about accommodation arrangements.

Al, his new boss, settled him into the office, introducing Edward to the Chief Designer and the design-team leaders. He suggested that they should take a look at the works, to familiarise Edward with the products, personnel and sources of vital, technical information. They strolled through the factory, visiting the various workshops and assembly halls. Construction and manufacture techniques impressed Edward, though they were quite unlike aircraft work, being mostly fabrication in heavy, steel plate. Several forms of welding were in use, whilst automated plate-metal cutting by flame torches and semi-automated fabrication were particular techniques in component manufacture. An extensive machine shop produced many parts for the various special components and hydraulic units used on the trucks and tractors.

Motive power for the tractors and forklift-trucks was, for the most part, electrical, providing for safe and relatively quiet operation indoors – in factories, warehouses and deep-freeze rooms, for example. Sizes ranged from small, pedestrian-operated vehicles, to the larger, rider-operated machines. Some had the capacity to lift several tons. Electric motors to drive the vehicles and batteries to power them, came from outside suppliers, together with specialised wheels, with solid rubber or neoprene tyres, flexible hydraulic pipes, multi-link lifting chains, lights and the thousand-and-one nuts, bolts and washers needed for assembly. Al seemed to revel in explaining things to him and his obvious enthusiasm for the products began to rub off on Edward. It was an interesting and instructive tour.

When they returned to the design office, Edward took his place at a desk next to his boss and studied several manuals relating to the various trucks produced at the factory. These included the Operating and Sales manuals, the production and up-dating of which was Al's responsibility. Edward's job would be the preparation of updated illustrative material, under the boss-man's direction and supervision.

Al suggested that a good starting point would be performance graphs, illustrating such things as weight ratios, battery capacities and wheel-traction on various surfaces, related to types of tyres, etc. Turning circle charts, lifting heights, overall dimension charts and the like, would all come to Edward's desk for his action and preparation. New models were on the drawing boards and, in liaison with the designers, he would need to extract relevant information for conversion into easily-read charts and graphs. The Sales Manual was of primary importance to the vending department for their extraction of the latest technical information, if they were to sell the product successfully.

40

Edward's introduction to the factory that morning was mind-boggling but Al, sensing his preoccupation, reassured him and told him not to worry, things would work out fine after he'd settled in. Meanwhile, the personnel department had arranged accommodation for him in the centre of old Basingstoke, at a hostel for temporary workers. The place was full of road-building people, employed by Tarmac, then working on the motorway construction from London to the south and west. They were a noisy bunch, but worthy fellows and friendly.

Edward spent three uncomfortable nights at the Hostel. Uncomfortable because the weather was extremely cold and the gas heating in his small, bare room had an enormous appetite for shillings. His cash-funding didn't run to it and so he huddled, shivering in his bed, clad in pyjamas, a thick sweater and a dressing-gown, under the single blanket and his heavy, motorcycling topcoat. However, the abundant food at mealtimes made up for the discomfort, and Edward ate well in the dining hall alongside the rough and ready, duffle-coated road men.

A few days later, along with another newly recruited chap, the company placed him in private lodgings on the edge of town. Their landlady was very rigid about mealtimes and she insisted on locking the front door at ten o'clock, lights-out at ten-fifteen after a cup of cocoa and an early call at seven-thirty next morning. Her tiny living room had the comforts of home, apart from the folded, brown paper coverings to protect her chair-cushions from the rumps of her lodgers and the trail of neatly placed newspapers, leading from the front door, to protect her carpets. At least, the place was warmer than the hostel.

Edward's lodging companion, Joe, came from the North of England. He was somewhat older and after work he busied himself with the task of house-hunting. He was a good-natured man who worked in the same office, with a battery of filing cabinets and a desk, next to the two illustrators. The father of two teenage children, he was eager to find a house to buy, so that he could move his wife and family down to Hampshire.

At weekends Edward motored home to Yeovil, setting out after work on Friday evenings and returning early Monday morning in time to start work at nine. The landlady wanted her lodgers out of the house during the daytime on Saturdays and Sundays anyway, so it suited Edward better to go home to see his folks rather than kick around in Basingstoke from breakfast until suppertime. Accordingly and in addition, his weekly expenses saw a reduction against five days' board and lodging instead of seven.

Edward, always pleased to see his family, was welcomed by his parents each weekend with expressions of deep interest in his new exploits away from home. As parents tend to do, they felt concerned for their middle son's welfare, even as they did for Freddy, who'd been working in West London for several years by then. Freddy visited Yeovil from time to time, when work permitted. Sometimes the brothers' visits to Yeovil, coincided. Such occasions were happy affairs, especially for their mother. She so enjoyed having the family together, if only for a couple of days once in a while.

After a few weeks at Basingstoke, Edward's lodging-companion, Joe, found a suitable house to buy. Situated near the village of Worting some miles West of the town and its industrial zone, the bungalow lay on a small, new estate. Joe took Edward along to see it one evening, to get his opinion. Edward thought it a compliment really, for Joe was twenty years his senior and had far more experience in real-estate dealing. The property was one of ten bungalows, recently built on a small housing development close to the mainline railway serving Southampton and the West Country.

Apart from the trains that passed frequently, its setting in the countryside offered peace and tranquillity. The nearby village, with a couple of small shops, a post office, a pub and a police station, straddled the North-South line of what was once a Roman road. The locality was quiet and rural. Joe reiterated his liking for the place, and since both of his near-grown-up children had finished their education, schooling wasn't part of the equation. They had a good look around and Edward could only agree over its suitability to Joe's needs. He decided to buy the place and bring his family down from the North, promptly.

Soon after he'd arranged the purchase of the bungalow, Edward's colleague moved out of the lodgings, camping out in his new house to prepare for the arrival of his family and furnishings. Joe explained that his son, the eldest of his two children, had found a job in London. His daughter would look for something locally after settling into the new home.

Edward remained at the digs and after Joe had moved out, their humorous winter evenings spent in front of the landlady's tele were a thing of the past. She spent most of her time in the kitchen, watching a small TV in there and with Spring coming on, Edward didn't fancy sitting on his own after work, indoors until lock-up and lights-out time. He spent his evenings driving his bike around the district, getting to know the county.

It was his custom to buy petrol for the *Norton* at a small filling-station, not far from the lodgings. One evening he drove into the forecourt and drew up at his usual pump for a few shillings-worth of petrol. Normally it was the owner or his mechanic who served, but this time an attractive girl attended to Edward's needs. She was new at the station, aged about seventeen and very pretty but, nevertheless, she seemed well versed in the workings of the petrol pump and the bike's tank-cap.

The slender girl smiled and chatted as she filled the tank, flicking her long brown hair back from her brow with a tiny toss of the head (an enchanting mannerism) as she returned the pipe-nozzle to its resting place on the pump. She was cheerful and friendly, making some admiring remarks about the bike. Edward needed some oil too, so he walked with her to the office, where he could select a small can of a suitable brand. They chatted for a few moments more at the doorway before he cranked-up the *Norton* to take his leave. The thought struck him that he'd definitely continue to use this station. He motored slowly back to his digs in a very happy frame of mind. Next day at work, his ex-lodging companion, Joe, commented that his daughter had found a temporary job at a petrol station, serving at the pumps.

Putting two and two together, Edward grinned and explained that he'd already met her, just the evening before, assuming, of course, that they were talking about the same gas-station. They chuckled over the co-incidence and Joe suggested that Edward should pop over to the new home and meet his wife too, after they had properly settled in. "I've got my son's motorbike in the shed at home," he added, "it's a *Royal Enfield* three-fifty, but it's not running right. Perhaps you could have a look at it when you visit."

That day Edward's boss told him that he was due for a spot of leave. Al wanted to take a two-week break, starting at the coming weekend. With that in mind, they spent some time preparing a list of work that Edward should press-on with during Al's absence. They agreed that if Edward should finish all the set tasks before Al's return, then he could start the preparation work for some complex cut-away and exploded illustrations that Edward had asked about earlier. This was what Edward had been waiting for. At last, finally, he could get his teeth into some genuine illustrative work. The idea made him very happy.

He popped into the gas-station again at Friday lunchtime to fill the bike's tank in readiness for his evening trip to Yeovil. Joe's smiling daughter came out to serve him as he pulled onto the forecourt. While she filled the tank, Edward mentioned that her father and he worked together, in the same office at the factory.

"Yes, I know. Dad told me," she laughed. "Do you like the job? You're an illustrator aren't you, Ted isn't it?" Urged on by the new intimacy of her cheerful demeanour Edward replied. "Yes, the job's quite interesting but I'm an aeroplane-man really, I took this on to get some experience and to develop my illustrative skills. What's your name, your father didn't say?"

"Marion." She replied with a smiling, sidelong glance, before light-heartedly bantering on about her anticipation of a couple of days off. It had occurred to Edward, as the petrol flowed, that her style of dress was hardly suitable for serving at the pumps that day, it looked more as if she was ready to travel. Marion seemed to read his thoughts. "I'm going up to London by train this afternoon, for the weekend. Lucky for me I don't have to work the Saturdays and Sundays," she explained. "What do you do with your weekends?"

"I usually go home to my parents' place in Somerset. I'm off tonight, after work," he responded, "I'll be back on Monday morning."

Edward paid for the petrol and snapped home the tank cap.

"See you next week then," said the girl as he kicked the *Norton* into life again. "Have a good trip."

"Thanks, you too; see you Monday."

It rained heavily as he drove to Yeovil that evening, and he got thoroughly soaked. But he didn't care, it's one of the joys of motor-cycling. Coupled with this, he was happy at work and pleased with the new friendship he'd struck up with Joe's daughter.

The following week on one of his almost daily visits to the gas station, Marion asked him if he would like to run her home on the *Norton*, after she'd locked up the premises. Her suggestion delighted him, and he readily agreed. Edward had nothing better to do – indeed Marion had pre-empted the offer he'd been contemplating.

Marion went on to explain her presumption in asking. "Dad suggested that I could ask you over tonight, now that we are all settled in at home." Joe had taken a couple of week's leave, like Al, and Edward hadn't seen him since the weekend before. It didn't take long for him and Marion to clear away the loose equipment lying about on the forecourt. Edward helped her willingly and soon enough they had the place firmly secured under lock and key. Her attire was more appropriate than when last they met. She wore blue-jeans and a light sweater, so hopping on the bike behind didn't present any problems. She settled into the seat with her hands on Edward's shoulders and they took off for Worting.

In those days crash helmets were not obligatory, though Edward always wore his – well, nearly always anyway. He made a mental note to buy another for future occasions, if or when his new friend rode with him again: an immodest thought perhaps. Marion shouted into his ear as they gathered speed and she moved her hands from his shoulder to his waist. Edward didn't catch the words but he fancied they were whoops of joy. On the corners Marion leaned with the bike. She was clearly not a first-timer on the back of a motor cycle.

With her hands at Edward's waist and the press of her slim body close to his, the proximity reminded him that this was the nearest he'd been to a girl since Valerie. He liked the feeling.

The village of Worting was to the left of the road upon which they travelled and at the signpost they turned off, passing through a tiny, one-street village, before turning left again into the little estate of bungalows were Joe and his family lived. Marion dismounted and leaned around to see Edward's face. Grinning from ear to ear and breathless, she tossed her head in a vain attempt to untangle the strands of windswept hair. "Wow Ted, I do like your bike. Can I have a go on it one day?" She exclaimed pushing the fingers of both hands through her windblown mane.

"I should think so; but only if you promise to be very careful." He answered, amused at her enthusiasm.

"Oh, that's great."

Joe came out to greet them as Edward wheeled the bike through the garden gate. "Come on in Ted, and meet my wife. D' you fancy a cuppa?"

Joe's wife welcomed him into the living room and after gushing formalities, she rushed off to the kitchen to put the kettle on for tea. Marion threw herself onto the deep cushioned sofa to watch TV as her Dad offered to show Edward around their new home.

It was certainly different from the first visit Edward had made, when Joe was prospecting for a place to buy. His wife had turned the bungalow into a real home. She'd clearly enjoyed the task. Bright, flowered curtains hung at the windows and thick, beige carpeting (all new, Joe explained) stretched from wall to wall, complementing the neatly placed furnishings. Joe took him through the French windows to the back, where he'd already started to arrange the garden, during his first week's leave. They sauntered around to see the various projects he'd started on, as Joe described future plans with pride. He was like a child with a new toy, happy and contented.

"After we've had a cuppa," he said, as he'd completed the guided tour, "would you like to have a look at the boy's *Enfield*? It's in the shed at the side of the house."

"OK, my pleasure." Edward's enthusiastic response was born out of his pleasure at being made so welcome, the good working relationship that Joe and he shared at the office, his friendly and often humorous companionship under the strict rules of their landlady and, not least, his growing attraction to Joe's daughter.

The *Royal Enfield* stood in the shed, collecting dust. It was a three-fifty Bullet, probably five-years old and by the look of it, still in good condition. Joe said that it ran erratically and sometimes stopped completely. No amount of cranking would start it, if this happened, until the engine cooled off. Since Edward needed some additional tools, he promised to have look at it another day. There was no time that evening anyway. It was close to nine o'clock and he had to get back to his digs, or find himself locked out.

"I should be going now or I'll be sleeping out. It's getting on for curfew time," joked Edward. They laughed over his reminder. Joe well remembered the rule of law under their landlady's roof. They went inside again and Edward made his farewells to Joe's family. They all followed him out to the garden gate to see him off with waves and smiles.

"See you tomorrow, at the petrol station." Marion called, as Edward pulled away on the *Norton*.

Besides his interest in the *Enfield*, Marion attracted him considerably. She reminded him in many ways, of a foal – delicate in her movements, slender and long legged and her unruly sienna-brown hair, like a mane was always floating across her eyes to be flicked away with a light toss of her head. She was very attractive and had a captivating way of eyeing others from beneath her dark eyebrows and lashes. She liked to talk about simple things like pop music, motorcycles and the movies, but Edward suspected that she had a tomboyish character beneath the surface.

He liked her, and she always seemed pleased to see him.

However, when it came down to dating, Marion invariably declined invitations with indifference or plausible excuses that dampened persistence on Edward's part. Motorcycles seemed to be of common interest to them and Marion talked incessantly about the latest machines as they came onto the market. Conversations between them, on the forecourt, often stopped abruptly, cut short for a moment when a new type motorcycle noisily passed the petrol station, gaining her entire interest.

The petrol station was their meeting place since Marion's job there placed her pump-duty times in the mornings and evenings. Afternoons were hers. It occurred to Edward that it was perhaps her hours of work that prevented her from accepting his invitations to go out together.

Meanwhile, his career ambitions continued to lurk in the back of his mind, and he relentlessly pressed on with the search for job opportunities, to get him back into aviation.

That was not to say that he wasn't enjoying his illustrating job. His work at the fork-lift truck company continued to provide interest and variety and his abilities grew with each new project that he undertook. While his boss was on leave, he did manage to complete the tasks left for him, and during the second week he developed some preliminary cut-away and exploded illustrations, as planned. This gave him enormous satisfaction and when Al returned to the office, he expressed his delight with the results of Edward's work on the projects.

Joe's appeal for Edward to have a go at mending his son's motorbike added further interest outside work and, moreover, offered him an opportunity to improve his disposition with Marion. As a consequence, he was reasonably happy with the new life away from home, for the time being at least.

Part 16

The Keys to a New Life

41

Edward got back to the office on Monday morning, after the weekend with his parents in Yeovil. He'd picked up a few useful tools from home, planning to make another trip out to Worting with the idea of working on the *Enfield.* Joe was still on leave for a further week, and so, at lunchtime, Edward popped into the petrol station to arrange a suitable time with Marion. He didn't want to barge in, unannounced, for he knew Joe and his wife would still be busy putting final touches to their new home.

Marion said that she was sure it would be all right to visit at any time, "Why not this evening?" Edward suggested that he could pick her up after work, run her home, and take a look at her brother's bike. She agreed, reminding him with a grin that she still wanted to have a go on the *Norton.* With the arrangement made, he told his landlady that he wouldn't be in for the evening meal and directly after work he drove to the gas station to collect Marion. Their arrival together quite surprised her parents and they proposed that Edward have something to eat with them. Joe's wife insisted, saying that the evening meal was almost ready and that he was more than welcome. Her clear insistence made it difficult to refuse.

Joe and Edward went to the shed to make some preliminary checks around the *Enfield.* Marion joined them to watch. On starting the engine, it was clear that the erratic running could only be one of three things, the fuel, the ignition or the valve timing. They agreed that after the meal they'd tackle the easiest possibilities first: fuel-supply and spark plug.

Marion decided to wash her hair after dinner, but Joe lent a hand with the *Enfield.* It didn't take long to discover that the fuel was clean, the lines were clear and the carburettor was functioning properly. However, the spark-plug was in an awful state. It needed considerable force to turn it. Removal was difficult and Edward had to use care for fear of damaging the threads in the cylinder-head. Finally the plug came out, but its condition plainly showed why it was so tight, and why the engine didn't perform properly.

"This looks like the original plug Joe; has it ever been changed?" Edward asked, holding the plug up for Joe to see.

"I've no idea, lad, my boy isn't very technical. He just likes to ride the bike and only cleans it when he feels like it. The bike hasn't been used for a few months," he said, shaking his head.

Edward remembered that he had a couple of spares with the *Norton*. "Well, I have a couple of new spark plugs that I keep with my bike, as spares. I just hope that they're the same type." He fished around in the *Norton*'s tiny toolbox and found the plugs. They weren't exactly the same by numbering but the thread size and the reach, were. They decided to try one of them anyway, after setting the gap. The *Enfield* started first kick and it ran smoothly enough for a single cylinder machine. Edward blipped the twist-grip, revving up the engine, then left it to idle awhile, until it warmed up. Joe smiled and patted Edward on the back.

"You want to give it a try Joe?" Edward asked. "Take it up the road and see how it feels now."

Joe shook his head. "No lad, you do it, I'm not a biker any more you know." At the sound of the smooth running *Enfield*, Marion came rushing out with a towel wrapped around her head.

"You fixed it then," she exclaimed, "are you going to give it a try up the road?"

"Yes," returned Edward, not without a little triumph in his voice, "it's running well but I'll take it for a whirl anyway."

He took the *Enfield* for a short spin into the village and back and then down the straight, narrow road, that followed the ancient Roman route. The bike performed well and showed no sign of the erratic running displayed when they started it earlier that evening. Content with the results of their work, Edward turned again and rode the bike back to the bungalow. Joe was as pleased as punch and asked Edward if he could pay him for his trouble. Edward told him that it was no trouble and payment was unthinkable, especially as they had given him a fine meal and their kind hospitality. Joe conceded and together they put the *Enfield* back in the shed and found a piece of old bed sheet to cover it with. Joe said that he'd give the bike a clean sometime, before his son came home again.

Meanwhile, Marion had climbed onto the *Norton* in the driveway, and she sat there wearing a hopeful look on her face. She smiled sheepishly as Edward approached and asked if they could go for a spin and let her have a drive. Joe shrugged and said that he had a few jobs to do in the garden before the sun went down. Edward looked at his watch but Marion's look was so imploring he couldn't refuse. "OK, let's do it, d'you want me to take it up to the

Roman road then, and you can have a go on it there?" She nodded with a whoop of joy, moving back onto the rear seat to let Edward take the front and they set off for their little excursion. Edward thought they should use the Roman road because it was straight, and although narrow, it was smooth and little used by heavy traffic.

On the farther side of the village, he pulled over to the verge and stopped the bike. They dismounted and after Edward had shown her the controls, he invited Marion to take over. She swung a leg over the seat to sit in the driver's position and he remounted to sit behind her. She already knew about the brakes, the clutch and the twist-grip throttle, but she asked about the gears.

The foot changed gears on the *Norton* were one notch up from neutral to first, then one notch down (passing through the neutral position) for second, down a notch again for third and down again for fourth. One up and three down they called it. Some bikes were different, like Triumphs and some *BSA*'s. Their gears were one down and three up.

Marion put her weight on the kick-lever to start the engine. Edward felt a trifle nervous but he kept his feet up on the passenger's foot-rests, and let Marion take all the weight on her own legs. The *Norton* was a heavy bike but she managed to maintain an equilibrium as she lifted her right foot to engage first gear. Then they took off. It was a smooth, well-controlled acceleration and Marion slipped the machine into second gear without any jerks. They gathered speed down the narrow road and as Edward looked over Marion's shoulder, he saw a car in the distance, approaching from the other direction. The road was straight and one could see a fair distance down it. As they drew nearer, Edward realised that it was a police car. This worried him because he knew that Marion didn't have a driving license, provisional or otherwise, neither of them was wearing a crash helmet and they had no "L" plates displayed.

"Now we're for it," he thought, but he didn't say anything to Marion. She was by then fully engrossed in driving. To Edward's astonishment, as the police car passed, going in the opposite direction, Marion raised a hand to the occupants, shouting something as she did so. The two policemen in the car returned the wave in a very friendly manner, then they'd passed and Marion sped the bike on towards the end of the straight. She slowed the machine, changing down through the gears quite professionally, and brought the bike to a halt at the side of the road. She turned her head with a grin, her eyes bright with excitement.

"That was super!" She said. "Can I drive it back?"

"Yes," Edward answered, "but what about the police in the patrol car?"

"Oh, don't worry about them. I know them, they're the local

police from the village and they've seen me driving a motorbike before." Marion explained.

"Oh, your brother's bike?" Edward ventured. "The *Enfield*?"

"No, silly, my boyfriend's. He comes over from London sometimes and lets me drive his bike all around the district." Her response answered the questions in his mind about her clear understanding of how to handle a motorcycle but it also explained why she'd consistently evaded Edward's petitions to date. It rather took the wind out of his amorous sails. "Let's go back then," he said, feeling more than a little put-off. When they got back to Joe's house, Edward realised that it was time to go.

Apart from his bruised sentiments, the time was getting on. Perhaps he'd over-reacted. Perhaps he was too sensitive, or just plain naive, but Marion's reference to her boyfriend, with a possessive inference, as opposed to simply a boy with whom she was friendly, had rather upset him. He felt a tiny bit absurd really, but her statement smacked of permanence and seemed to close the door on any further advances on Edward's part. He shrugged mentally, hiding the self-analysis behind a smiling exterior, and made his exit-stage-left with a joke about sleeping in his landlady's garden if he didn't make a move. On that note he wished everyone a good night and drove the *Norton* thoughtfully back to his digs.

The following weekend in Yeovil, Edward picked up a copy of his usual aviation magazine and scanned the Situations Vacant columns. His eyes lit on a small-ad that seemed intriguing, although it said very little and referred only to an urgent need for aircraft fitters, based at home or overseas. The advertiser gave only a telephone number, so he decided to try it after the weekend.

He called the number on Monday and within seconds found himself talking to a loud voice with a commanding ring to it. Edward introduced himself and expressed his interest in the advert, offering a little information about his background and experience.

"Where are you calling from?" the voice asked.

"I'm in Basingstoke."

"Oh, good, that's not too far. Can you come over to the Isle of Wight sometime this week, for an interview?"

"Yes, tomorrow if you like."

"That's splendid," exclaimed the voice, "if you can get to Ryde, by the ferry from Portsmouth, there's a good one at Ten o'clock, the train from the pier will take you to Brading. I'll arrange for someone to pick you up from there, and bring you to the office."

"OK, I'll get down to Portsmouth in good time."

"Good. Look out for a dark-green Land Rover at Brading."

"Yes, I've got that."

"Right, see you tomorrow then. Good-bye."

The conversation told Edward nothing about the job, yet it was clear to him that an adventure was about to begin; he could feel it in his bones. The invitation to interview left him to use his own initiative in making the trip. Perhaps this was a test. He asked his boss for a day's leave of absence and upon his agreement Edward made a list of things to remember. He spent the rest of the day unable to concentrate clearly on his work. His excitement was too great. During the evening he took out his maps to plan the route he would take next morning. An early start was clearly necessary and he advised his landlady before turning in for an early night.

The journey was very straightforward and he made good time getting down to Portsmouth. He was early but that gave him time to find a place to park the *Norton*, collect his wits and look for the ferry departure quay. He didn't know Portsmouth, not having ever been before, but it didn't take him long to find the ferry-quay at Southsea and a place to park the bike.. The ten o'clock boat was about to take on passengers, so he bought a ticket and boarded with ample time in hand, before its departure.

Edward had never been to the Isle of Wight either, although he knew that his parents had spent their honeymoon at Shanklin, on the South-eastern coastline. The maps showed him that the seaside town of Ryde lay on the North-eastern side, almost opposite Portsmouth, some forty-five minutes by ferry across the waters of Spithead.

From the deck he could see the hazy blue-purple outline of the island stretching Westward, it's central, undulating down-lands vaguely visible in the early morning haze of Summer. Whilst he occupied himself in thought and taking in the maritime view, the ferry quickly filled up, its decks crowded with chattering, laughing holiday makers. The ship's siren blared and quay-hands cast off the rope cables.

He breathed deeply of the salty air, borne on the stiff sea-breeze that wafted strongly across the decks as the vessel moved out slowly into the calm waters of the estuary. Below decks, the throbbing engines drove the ferry smoothly across Spithead, lurching gently as it crossed the dissipating bow-waves and wakes of larger ships on route through the Solent to Southampton's great port. The swirling splash of parted sea at the ship's bow, and the churning rumble of its own wake at the stern harmonised with the deep thrum of the ferry's diesel exhausts at the funnel-head.

The plaintive cries of seagulls joined the sea-borne sounds, and Edward watched with amusement as the grey-white monarchs of the coastal-airs, hovered and wheeled alongside, in the stiff breeze, dipping and flapping to snatch morsels of bread from the fingers of wary passengers. It was a good way to start a holiday, he thought, and a relaxing way to go for an interview.

As the ferry tied up against the docking-pier at Ryde, he could see that his connecting train was standing at the station. The railway reached right to the head of the pier so that transferring passengers, with destinations on the other side of the island, had little to worry about, and only a short walk to make with their luggage. The narrow-gauge train's ultimate destination was Shanklin, halting at Brading and Sandown along the way. Edward remembered the Manx railway that he'd travelled on all those years ago, as the little tank-engine moved off with a piercing whistle, and tugging jerkily at its carriages. His excitement grew and new sensations filled his thoughts. He felt at ease and relaxed, yet excited at the prospect of what he might discover that day.

Over the years, he'd discovered that his intuition was very reliable and that day it told him something good was about to unfold. His thoughts focused as the train-wheels clicked rhythmically over the rail gaps and the carriage swayed through the gentle bends. He was on his way to an interview for a job, about which he knew nothing. The military voice on the phone hadn't mentioned anything to answer any of Kipling's five famous questions: Why, What, Where, When and How. All he knew was that jobs where available for aircraft fitters, at home or abroad, just like the advert said, nothing more, nothing less.

"Brading! Brading!" The shouted station-name brought him back to his senses as the train came to jerking rest at a tiny station.

Edward alighted along with a small crowd of noisy people clad in sailing gear, red jeans, dark blue sweaters and rope-soled footwear – holiday-makers out for a couple of weeks sailing; bobbing about in their tiny boats on the briny sea by day, and sampling beer over the island's famous lobster or crab cuisine of the many pubs and restaurants, by night.

The ancient Roman town of Brading nestles with its steep streets close to a tiny stream (hardly a river) that drains the down-lands to the West. The stream empties into Bembridge harbour, a popular and well-used haven for the sailing and yachting fraternity. Edward didn't know this at the time but later it answered the questions in his mind about the nautical flavour of dress that his train-companions wore. Whilst the happy sailors noisily gathered their luggage, Edward had his return ticket clipped as he stepped through the station gate into a small parking area.

He cast his eyes about the yard, looking for the promised pick-up vehicle and of several private cars and a couple of taxis parked in the bright sunlight, the green Land Rover was easy to spot. It stood in the shade of a tree ...very thoughtful of the driver. A young man in white overalls got out as Edward approached. He was about the same age as Edward; he had blond, curly hair worn with long side-burns and his blue eyes brimmed with humour.

"Is it Mr. Weinel?"

"Yes."

"I'll run you into Bembridge then," said the lad, with a grin and holding the cab-door open for Edward, "was it a good crossing?"

"Yes, smooth enough."

"You didn't bring your motor bike with you, did you?" The lad asked before climbing into the driver's seat, eyeing the crash-helmet that Edward carried by its strap, before glancing towards the station-platform as if expecting the bike to follow at a whistle, like an obedient dog.

"No, I left my bike at Portsmouth. I'll pick it up when I go back, later." They both laughed nervously and the young man hopped into the driver's seat looking somewhat relieved. Perhaps he thought that Edward had driven up the railway track, behind the train, and they'd have to load the bike in the back of the Land Rover. Who knows? He drove at leisurely pace out of the village, turning left to take them Eastwards on the road to Bembridge.

The young man said no more and Edward was content to watch the green countryside drift by. Brading marked the steep Eastern ending of the downs. The farm-land and wild heath-land they passed on their way, swooped and dipped in gentle hillocks, winding the narrow road over and between green fields, knolls and combs.

The steep, green, humped rise of Culver Cliff rose up on their right, standing on the South-eastern coastline, breaking at its rounded peek on the seaward side, to plunge, white and near vertical, into the English Channel on the southern face. They rounded another curve, descending then, along the landward foot of Culver. The sea at Spithead was visible again in the distance and a windmill on a wooded rise, beyond a broad field with a wind-sock and a hangar, stood at its roadside edge.

"This is the airport," unnecessarily announced Edward's long-silent driver, "and that's the hangar." He pointed with his finger to emphasise another obvious fact. Passing the hangar, they turned sharp-left into the open, roofless forecourt of a petrol station where four pumps lined up in the middle of a paved area, in front of a small hut painted with yellow and red squares. The hut clearly served, not only as a place for the petrol pump attendant to sit, but also as a tiny, airfield control-office. Next to the hut, on the airfield side, a small grassy area with a couple of long benches seemed to provide a pleasant place for visitors to sit – perhaps to wait for the arrival of an aeroplane, or just to sit and enjoy a pint of beer from the pub that Edward could then see, nestled alongside the hangar.

Prettily fenced with knee-high, white painted, wooden palings, the grassy area took in a beautiful view at the Eastern tip of the island, from the aerodrome, across the fields and landscape of St. Helens, Nettlestone and Seaview beyond, rolling away, downhill, Northwards to the sea at Spithead.

The driver parked the Land Rover and offered to take Edward to the offices, nudging with his elbow as they walked across to a side door in the hangar, next to the pub. With a theatrical wink, he raised his arm in a tipping motion, his hand gripping an invisible pint.

"It's open," he said, with a laugh, "I'll be supping a couple at lunchtime." Edward looked at his watch and noted with surprise that the time stood at noon but the pub's sign didn't surprise him, however. The Propeller Inn seemed like an aptly fitting name in view of its surroundings, sited on a small airfield. The place reminded him of the flying club at Liverpool. It carried the same atmosphere and a friendly aspect.

With no less than five brass plaques next the side door into the hangar, it was clear that more than just one company registered at the works. Each bore a different name. The driver explained that they were all subsidiaries of the same company.

Inside the hangar, several aeroplanes occupied the space just inside the big sliding doors that stood open to the field. What Edward took for a hovercraft, painted red and circular in shape, lay in one corner, close by. The sound of drilling and hammering, filled the lofty span of the workplace, as a group of men in overalls worked intensely around another type of craft, in the centre of the floor. There were benches next to a row of ground-floor workshops, in an open, lean-to area to one side, where more people worked.

42

The first floor offices lay along the back wall, facing the airfield. A staircase in one corner gave access to them along a balcony overlooking the hangar floor. Edward's young driver took him to the one marked General Office, at the head of the stairs, and seated him with a couple of typists for company. With another wink at Edward, he parked his backside on a desk-corner and made his announcement to the girls. "Mr. A will want to see this chap, he's expected. I just picked him up from Brading."

"Oh yes, it’s Mr. Weinel isn't it," one of the girls answered, "You're expected, I'll just see if Mr. Ayles is free." She hurried out of the office and Edward's driver dropped into deep conversation with the other girl.

Moments later, a big man in his shirt-sleeves swept into the General Office like a whirlwind. He was tall and heavily built; his ruddy complexion and silvering hair put his age at around Fifty. He grasped Edward's hand in a vice-like grip and greeted him with a loud voice. Edward recognised him as the man with whom he'd spoken on the phone, the day before. The big man thanked the girl who brought him and ushered Edward out on to the long balcony overlooking the hangar's work-space below.

"Come along to my office," he said, marching quickly ahead, "I'll talk to you first and tell you all about what we're looking for. Then, if you have the right background, and if you're still interested, I'll hand you over to the our operation manager."

In his small office, the big man seated himself behind a desk in front of a large map pinned to the wall and displaying North Eastern Africa. He waved Edward to a chair and leaned back ponderously in his own. An affected smile belied a rather humourless, piercing, blue-eyed gaze he directed at his visitor. With elbows resting on the arms of his chair, he put the finger tips of his large hands together, tapping his lower lip precisely with the forefingers, before speaking again.

"We're looking for aircraft fitters and air-strip supervisors for a season in the Sudan." He announced. "We operate aircraft to spray cotton plantations with insecticides, starting in August and ending in December or January – a six to seven month season." He swung his pivoted chair to look at the map behind him and pointed to a vast area marked in pink. From where Edward sat, he could see that it lay south of Egypt, with a stretch of coastal strip on the shores of the Red Sea. His interviewer tapped the map with his finger at a point where two rivers joined into one, before continuing with his lecture.

"Khartoum," he said. "We're taking four aircraft out at the end of next month, brand new *Piper Supercubs*; we're in the process of modifying them now." He swung his chair again to face Edward and ask him if he had his CV handy. Of course Edward had, and passed it over for scrutiny. "This looks good, I must say. I see you have some experience with light aircraft; that's useful." The big man seemed pleased with that and levelling a very direct gaze at Edward, he asked him if he'd ever been abroad before.

"No," Edward answered, "but I'd like to. I've been looking for the opportunity."

"It's living rough you know, once we're out of the city and on the job in the field ...long hours, dawn till dusk, mosquitoes at night and flies in the day-time, but we're quite well organised. It's a bit like camping." Mr. Ayles gave him a sidelong, questioning glance. "You haven't been in the army, have you?"

"No, but I've had a bit of camping experience and I know about living in tents." Edward said, in response to this unhidden test of his resolve.

"Good, excellent. When would you be able to start?" The big man positively beamed.

"In a week. Will that be OK?"

"Oh good, the sooner the better. You see, if we take you on, with your technical background I think we'd like you to help out with the modifications and flight preparations on the *Pipers*, before we fly them out to the Sudan." He pronounced the country's name with strongly accented emphasis: Soo-dan.

"Fine, I can give my employers a week's notice as soon as I get back."

The big man stood up and handed the documents back to Edward. "Well, I'll take you to see our operations chap now. He's the manager of the Sudan operation and ultimately it'll be his decision whether to take you on or not."

The ops man, short with fair, thinning hair and glasses, sat in the next office. He smiled and stood up from his desk as Mr. Ayles and Edward entered. "Les, Mr. Weinel has come about the airstrip supervisor's job in the Sudan. I've had a chat with him and his CV looks very promising. He can start next week if you think he's suitable." The big man gushed, adding more favourable remarks, intended to be convincing that this applicant was eminently suitable. Edward began to wonder what kind of interview this was turning out to be. The ops man shook Edward's hand and offered him a seat as he thanked Mr. Ayles

"OK Peter, I'll have a look at his CV too, and give him some more background on the job." Then, directing himself to Edward as the big, affable man left the office, "Peter is our chief pilot. Now, let's have a look at your experience." The ops manager spoke with an unusual accent, with origins in Lincolnshire, Edward suspected. His dapper appearance and straight-backed stance gave a strong impression that he too, had a military background.

The CV seemed to please him and he asked a few technical questions before explaining the company's Sudan operations in more detail. His own particular speciality was the management of aerial crop-spraying work. He'd been in that line for a number of years, mostly in UK but with the seasonal excursion every year, to the Sudan. For many years they (the company) had used *Tiger Moth* aircraft, but during the previous year they'd ended badly with all the planes written off in accidents. This year, with the purchase of new aircraft they expected better things.

He explained further, that the spraying contracts for the coming season were small. They were spread out among three basic areas of the cotton-growing region, between the Blue Nile and the White Nile, south of the capital, Khartoum. He said that the company needed air-strip supervisors for ground support at each of the three locations, organising the ground tanker-units, chemical mixing and aircraft loading. Clearly, applicants having aircraft experience would be more useful; being able to assist with the periodic aircraft maintenance, as well.

"How do you feel about assisting with preparation of the new aircraft, before they're flown out to the Sudan?" he asked. "We have four and they must be ready by the end of next month."

"Fine, I told Mr. Ayles that I could probably start in a week's time," Edward said, "I want to get back into aviation again and I've been looking for an opportunity to work abroad."

"OK, you could spend a few weeks here, helping with the aircraft preparations, that'll give you the opportunity to familiarise yourself with the *Pipers* and also give us time to get your visa documentation organised. Do you have a passport?"

"No."

"That's OK, we can arrange that, too." The ops man wrote a few notes and with a smile he told Edward that the job was his.

"The few weeks here, will give you the opportunity to learn about the rotary spray gear too," he said, as they shook hands over the deal. "Its specialised equipment of our own design, made here on the island."

Edward's thoughts were in turmoil, everything was happening so quickly. He tried to focus his attention on some of his own questions as he suddenly realised that there were preparations of his own to make.

"What sort of clothing will I need?"

"The Sudan is quite hot in the summer so you'll need tropical kit, cotton shorts and shirts are best. Get Khaki drill because the job is quite a dirty one. Three or four pairs should be enough. Take plenty of underwear and a couple of towels. Long trousers are useful for the evenings but don't take more than one suitcase because you'll be moving about a lot."

"Will I need vaccinations?"

"Yes, all that can be done by your own doctor, or we can arrange it from here. I'll give you a list, there's only three, nothing dramatic."

"Will I need my driving licence – though it's only for motor cycles?"

"Take it anyway, a local one can be obtained on presentation of a British licence."

Edward asked a few more questions and within fifteen minutes the interview was over, on the promise of a contract when he joined the company in a week's time. They shook hands again, the process completed.

Much to his own astonishment, Edward found himself wondering what had happened. There he was, standing on the Isle of Wight, accepted for a new job that was about to propel him back into the aviation industry, working with aeroplanes again and, to boot, with an overseas contract in sight. What a fantastic stroke of luck!

He didn't have a great deal of time to get organised, to give a week's notice, to get home and explain to his people, to buy the things he needed, to attend a medical and get his vaccinations. It seemed like an awful lot to manage in such a short space of time. On the other hand, if he was to spend a month or so working at Bembridge, there really was plenty of time to sort out the finer points. He'd landed on his feet, and he felt rather like a high-jumper successfully beating his own record. It was mid-June 1961; a time to remember and the beginning of a new life that promised to give impetus to his modest ambitions. Little did Edward realise just how profound was the chance-sighting of that tiny advert in a magazine; indeed, it was the key to the door of his whole future life.

Time was getting on and the ops manager suggested that they

go down to the pub for a drink. The place was full and they had to squeeze through the crowd to reach the bar. Mr. Ayles saw them coming and hailed them over to a space next to him. Edward wasn't a drinking man then but he accepted a brown ale, for his mouth was dry with excitement, and, moreover, he felt that a small celebration over his eminently successful interview was quite in order.

Peter (Mr. Ayles) asked if the interview was OK and seemed very pleased when Les (the ops man) affirmed that Edward would start with the company in a week's time. Peter beamed and craned to sweep his china-blue eyes about the bar until he found his target. "Let me introduce you to Trevor. He's involved with the *Piper* mods and he's coming to the Sudan with us. You'll be working with him."

At Peter's hail, a younger man squeezed through the crowd to join them and meaningfully set down his empty pint pot on the bar.

"Hello Peter, Les, what's on then, who's round is it?" he asked in a Hampshire accent with a mischievous grin as he winked at Edward.

"I want to introduce you to Mr. Weinel," smiled the big man, then to Edward, "what's you first name?"

Trevor watched his pint pot intently as the barmaid refilled it, before turning his eyes to rejoin the conversation with a broad smile and the freshly charged tankard clasped in his fist. "Ted," answered Edward and shook hands with the vigilant Trevor.

"Ted's joining us next week and he'll be helping you with the *Piper* mods. You'll have to show him the spray gear, too, and train him up a bit," explained Peter.

"Fair enough," said Trevor, turning to Les, nudging him with his elbow, "where are we going this year Les? No more Bonziga I hope."

They all laughed as Les said that he didn't think Bonziga was on the books this time. Of course Edward hadn't clue about what they were talking about, nor what they found so amusing, but it seemed to him that the relationship among the three fellows was team-like and extraordinarily familiar, in spite of the clear differences in their positions with the company. It was refreshing and again, reminded him of his flying-club days.

He declined an invitation to lunch in favour of heading back to Basingstoke. Les arranged transport to take him to Ryde for the ferry. The driver was going that way on an errand anyway, and it would be easier than messing about with the train. Edward got back to Basingstoke in good time for the evening meal at his digs and he explained to his landlady that he was about to give notice to his employers. Subject to which he would then give her a week's notice, too. Although surprised, she accepted the situation without complaint.

Edward was sure that the Fork-lift Truck Company would send someone else to take his place as a lodger, since they were expanding and always had more people taking up new positions. Edward's would be no exception he supposed. They would have to replace him. Next day he gave notice at work. His boss was a bit disappointed but expressed his understanding at Edward's desire to travel. He had spent a few years in the Royal Navy and consequently he'd travelled considerably, himself.

Close to the petrol station where Marion worked, there were a number of small shops. One sold books. On his way to buy petrol, the day after his visit to the Isle of Wight, Edward called in to see if the shop had any reading material on the Sudan. As it happened, they did – a rather large book containing geographical, political and cultural descriptions of the country. He bought it so that at least he would have some idea of what to expect when the time came.

He told Marion of his success at interview and that he would be leaving Basingstoke within the week. She showed no particular interest other than to congratulate him on his good fortune, but how she felt about it mattered little to Edward anyhow. He was full of excitement about starting the new job, getting back to aeroplane work and the prospect of going overseas. In any case, they were just good friends, as they say, after she had successfully, if unwittingly, switched off Edward's ardour.

When the weekend came, Edward rushed off to Yeovil as usual, but this time with his special news for the family. Edith expressed concern, naturally enough, but his father seemed rather pleased. They had a long talk about the job and Fred seemed to think that a short contract of six months would likely give his son a good idea of life away from UK and family. However, he warned Edward about the dangers of falling into the wrong company, and of drinking and gambling: two easy traps to fall into. Not that Fred was a prude, on the contrary, but having been overseas himself, he well understood the ramifications for a first-timer. He asked his son to write home to Edith at least once a week, to which Edward gave him his solemn promise that he would and he kept that promise, without failing, over the ensuing years.

Edward set himself a target, though looking back he supposed it was a bit silly. He would try to achieve at least three years in overseas work. His brother Fred had spent three years away from home during his time with the army, and Edward wanted to do the same, or better. In retrospect, perhaps this petty ambition came from experience in his schooldays of always being compared to his elder brother's achievements, by teachers at Primary School. He wanted to be more than a patch on Freddy's reputation, though with all respect to him, of course, such comparisons were never, ever, of his brother's making. Perhaps Edward just wanted to prove a point to himself, if no one else.

With just over a month to serve at Bembridge before going to Africa, he planned to drop his usual weekend trips to Yeovil, except for the last weekend before leaving England. It would be like a dummy-run at being away for an extended period. That way his final departure, when it came, would perhaps be less of a wrench for his mother. He knew that she wasn't so keen on the idea of his going away to foreign parts, though she didn't say so outright. Apart from that, there was much to do at Bembridge and he had a lot to learn about the spray-gear. His weekends would be therefore be well used in that regard. However, he needed to lay-up the *Norton* for the duration of his absence abroad, so the last weekend at home would be essentially for that and for, what he knew would be difficult, farewells to family.

His last day at Basingstoke was a Wednesday. Edward decided to travel down to the Isle of Wight directly, the following day, on Thursday morning. That way he would have time to settle in and to deal with the initial formalities of his new employment. Accommodation came high on his list of priorities, though he expected the company to have some pointers in that direction. Accordingly, on Wednesday afternoon, he made his farewells to the bosses and companions at the office. Joe invited him over to Worting that evening saying that his wife would like to say Good-bye, too.

"She's rather taken with you, Ted and there were moments when she thought you might get to know our daughter better – put the brakes on her a bit, you know. Marion's a bit wild sometimes."

"OK, I'll take a run over later on."

When he arrived at their house, Joe's wife made a point of telling Edward how upset she was, not with him but with her daughter. Apparently Marion's 'so-called-boy-friend' had arrived that afternoon, quite unannounced, expecting to stay-over for the weekend. "With not so much as a by-your-leave," complained Joe's wife, "and then, when Marion came home from work, they just upped and went off on his motor bike. I just don't know what's come over the girl. Kids don't seem to have any respect or manners nowadays. It's just too bad that she couldn't be here on your last visit with us, it's just too bad." She sniffed and moved away towards the kitchen, saying that she'd make some tea. Joe shrugged with a thin smile, and he and Edward sat silently for a moment until his wife returned to join them. "Kettle's on, we'll have some tea in a minute," she said, smiling now. She came to sit next to Edward on the sofa, patting his knee as she did so. "It's a pity you couldn't have got to know Marion a bit more, Ted. I don't like that boyfriend of hers. He's not good for her I'm sure, but now you're off to Africa." She sighed resignedly.

Edward kept his thoughts to himself. Never having met Marion's boy-friend, he could neither comment nor commiserate and anyway Marion had put him down as a friend of her Dad's rather than someone who might have an interest in her. He certainly wasn't going to make an issue of it though. It had never occurred to him that Marion's mother held him in such esteem. Joe's wife spoke again, breaking into his thoughts. "I expect they'll be back before you have go, Ted," she said, as if everything hinged on it.

From where he sat on the sofa, Edward could see the garden gate through the front windows. He'd parked his bike on the road outside, and he concluded that if Marion and her friend did come back before he left, he would be the first to see them. He decided to make the sighting his cue to leave; it would be better that way. He had no desire to impose; the status-quo was not of his making.

The three of them had tea as they chatted over the many questions about Edward's new job in the Sudan. Neither of them had been overseas themselves and Joe's wife couldn't get over the fact that Edward should, willingly, want to work in Africa. Joe changed the subject.

"You'll be missed in the office you know."

The sentiment was touching, but Edward really had no regrets about changing his job. Quite honestly, he still had some difficulty in realising that luck had provided him with the two things that he most wanted in his career – a return to aircraft maintenance, and the opportunity to go abroad.

Joe got up and switched on the TV to catch the news whilst his wife took the tea things out to the kitchen. Edward noticed movement in the street outside and caught sight of a motor cycle, with two people astride, coasting silently into the driveway through the open front gates. Marion and her friend had returned. He took his self-made cue and, standing up to shake Joe's hand, he said that it was time to be off. Joe's wife came back from the kitchen, and although she was a little surprised at Edward's sudden intent, she gave him a little hug. She said how pleased she'd been to meet him, and to look after himself in his new venture. Neither she nor Joe had heard their daughter's return.

As they stepped out of the front door uttering last farewells, Edward imagined that Marion and her friend had entered by the kitchen door at the back. Edward was half way up the front garden path with her parents as Marion came out onto the front doorstep.

"Oh, you're off then, Ted."

Startled at the unexpected voice of their daughter behind them, Joe and his wife turned in dismay.

"Yes, I'm off now. I have an early start tomorrow and I don't want to spend my last night in Basingstoke, sleeping in the landlady's yard." Edward joked. Looking a bit sheepish, Marion came to where they stood at the gate. She'd clearly forgotten that her father had invited Edward over for tea that evening, to say good-bye.

"Well I hope you have a good journey and good luck in the new job. Will you be coming back this way sometime?" she said, tossing tousled hair from her brow.

"I can't say really: maybe." Edward bit back on the desire to ask why; that would have been a bitter retort.

"Can I write to you?" An odd question for someone who'd burned Edward's fingers and perhaps (a vain thought on his part) burned her own boats against his further advances in an immature relationship. Edward kicked himself mentally for such a cynical thought, rebuking himself, and answering with civility.

"If you like, I gave my home address to your Dad. I haven't got an address in Africa yet."

"OK, I'll write then, Good-bye; sorry I wasn't here when you arrived. I think I'll probably miss you, too."

On that note Edward kicked the *Norton* into life, and with a final wave he pushed off to the sound of gently muttering exhaust pipes. He didn't look back. With an enormous amount in his future, he looked forward.

Part 17

Gone was the Boy

43

Edward arrived at Portsmouth accompanied by warm sunshine and blue skies. A week had passed since his first visit to the Isle of Wight for interview but it seemed like only yesterday as he boarded the ferry. On this occasion he took the car-ferry to the estuary of Fishbourne, some little way westward along the coast from Ryde. He'd loaded up his motorbike with two pannier-bags and a suitcase strapped across the pillion seat. Heavily laden though it was, the *Norton* had carried him, and his gear, faultlessly on the journey south, from Basingstoke.

Edward reported to the company's General Office on arrival at Bembridge, and the girls handed him over to the Works Manager. Edward hadn't met him before, but his welcome was affable and friendly. He'd anticipated the need for temporary accommodation, and explained that the company could offer the use of a caravan on the airfield. The idea appealed and together they went to look it over.

The caravan lay out of sight from the road, on the airfield side of a thick hawthorn hedge, next to the petrol pump attendant's red and yellow painted hut. Parked permanently, with connections to main water supplies, drains and electricity, the temporary quarter was only a hundred yards from the hangar. It was the perfect arrangement. Not having to drive to and from work each day was clearly to Edward's advantage, and he was quite capable of catering for himself. He accepted the offer.

The Works Manager told Edward to get settled in and to start work on the following day. This gave him the opportunity to get his bearings and to do a little shopping for food in the village. It all seemed too good to be true. Accommodation without charge next to his new place of work (for a month or so anyway), no driving between digs in the village and the airfield, a petrol station right next door, and a pub where he could get lunch each day, not to mention a pint or two in the evenings, should the fancy take him. Who could have asked for anything more?

Next day Edward woke early and by the time he'd taken a light breakfast, the sound of cars parking on the other side of the hedgerow brought him to the door of his caravan. The company's workers had begun to arrive. Among them, Edward saw Trevor and hailed him as he locked up his temporary home.

"You made it then," said Trevor, detaching himself from the others and beaming all over his face, "are you ready for the fray?"

"Yes, I arrived yesterday." They shook hands and fell into step towards the hangar.

"You're in the caravan then?"

"Yes, it's a bit of luck really; living on site saves me chasing around to and from digs and work."

The works manager arrived in the company Land Rover, alighting with keys in hand to unlock the side door of the hangar. More cars arrived as the small crowd of fellows bantered and chatted, moving at a leisurely pace towards the open doorway. Inside the hangar people dispersed, to prepare for the day’s work. Edward noticed that there was no time-clock to punch. That pleased him. The atmosphere of the workplace seemed very unhurried and easy-going. Trevor led the way, joking with a couple of fellows at the door as they crowded through, then to Edward he said, "Come on then Ted, I'll show you where things are."

He took Edward into the workshop area at one side of the hangar, where benches and machinery filled the space beneath the wooden beams of a mezzanine. Trevor introduced some of the other chaps as they passed inside and made his way to a small bench at the farthest end.

"We'll have a cup of tea before we start," he said, stepping into a white overall and raising his voice to catch the attention of someone he addressed as Norman, "put another cup on mate, please." A short man with a limp, dressed in an old fashioned bib-and-brace overall and a cloth cap, responded with a wave, as he busied himself with a large kettle and an equally large teapot at the far end of the workshop. Nobody seemed eager to start work and the men stood or sat, gossiping, as they waited for tea.

"We've got three jobs on at the moment," said Trevor, "there's the modifications on the *Piper* to finish, atomisers to clean, repair and assemble, and eventually we'll have to go over to Portsmouth airport to prepare the other three *Pipers* for the flight out to Khartoum."

The limping man passed among the men carrying a tray filled with large mugs of steaming tea. Trevor introduced Edward to Norman as he took the last cup, explaining that the newcomer would be a regular customer for a month or so.

"You'll be off to Africa then, Nipper." Norman directed his comment to Edward in a broad Hampshire accent.

"Yes, I joined yesterday to help with the *Pipers*, before we go."

Norman nodded and said that the tea, first thing, mid-morning and mid-afternoon, would be five-bob a week. "Pay me when you're ready, Nipper, no hurry."

Trevor took Edward on a guided tour of the hangar, showing him where to find the essential facilities before leading on outside to have a look at a small wind tunnel, used for testing the rotary atomisers. The excursion ended next to a beautiful little white and red aeroplane, parked next to the frontage of the hangar and tethered with ropes to picketing irons in the grass.

The single-engine, high-wing monoplane glistened in the early morning light. Overnight dew on its top surfaces captured a million images of the sun that now climbed into a clear, blue sky. Edward wasn't familiar with the type, a *Piper Supercub* but, having an enclosed, glazed cabin, it was not unlike the *Austers* he'd worked on years ago. Trevor opened the split-door on the right-hand side of the cockpit, lifting the hinged glazed portion to a clip on the underside of the wing and lowering the other half to hang down at the side. The wide opening allowed access to both seats, set in tandem, unlike the *Austers*, with their two seats abreast or, sometimes, two abreast and a third sideways-facing seat behind.

They stooped down to examine a bulbous, fibreglass belly-tank, secured with metal straps to the underside of the fuselage beneath the cabin, just aft of the undercarriage.

"I fitted this yesterday," Trevor explained, "there's not much ground clearance is there?" He was right. The long grass of the field brushed the underside of the tank. There wasn't much more than nine inches between the lowest point of the tank's curvature and the ground. "We have to rig the dump valve cable eventually and the pipe-work to the pump up front."

They moved forward and Trevor showed Edward the pump assembly, also mounted on the underside of the fuselage, but just forward of the landing gear. He spun the blades of the wind-driven pump and tugged at a Bowden-cable protruding from a rubber grommet in the aircraft skin.

"We have to rig the fan brake, too."

Stepping back, Trevor drew attention to a pair of short struts hanging from the rear spar-line on one of the wings. "These are for mounting the spray gear on: the rotary atomisers," he explained, "there's two on each wing. This side's finished and we'll start on the other side today. You can get on with that in a minute if you like, opening up the skin where it's marked – how's your fabric-work,

Ted?" Edward nodded, though he knew he was probably a bit rusty since he hadn't worked on light aeroplanes in almost six years, but he didn't think it would take long to get back into the swing of things. They moved around to the other wing, to see the marked places that would give access, through the fabric skin, to the wing structure inside. Trevor grinned. "There's a lot to do, Ted, and when we've finished all this – the spray gear, struts and the pump's fan-blades have to come off again for the ferry flight to Khartoum. The belly tank doubles as a long-range fuel-tank for the trip, so there's a wobble-pump and piping to be fitted, as well." He stood back, obviously proud of his work so far. "After we've worked out where the overload fuel piping will go, and when it's all fitted up, the drawing office people will take the details for the approval drawings needed, to make the mod-kit official."

"It's lucky for us that the other three aircraft are being modified at Portsmouth, otherwise we'd be here until Christmas." Edward remarked, as they walked back to the hangar. Trevor laughed and reminded his new companion that they'd have to go over there too, to load and prepare all four aircraft for the ferry flight.

Like many islanders, Trevor had a Naval background. He'd served in the Fleet Air Arm during the time of Sea Fury and Wyvern aircraft, aboard Ark Royal in the Mediterranean, during the Canal Crisis of the fifties. He'd served in Far Eastern, Australian and Asian waters, and he'd seen action several times.

Although Trevor was his senior by some ten years or so, Edward's first impressions of him were good. With a highly developed sense of mischievous humour, he seemed adaptable, tolerant and self-reliant. Above all, he enjoyed working with aeroplanes. His enthusiasm for technical work was so obvious that Edward already felt like part of a team.

They collected up some tools from Trevor's bench in the workshop and set about the work to finish the wing modifications on the *Piper*. They worked outside on the grass field, where the aircraft stood, for the weather was perfect, with sunshine and soft breezes. They chatted ceaselessly as they laboured at the task. Indeed, it was a perfect way to get to know each other. Trevor talked about his Navy days and his early experiences with the company's crop spraying activities up and down the country, mostly in Lincolnshire and Norfolk, and of his several overseas excursions in the Sudan. Edward spoke of his civilian apprenticeship and of his earlier background in light aircraft. It soon became clear that they shared a common love for light-aviation.

The day wore on and morning tea-break came and went. Whilst they worked, Trevor spoke of an apparent problem related to the Sudan operation. Other operators with earlier models of the *Supercub* had fitted additional oil coolers to reduce engine-oil temperatures in the tropics although *Pipers* had suggested that the later version of the type wouldn't need such a modification.

The spray-company's management team had expressed concern, however, but if they wanted extra coolers fitted, then they'd need more down-time for embodiment of the modification. Concerned or not, time they didn't have. Trevor opened the engine cowlings to show Edward the implications of such a mod. Connection to the circulated oil system presented no problem, nor did the support fittings that were all part of the mod kit, but cutting the nose cowl to allow the cooler to protrude, like an extra chin beneath the propeller boss, and re-arrangement of the air-ducting, would clearly involve much more time-consuming work.

Edward pointed out the two rigid-tube, oil-pipes, evident in the engine's two air intakes, either side of the propeller boss: the intakes allow cooling air to pass around the engine cylinders. "If we could increase the surface area of these two oil pipes, that would provide an additional degree of cooling to the circulating oil, wouldn't it?"

"Yes, I should think so, but how?"

"Fins, like those on the cylinders would do the trick, if they were big enough." An idea had struck Edward but he decided to give it some thought before pursuing it further.

At lunchtime, Trevor took off in his growling, souped-up, pale-blue and grey, *Zephyr Six* to take lunch with his family, at home in Ryde. For Edward's part, he retired to the dining room next to the bar in the pub and later, relaxing on the grassy area outside, he chatted with some of his new work-companions. They were full of questions, but they answered a few of Edward's too, for he was curious about the hovercraft in the hangar.

"Cushion-craft is what they are, lad." Corrected one fellow, a sheet-metal worker, deeply involved in skinning of the new craft under construction in the hangar. "The red one, up in the corner of the hangar, is a prototype, the first we built, but the new one is different in its design. The principles are the same though."

Edward shook himself mentally at the diverse activity in which his new employers seemed to have engagement: experimental cushion-craft, aerial crop spraying, rotary spray gear and, on top of this, somebody mentioned plans for the building of an aircraft too. Could all this be happening at this singularly, tiny facility at Bembridge? It seemed incredible to Edward, that so much could be going on in such idyllic surroundings.

When Trevor returned, he and Edward pressed on with the job on the *Piper* again. By evening time they'd completed all the structural work in the wing, leaving only pipe-work and hydraulic lines to fit, secure and connect. They both agreed that it was a good day's work and the bond of friendship between them had grown.

That evening Edward gave some thought to his idea concerning the oil cooling question. It seemed to him that without a recognised modification kit, nor the time to fit it, they might save time and cost simply by adding cooling fins to the oil pipes. The pipes, already exposed directly to the air stream at the front of the engine, presented four to five inches of accessible, straight length. All they needed was greater surface area.

It occurred to Edward that perhaps they could achieve this if they made a pair of finned, length-wise-split, tubular, copper sleeves, of the same inside diameter size as the outside diameter of the existing oil pipes on the engine. They could mount them to enclose the oil pipes, securing them tightly in place with suitable clips, like a couple of jubilees. To this end, Edward made some sketches.

Next day he and Trevor were back at work on the *Piper*, fixing pipes and hydraulic lines to the wing struts, in preparation for attachment to the rotary spray gear. Trevor collected a small wobble pump from the store and explained how they would mount it in the cockpit.

The pilot would use the pump to move fuel up from the belly tank, replenishing the wing tanks in flight, as they emptied during the many stages of the ferry flight to Khartoum. It was essential to locate it within comfortable and easy reach of the pilot's hand, but it was also important to minimise the extent of fuel-carrying pipe work inside the cockpit. They decided to mount the pump on the right, next to the front seat, at approximately the closest position to the belly-tank's proposed fuel outlet. This satisfied both requirements. To raise the position of the pump's handle close to the pilot's hand, they made a light-alloy, box-shaped platform, large enough to support the apparatus, but small enough so as not to impede entry and exiting of the cockpit. They made up the necessary hard-pipes and fitting in preparation for installation after the chief pilot had completed spraying air-tests.

So passed each day until, towards the end of that week, the *Piper* was ready for testing in its crop spraying role. The Sorensen belly tank carried sixty-four gallons or six hundred and forty pounds of water. Its weight together with the four atomisers, their struts, and pipe work, plus the pump assembly and fittings, all added up to a fair load for the little aircraft to carry.

44

The Chief pilot, being a big man, probably weighed-in at around fourteen stone. Since he was to perform the air test, it was possible to estimate that the *Piper* would be close to its maximum all-up weight capacity, in spraying configuration.

Peter came out to have a look around the aircraft. He beamed widely and joked about his size and smallness of the *Pipe*r's cockpit. Trevor had told Edward earlier that Peter had many years experience with light aircraft, in particular he was once a Senior Flying Instructor at some high-class flying club, before turning to the crop spraying business. He was to be Chief pilot in the Sudan operations.

"Right," Peter pronounced, after his inspection of the *Piper*, "I'll take just half-a-tank of water to begin with. I don't want to overdo it. Must get the feel of it first."

Trevor and Edward filled the *Sorensen* belly-tank to the half-way mark with the hose they'd trailed across the grass from the nearest tap. The undercarriage sagged a little. When Peter got in, the landing gear sagged a bit more, but a quick check of the ground clearance under the lowest part of the belly-tank showed that there was still a good six inches of space. Trevor nodded in satisfaction. "Once the tail's up, on take-off, the clearance will be much more of course," he said.

After starting the engine and warming up, Peter taxied gingerly out onto the airfield, moving slowly across to the farthest end of the grass landing strip. With the under-slung belly-tank, the *Piper* looked as if it was with child and about to give birth, waddling slowly across the grassy field. The take-off run was quite unspectacular. The sweet note of the *Lycoming* engine, muffled by the fine, green turf, drifted across the field as the laden aircraft lifted off smoothly.

Peter did a couple of circuits around the district, occasionally squirting a few bursts of water through the atomisers. Then, turning into wind over the village, he brought the little aeroplane towards the airfield, swooping down the hillside, over the telephone line, to pass the perimeter hedge at low level, Peter hunched over the controls in his tiny cockpit. He opened the spray-valve and performed a mock spraying run across the field, tail high at eighty knots, and at next to zero feet. The plume of swirling water-mist, trailing out behind, sank in a broad swath into the grassy, gentle slopes of the field, in a perfect spray-run. Then, pulling up steeply and turning, Peter repeated the operation in the downwind direction.

Trevor and Edward had already tested the belly tank's dump valve on the ground, releasing sixty-four gallons of water in an explosive gush, through a large, capped orifice at the lowest point of the tank. Peter wanted to prove it in the air and upon his return from the spray trials he asked for a full tank. It was important to know how quickly a dumped-load would take to clear the tank in flight. When used in an emergency, such as engine failure over the crop, the pilot had to be sure of his capacity to lighten the aircraft rapidly.

Peter took off again with his full load, climbed to get the feel of the near-maximum-all-up condition then, swooping in over the village again, he pulled the dump release over the field. The two fitters watched in awe as the sixty-four gallons of water spewed out in a wind-swept plume, and the *Piper* lifted skywards as the load-weight diminished quickly. They counted four seconds.

Peter came back, smiling widely over the results of his spray-tank, dump-test. "I got a good thump in the backside when I pulled the dump handle," he gushed, with a stilted laugh. "Goes up like a lift, and flies like a bird."

"Do the atomiser brakes work OK?" Trevor asked. "We had a bit of trouble with bleeding the hydraulic lines, they seemed fine on the ground but do they hold in flight?"

"Yes, perfectly. No problems there, Trevor."

Peter asked about the pump filter – its accessibility and the ease of its cleaning. Trevor reassured him and they stooped down next to the engine cowls to demonstrate. Satisfied, Peter wanted the *Piper* left in its spraying configuration for the time being. He explained with a smile that some of the other pilots had asked if they might have a go with the new toy, before removal of the struts and atomisers for the ferry flight.

Edward's interest was in the oil temperature and he questioned Peter about the figure he had recorded, on this hot summer's day.

"Normal," he said, "well within the green sector." Then, reflecting on Edward's question for a moment, he added. "*Pipers* say that this new model of the *Supercub* shouldn't need an oil cooler in the tropics, but I'm a bit worried all the same." Of course, the temperatures of the English summers were considerably lower than those expected in the Sudan but at least his recorded figure was a starting point. The object of Edward's idea was to achieve an overall reduction in oil temperature, if only by a few degrees.

Later, he again broached the subject of his oil cooling idea with Trevor. He showed the sketches too, and Trevor seemed impressed, though more by the quality of the drawings than by the theory behind them. "Well if you think you can make them, Ted, let's try it, there's nothing to be lost."

Edward spoke with one of the workshop chaps who was responsible for the welding equipment, to find out if he could use the gas torches. The fellow agreed, and set aside one of the hearths for Edward's use whenever he was ready. Making the five-inch, tubular bodies and the annular fin-pieces to slide over them presented no problems. Brazing the fins to the tubes wasn't difficult either but Edward ran into a problem when it came to cutting the assembled pieces lengthwise, to split them. The friendly welder came up with a method and by the end of the day they had the first set of cooler-fittings finished.

The devices fitted perfectly when offered up to *Piper*'s engine. They were snug – the inside surfaces of the split tubes making good contact with the straight section of the oil pipes. Enough free length at each end provided the plain surface necessary to accommodate a couple of Jubilee clips with which to secure the coolers in place. Air-tests the following day would prove the idea.

That evening Edward spent an hour or two riding the *Norton* through the narrow lanes, exploring the locality farther afield than Bembridge. He drove around the harbour and estuary to St. Helens, with its broad village-green, its Victorian houses and its many boat yards. A ferry plied across the narrow entrance to the harbour, between the sandy beaches and dunes of Bembridge and the shore of St. Helens. The boat was rather like a miniature tank-landing craft, with a drop-down ramp door at its bow, but it carried holidaymakers from Bembridge to a large pub on the other side, appropriately named the ‘Ferry Boat Inn’. From the St. Helens side, the ferry carried those who would return or those who would prefer to savour the beer in Bembridge, where there were several pubs to choose between.

‘The Pilot Boat’ Inn was a favourite. Built at the apex of a sharp corner of two intersecting roads, and overlooking the harbour, the pub resembled a ship, its prow being on the corner where the two roads met, and the sweep of its ‘hull’ sporting portholes instead of windows, deck rails and lifebelts. The pub had two bars, the Fo'c'sle at the bow (on the corner) and the Saloon farther along the hull's side. Lobster lunches were a speciality, a favourite among the holiday visitors.

Edward sat on his bike for quite some time, watching the landing-craft-ferry and taking in the scenery. Popular though it was with the sailing fraternity and holidaymakers, the village of Bembridge was, in many ways, like a village of bygone days: calm, tranquil and gentle. When he got back, he parked the bike in front of his caravan quarters and popped into 'The Propeller' for a pint of bitter to round off his evening before turning in; just one, for he wasn't much of a drinking man in those days.

He'd retired for the night, when the sound of a powerful engine brought him to the door of the caravan. Although the summer sun had set, there was still light enough to see. The hangar doors were open and additional light streamed out, illuminating the paved hard-standing and the grassy field beyond. Activity on the new cushion craft had reached a point where overtime was necessary and those involved worked for a couple of hours each evening and at weekends on Saturday. There was clearly some urgency in the work's completion.

Amplified by the confines of the hangar, the throaty roar of the engine increased to a crescendo. The new, white-painted cushion craft slowly emerged, swathed in a cloud of dust and surrounded by the workmen. Tilting to slide gently down the shallow gradient of the pavement, and straightening to level again as it moved out onto the turf, the new craft hovered upon its cushion of accelerated air. A storm of grass clippings burst in plumes from beneath the craft, sending the attendant group scurrying with backs turned against the swirling, green blizzard. As the pilot sought to control the directional vanes in the cushion of air upon which the vehicle floated, the craft moved farther out into the field. Swaying and swerving, this way and that, the craft twirled and bucked as if in some elegant ballet for machines. The pilot moved his charge backwards and forwards, sideways and in circles about its own axis, experimenting with the controls. It was indeed a sight to behold.

Edward shook his head and yet again he wondered at his luck in finding the opportunity to work with such a diverse company. He returned to his bed to dream of an exciting future. He had returned to aircraft again, after so long away, and he was about to go abroad. What else lay in store for him, he wondered, as the cushion craft's engine ceased to bellow and the hangar doors closed with a distant rumble.

Edward woke early next morning to another fine day. Birds in the hedgerow, raised their song to greet the rising sun, their tinkling melodies and twittering choruses drifting though the caravan's open windows with the light, morning-airs of Summer. His eggs and bacon breakfast never tasted so good, as the smell of coffee put an edge on his appetite. He thought that there was something very pleasant about living in a caravan, right next door to an airfield such as this. It reminded Edward of the caravan holidays he'd spent with his family in Flintshire, all those years ago, only a few steps over the sand-dunes to the Gronnant beach. For once, in recent times, Edward had a clear mind. He felt at ease and in harmony with the world around him. Contentment, excitement, anticipation, joy and incredible happiness filled him. He no longer felt herded. He was a free agent, released from the oppression of the time-clock of factory life. He'd competed independently for a new niche in life and was about to become part of an enthusiastic, benevolent and appreciative team.

Trevor and he prepared the *Piper* for more flying later in the morning and Edward retrieved the two oil coolers from the paint-shop, in readiness for the trials. Peter arrived in the company of a shorter man with an Australian accent and hunched shoulders. Peter introduced him to Edward as Jim, and they shook hands. In spite of the rather informal way he had been presented, he was a director, responsible for the crop-spraying side of the company, with many years of flying experience.

"You're off to the Sudan then," Jim said, scrutinising Edward with sharp blue eyes, "you'll find it hard work, but good experience."

Peter and Jim walked around the aircraft, looking at the spray gear and belly tank, nudging and shaking things as if unsure of their security.

Trevor nudged his companion and grinned. "Jim's been to the Sudan a few times when we flew *Tigers.* He crashed a couple last year," he said in an undertone, "he worries a bit, about technical things, but he's a great pilot."

"Is he going to the Sudan this year?"

"No, he's in charge of all the overseas operations now, as a director, and for the spray gear development, manufacture and sales, too."

The two pilots finished their walk-around in discussion, joining the two fitters again with complimentary comments about the installation and the good progress, though Jim seemed only marginally impressed, saying that the standard mod-kits cost a fortune.

The moment looked right to mention the oil-cooling idea. Edward suggested that Jim might like to take accurate oil temperature readings on the first flight, then they'd fit the coolers and he could do a second flight for comparisons. Jim agreed and said that he'd fly the *Piper* later. Edward showed him the two finished pieces, and explained where and how they would fit them.

Jim looked at the coolers in doubt, shaking his head enigmatically.

"Well, if they do the trick and save us cost, I'll be pleased," he said as the two pilots moved off back to the hangar.

Trevor nudged Edward again. "Jim's a bit tight with the money, too, he's like a frog at forty fathoms sometimes," he said quietly, from behind his hand, hiding the mischievous grin on his face.

Much to Edward's satisfaction, the oil cooler trials were a success. By comparative readings, they established a temperature reduction of some five degrees. On that basis, Jim asked Edward to make three more sets, one for each of the other aeroplanes. Later, he also made detailed sketches for the design department people, from which they might seek application for approval as an official modification.

During the following days Trevor and Edward prepared the overload-fuel wobble pumps, pipe work and ancillary equipment ready for installation in the three *Pipers* at Portsmouth. They stripped the spray gear, support struts and pump assembly from the 'trials' aircraft and prepared the belly tank for its temporary, fuel carrying role.

Fitting the long range equipment presented no problems. They found no leaks after filling the Sorensen and, to their satisfaction, they proved the integrity of the pipe work without further adjustments. Later, they established that the wobble pump had sufficient capacity to deliver fuel, from the belly tank to the wing tanks, at more than twice the maximum rate of engine consumption, with very little effort; a very important consideration and with a result better than expected. The pilots wouldn't want to be flying over the Mediterranean only realise that they couldn't pump up sufficient juice, quick enough to meet the engines' needs.

"They'd end up in the drink with sweat on their brows and blisters on their right palms." Trevor joked. "It'll give the pilots something to occupy themselves with, as they fly the long stretches."

The final test proved how much unusable fuel would remain in the tank. This was interesting owing to the shape of the tank. However, worries on that score were unnecessary. In both 'tail-up' and 'tail-down' attitudes, the unusable fuel quantity appeared to be about the same, just two gallons.

Time passed quickly and Trevor began to talk about personal preparations for the coming six-month season in the Sudan. He spoke about the spraying and told Edward much about the places he would likely see and the people he'd meet and work with.

It occurred to Edward that he should make plans for his final visit to Yeovil, make some final purchases for clothing and footwear, lay-up the bike and most importantly, reassure his mother that his proposed overseas trip was not something she should worry over. After all, it was only a short contract and although Edward would miss Christmas in UK, he'd be home early in the new year.

Trevor and Edward spent several days preparing atomisers for the other aircraft; stripping, cleaning and reassembling each unit, replacing worn parts as necessary. Trevor assured his companion that the experience would be invaluable, for Edward's part in the Sudan operations.

"Even though the pilots will take a final load of water each day, to flush out and clean the system, the atomisers will need to be stripped regularly." Trevor told him. "Some of the chemicals leave a residue that builds up in the atomiser gauze: especially Endrin and Ecatin."

Since the weather remained fine and hot, they conducted this work outside, in an open area close to the wind tunnel, on the farther side of the hangar. It was just as well, for the atomisers smelled very strongly of the congealed chemical residue, left over from their previous use. Certainly, they wouldn't have been popular doing it in the workshops. Anyway, it was more convenient to work near the wind tunnel because the atomisers needed testing after reassembling. In the evening Edward made sketches and exploded drawings of the equipment to remind him of the essential parts and their assembly sequence. He added part numbers and descriptions to identify them.

His medical results came back. They were satisfactory and he suffered no ill effects from the vaccinations that he'd had the week before. His passport had also arrived, his very first, complete with entry visa for the Sudan. He'd signed the contract for overseas employment to start upon departure from London, Heathrow. Everything seemed to be working out very well.

The four planes in preparation for the coming season in the Sudan, bore Panamanian registration. Trevor explained that the company had bought them through a subsidiary in that country. For this reason the approved maintenance firm at Portsmouth would perform final inspections, before releasing the aeroplanes for aerial work, along with the necessary documents and certificates. With only one more week to go, Peter took the *Piper* over to Portsmouth to join the other aircraft at the contractor's.

The same day, just before lunch time, a visiting aircraft joined the circuit at Bembridge. The aeroplane circled once, made a long approach and landed without fuss in the middle of the airfield. The sound of its engine brought a few people to the hangar doors.

At first glimpse, the aircraft appeared to be a *Tiger Moth* but a glazed canopy over the normally open cockpits identified it as a *Jackaroo*, a modified *Tiger*, designed to carry four people instead of two. The *Jackaroo* was the recent brainchild of a firm at

Thruxton, and a number of these hybrid planes were already in use by the flight training centre at Southampton. The ATC student pilot of this one was on a solo cross-country navigational exercise, calling at Bembridge before his final leg back to base.

He taxied to the parking area in front of the hangar and whilst he sought out a responsible person to sign the proof of his visit to Bembridge, we gathered around to examine his unusual aircraft. Edward had never seen this version of the *Tiger Moth* before.

The accommodation of four seats, two by two, in the aircraft's widened fuselage, was a substantial modification both to the structure and the control system. The addition of a glazed canopy ensured a measure of comfort to the occupants, protecting them from the rigours of wind and weather, normally associated with open cockpit flying.

When the student returned to his plane, Trevor offered to swing the propeller for him but the proud cadet declined, saying that he could manage very well on his own, thank you very much. It was lunchtime anyway. Trevor shrugged and retired to the grassy area next to the red and yellow hut, where a small crowd of holidaymakers from the pub and a few fellows from the hangar had gathered to watch the *Jackaroo*'s departure.

The student, meanwhile, prepared his plane for starting. He must have set the throttle too wide, for as the engine caught, bursting into life as he swung the prop, the aircraft began to move forward. Before swinging the propeller, the student had foolishly removed the chocks, which had been placed in front of the brakeless wheels. Edward supposed that the student thought it would be easier and quicker to dispense with the formality of additional running about, to pull the chocks after start up, but it was a big mistake. The pilotless plane began to move away under power.

At least the cadet had the presence of mind to move quickly out-board to grab a wing-tip, before the plane gathered too much way. Having done that, he found himself in a serious predicament. He dared not let go of the wing for fear of the plane trundling off to bury itself in a hedge or a ditch. All he could do was to hang on and keep the plane circling about the wing tip he'd taken hold of, in desperation.

He had three choices. He could let go of the wing tip and watch the aircraft roll away, perhaps to destroy itself, or he could hold on, circling until the petrol ran out, or he could try to work his way along the wing, to scramble into the cockpit and switch off the mag switches. Fortunately, he didn't need to do any of these things since several of the hangar people rushed over to help. Someone clambered up into the cockpit and stopped the engine. The excitement was over.

A renewed offer to swing the prop received the cadet's abashed acceptance after the drama, and it was a red-faced student-pilot who seated himself in the cockpit to perform his end of the procedure.

At last he took his plane into the sky again and with a modest over-flight of the field, he waggled the *Jackaroo*'s wings before heading off towards his next map reference.

45

After lunch Les sent for Edward and Trevor, to discuss arrangements for their planned visits to the mainland. The company kept a couple of vehicles at Portsmouth, garaged in the town. They were to use one of these. Starting from the following day, they were to catch the early ferry from Ryde, pick up the car and spend all day at Portsmouth airport. The return in the evenings would depend upon progress with the packing of spray gear, spare parts and ferry flight kit into the four *Pipers*. Les anticipated that three days should be enough time to complete the task.

Edward decided to use the *Norton* as far as Ryde, leaving it in a secure parking slot at the foot of the pier ready for his return trip to Bembridge in the evening. Trevor, living in Ryde, walked from his home to meet up with Edward at the pier-head. The seven o'clock ferry wasn't the earliest of the day but it suited their purpose – to reach the mainland and Portsmouth airport in good time.

On the first day, there were few passengers and Trevor suggested a cup of coffee in the uncrowded cafeteria, to start them off in the right frame of mind. The island receded behind them as the ferry moved sedately over the smooth waters of Spithead. The roof tops of Gosport, Portsea and Hayling, glinted distantly in the morning sunlight as they approached Portsmouth harbour. Known affectionately as Pompey to residents, naval ratings and officers alike, the town was one of Trevor's old stamping grounds.

He led the way to a side-street garage where the company kept its mainland cars, chattering ceaselessly about his navy days, and liberty nights. He pointed out some half-remembered pubs that he'd rather not return to. "Too much money, too much beer and not enough sense," he said, "those were the days!"

Trevor signed for the car and drove the short distance to the airport. They found the four aeroplanes neatly parked in a line on the grass, outside one of the contractor's hangars. The contractors had finished all the spray gear mods and it only remained for the lads to fit the wobble pumps and pipe work for the long range fuel systems. After that, they could pack the gear and ferry-flight equipment securely into the stowage spaces and rear seats. Trevor went off to the hangar, to find the gear, sent over from the island the day before, whilst Edward started preparations for the first of the remaining wobble pump installations. They work through each day, breaking only briefly for lunch and commuting each morning and evening. During the last of the three days at Portsmouth they finished the task. The aeroplanes were ready to go.

Peter and Les, along with two other pilots who would also make the ferry-flight to Khartoum, flew over in the company's *Gemini* to view the progress and to secure the all-important documentation with the contractor. Pleased with their work, Peter explained that the *Pipers* would leave next day, with an early start.

Trevor's mischievous humour gave them a moment of mirth as he presented Peter with a peaked cap, displaying an anchor on the front. Such caps were commonplace, popular with holiday-yachtsmen, but since Peter was the Chief Pilot and Captain of the ferry flight, Trevor thought it was appropriate that he should wear a captain's cap. Peter laughed heartily and immediately donned his new acquisition, promising to wear it for the whole journey.

As it turned out, joke or not, the cap came in very handy when it commanded respect at certain places along the ferry-flight route into North Africa, but that's another story.

Early next day, the four, beautiful *Piper Supercub*s lined up for take-off. Loaded down with full fuel tanks and cabins packed with equipment, they heaved themselves into the air to climb away into the morning sky before turning southwards towards the coast of France. They were on the first leg of their long flight to Khartoum. The brilliant sunlight touched their wings briefly, gleaming across the surfaces, as the pilots closed up in loose formation with their leader, resplendent and in charge, with his brand new sailing cap perched upon his head.

The glorious summer of nineteen-sixty-one seemed never-ending. The weather remained warm and fine. Certainly, during Edward's very first month on the island anyway, it rained only a couple of times at night and over the whole of just one weekend. For the rest, it was dry, sunny and perfect for working outdoors, and for the island's holiday trade. Many of the local farms around Bembridge turned over some their land to campers, whose brightly coloured tents of all shapes and sizes, dotted the lower slopes of Culver. The country lanes carried the more energetic, on their walks to explore the locality and the wide variety of scenery. To Edward the Isle of Wight became like a second home. It remained close to his heart, a jewel among his memories, filled with happy times and the seat of a significant turning point in his younger life.

With the aeroplanes on their way, it was almost time for Trevor and Edward to make their trip to the Sudan. They had to be there before the *Pipers* arrived, to prepare for their immediate conversion into crop-spraying configuration. Friday came and Edward rushed off to Yeovil to spend the final weekend with his family. This time he drove across the island to Yarmouth, at the western end. Catching the ferry to Lymington, he anticipated that the trip would be quicker than using the Portsmouth route.

Of course, it pleased his folks to have him home again for the weekend, and although it was only a month ago, they made his visit seem as if he'd been away for years, bless them. It was a bit like the return of the prodigal, and inwardly Edward wished that his parents would play down the sentiments a little, keeping it to proportions more suited to the occasion. His next visit wouldn't be until after Christmas, in six months or so. However, it was their way of demonstrating the love and concerns they shared for their second son, who was about to spread his wings. Edward had written to his mother every week, honouring the promise that he'd made to his father. For this reason, his absence from Yeovil for the preceding month had been like a proving run, just as he'd wanted, before his forthcoming and much longer absence, abroad.

Les scheduled the two fitters to leave for Khartoum on the following Tuesday. This meant that Edward had to get back to the Isle of Wight on Monday. He made his last minute purchases in the town and spent Sunday bedding down the *Norton* for a long period of inactivity. He and his father made room for it in the garden shed, and covered it with old sheeting to keep the dust off.

Sunday lunch was a sad affair, without the usual family banter. They made an effort to keep up cheerful conversation, avoiding references to Edward's journey and the new job overseas. Edith kept her silence, trying hard not to show her emotions over how she felt about her son's decision to go abroad.

Edward's plan was to catch an early train from Yeovil Junction on Monday morning, one that would get him to Portsmouth in time for the ten o'clock ferry. He had already arranged for a taxi to take him to the station and, with his bag packed, he was ready. He asked his people not to insist on going to the station to see him off. He wanted to take his leave quickly next day, at the front door.

They had tea in the garden.

The evening sun was still warm and it was less formal than sitting as usual in the living room. Surely they all felt more at ease and they joked and laughed over silly things to keep their true emotions in check. As a family, they weren't very demonstrative but that didn't mean that they were without feelings of love. Quite the contrary, they were very close.

They tried to keep innermost, sad thoughts and emotional disquiet hidden from each other. For his part, and probably it was the same for his brother David and their Dad, self control was aimed at protecting Edith's sentimental feelings.

Trevor and Edward boarded the train at Portsmouth, bound for London. Their journey had begun. It was hard for Edward to believe that within twenty-four hours he would be in Khartoum. What was he to expect? Would it be awful? Would it be splendid? Would it be disappointing?

He wasn't familiar with London but Trevor explained that they would take a taxi from Waterloo to the Buckingham Palace Road air terminal, check their bags in, have a couple of beers and meet the others, before taking the airline-bus out to Heathrow. The *B.O.A.C. Britannia* flight would fly some twelve hours to reach Khartoum, with one stop at Rome. This would place their arrival at about three in the morning.

"Who are the others?" Edward asked.

"Well, there's a pair of entomologists and a couple of airstrip supervisors; that'll make six of us all together. I know the entomologists from last year but the others are new, like you."

They fell silent, each intent upon his own thoughts; Trevor's, of his wife and family back home in Ryde, and Edward's of his own family, probably relaxing after lunch and thinking about him, he hoped. Edward gazed from the window as the train rattled on, the countryside speeding by. They passed a solitary figure, lonely in his fields.

Edward felt a great sense of detachment and wondered what the farm-worker was thinking as he stopped his work for a moment, to lean upon his scythe and watch the train rush on to London. Did he, too, wonder what thoughts coursed through Edward's mind? Could he have seen Edward's face at the carriage window? Had he dreamed of adventure, as Edward had done, and of a way out of drudgery?

The older Edward recalled that he'd often had this feeling of detachment when travelling, whenever he saw people from the window of a train or a bus, or from an aeroplane at take-off or approaching for landing; they, going about their business in a normal, perhaps even, ordinary world and he speeding to or from distant parts. He always wondered who they were, where they were going, how they felt about the life they led. Were they really ordinary? Where they happy or sad? Did they seek change, or were they content with what they had? Did they look up to wonder where Edward was going or where he'd been. Did they feel the shadow of his spirit touch their thoughtful brows and did they hear his silent question when he asked of them, "Who are you?"

For a moment Edward felt a wave of panic, a burst of racing pulse, a surge of adrenaline. The train propelled them on. The adventure had begun. There was no stopping, no turning back.

Had he made the right decisions? The train lurched in a curve and roared through a wayside halt. Trevor nudged his thoughtful companion and stood up, bringing Edward back to reality. "Let's have a drink, come on, Ted." Edward's serenity returned like waking from a dream as they made their way to the bar in the restaurant car.

The cool quietness of the city air-terminal brought a sense of peace and tranquillity to Edward's pounding trains of thought, in contrast to the hustle and hullabaloo of Waterloo station and the mad race by taxi through the busy streets of London.

How he hated crowded places. He'd felt uncomfortable in the jostling frenzy of people at the station – thoughtless of others about them, and intent only on getting to their destination, at all costs. Pin-striped suits and bowler hats, umbrellas, briefcases and newspapers tucked under arms, hurrying on to drudgery, there to wait impatiently for the hour of five to strike.

Would they then repeat the madness of commuting-travel in reverse, to dine in silence with indifferent spouses in suburban or urban homes, their children, long in bed. Would they sit to watch the news on their television set that seemed, at times, to rule their after-work lives? Were they all in a rut, as Edward had been? He pictured an ants' nest, teeming with industry or the migration of Lemmings, rushing, helter-skelter to their doom. Imagination got the better of Edward again.

After checking-in with their luggage and tickets, they made their way to a waiting area near the bar. Trevor hailed and waved to a couple of chaps who craned about them as if looking for their fellow travellers.

It was like the finding of Livingstone – hearty handshakes, wide smiles and noisy banter. Trevor introduced Edward to Rene, a tall, lantern-jawed fellow with a mop of curly hair. He seemed to be the senior and about the same age as Trevor. The other was younger, close to Edward's age, short of stature with his shoulders lifted in an awkward, nervous stance. His name was Peter. Both were the entomologists Trevor had mentioned.

Peter rubbed his hands together in glorious anticipation. He wore a permanent grin on his face, his lank hair was in what looked like a permanent state of disarray. "Who's getting the beers in then?" he asked.

"You can, if you like," said Trevor with a laugh, "mine's a pint."

"Anyone else?" Peter's grinning eyes glanced at Rene and

Edward.

"I'll have a brown ale, thanks." Edward said, as Rene shook his head, declining the offer. Peter casually strode off to the bar, his hands thrust deeply into his trouser pockets, pushing his shoulders up even higher. Rene smiled and explained that Peter had supped a few on the train, coming down from Durham.

The lounge area began to fill with travellers and, as Peter returned with the beers, two more fellows detached themselves from the latest group to check in. They eyed Trevor's jolly group and approached to ask if they were Mr. Hewitt's team. Rene's response brought smiles and more hand-shaking. Peter gleefully rushed off to get more beer and the newcomers sat down to introduce themselves.

Glen, the taller of the two newcomers, spoke well and precisely with no discernible accent. He was about Edward's size, fair of hair and carried himself importantly. John was short and thickset and spoke with a slightly Norfolk accent.

Peter came back with the beers, and conversation reverted to humorous tales about last year's season in the Sudan. It crossed Edward's mind that his first trip overseas was beginning to take on the feeling of a holiday, and a beery one at that.

Whilst Trevor and the entomologists chuckled over tall stories, Edward asked the two new arrivals about themselves. Glen, it seemed, had been an Outward Bound boy. He joined the merchant Navy and had travelled extensively, circumnavigating the globe several times before he'd reach the age of twenty-three. More recently, he'd learned to fly and qualified for a private pilot's licence. He expected to build up some hours, flying communications in the Sudan, to help his cause towards a commercial license. John also had Outward Bound experience but he was currently studying agriculture. He thought the Sudan experience would aid his studies. He wanted to write a paper on pest control in agriculture.

The announcement of their flight number brought conversation to a halt, and they made their way down to the waiting bus that would take them to the airport. Edward hardly saw the passing city as the bus made its tortuous journey out of town. His thoughts centred upon the grand adventure that was about to begin. This was to be his first flight by international airline. He felt excited, having flown only in smaller aircraft like *Rapides, Ansons* and *Herons.* Flying to East Africa with BOAC would therefore be an entirely new experience for him.

They boarded at the rear doorway and passed into the long pencil-like fuselage. A pretty stewardess showed them to their seats and they settled into the soft cushioning. Trevor offered the seat next to the window from where Edward could see the starboard wing and its two engines jutting from the leading edge, the metal enclosures curving back over the top surface in smooth contours.

Edward found himself listening to the quietness of a sound-deadened atmosphere within the narrow confines of the cabin's interior. There were no echoes and the thick carpeting absorbed every footfall down the central aisle, muffling the sound. Below, on the concrete hard-standing, service trucks, technicians, loaders and tractors moved about with organised precision as they performed their final assignments. Raising his eyes, Edward could see scores of other aircraft lined up on the ramps. A forest of tails, with their coloured insignia, towered above white-topped fuselages.

A soft thud, felt rather than heard, announced that the flight attendants had secured the cabin doors. More thumps, emanating from beneath their feet, told them that the ground crews had closed the baggage compartment doors upon the luggage and freight, loaded in bays below the floor.

The aircraft came to life with vibrations and dull rumblings, increasing in amplitude as the flight-deck crew started the engines. Edward watched the huge propellers spinning in front of the engine nacelles on his side of the aircraft. Within minutes, and with a gentle lurch, the aircraft moved backwards, pushed by a tractor at the nose wheel, until clear of the of the ramp and its long line of aircraft.

A moment's pause, then they moved forwards under power, away from the buildings, along the taxi-ways. Edward's thoughts raced as the aircraft turned onto the end of the runway. It was years since he had flown, but this journey was like nothing he'd ever experienced before.

The propellers became almost invisible to sight as the pilots opened the throttles. Edward felt pressure at his back as the brakes were released and forward speed accelerated for take-off. The tubular enclosure tilted as the aircraft's nose lifted and lingered for a brief moment in the transition from a bumping, almost wallowing, high-speed dash on wheels, to smooth, gentle flight.

Edward heard the sound of hydraulics retracting the undercarriage, ending in a thump as the doors closed to cover the spinning wheels.

Then, as the aircraft climbed steeply, he watched the ground dropping away beneath the wings, revealing, like a living map, the outskirts of London, the green and golden fields of a countryside close to harvest time, distant hills and the patchwork quilt of rural countryside. It was teatime in leafy, sunlit, English gardens far below the aeroplane's droning, southward passage in the afternoon sky.

The whole of Edward's short life flashed before his mind's eye, as if he were close to violent death. His heart pounded and his thoughts spilled over with fresh emotions, his hopes and dreams fulfilled.

Joy and sadness clutched at his heart, as the whispering giant carried him away from those dearest, fairest islands of his youth. Gone was the boy. Would a man return.

Lightning Source UK Ltd.
Milton Keynes UK
UKOW06f0915240315

248413UK00009B/419/P

9 781908 135093